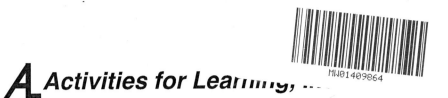

Activities for Learning, Inc.

RightStart™ Mathematics

by Joan A. Cotter, Ph.D.

Level D Lessons

For Home Educators

Special thanks to Sharalyn Colvin, who converted *RightStart™ Mathematics: Grade 3 Lessons* into *RightStart™ Mathematics: Level D For Home Educators.*

Note: Rather than use the designation, K-4, to indicate a grade, levels are used. Level A is kindergarten, Level B is first grade, and so forth.

Printed in the United States of America

www.RightStartMath.com

For more information:

info@RightStartMath.com

Supplies may be ordered from:

www.RightStartMath.com

order@RightStartMath.com

Activities for Learning
PO Box 468
321 Hill Street
Hazelton ND 58544-0468

888-775-6284 or 701-782-2000

701-782-2007 fax

ISBN 978-1-931980-13-5

August 2014

Home Educators

RIGHTSTART™ MATHEMATICS

by Joan A. Cotter, Ph.D.

The following are items needed to complete the RightStart™ Mathematics Level D Lessons:

STATUS	ITEM	CODE
REQUIRED	***Level D Lessons*** (Third Grade)	T-D
REQUIRED	***Level D Worksheets***	W-D
REQUIRED	***Math Card Games*** book	M4
REQUIRED; choice of abacus	Classic AL Abacus - 8-1/2" x 9-1/2" hardwood frame & beads	A-CL
	Standard AL Abacus - 7-1/2" x 9-1/2" plastic frame & beads	A-ST
	Junior AL Abacus - 5-1/4" x 6" plastic frame & beads	A-JR
RECOMMENDED	Place Value Cards	P
RECOMMENDED	Abacus Tiles	AT
REQUIRED	Thousand Cubes	AC
REQUIRED	Cards, Six Special Decks needed for Games	C
RECOMMENDED	Fraction Charts	F
REQUIRED	Basic Drawing Board Geometry Set	DS
REQUIRED	Colored Tiles, apx 200 in set	RH2
REQUIRED	Casio Calculator SL-450	R4
RECOMMENDED	Math Balance (Invicta)	R7
RECOMMENDED	Centimeter Cubes, 100 in set	R8
REQUIRED	4-in-1 Ruler	R10
REQUIRED	Folding Meter Stick	R15
RECOMMENDED	Plastic Coins	R5

Note: If a child has not previously worked with the AL abacus and is just starting RightStart™ Mathematics, *RightStart Mathematics Transition Lessons* are required before starting the *RightStart Mathematics Level D Lessons* .

TO ORDER OR FOR GENERAL INFORMATION:
Activities for Learning, Inc.
PO Box 468 • Hazelton, ND 58544-0468
888-RS5Math • 888-775-6284 • fax 701-782-2007
order@RightStartMath.com

AL Activities for Learning, Inc.
RightStartMath.com

05-08

RightStart™ MATHEMATICS: OBJECTIVES FOR LEVEL D

Name ______________________________

Teacher ______________________________ Year ________

	1ST QTR	2ND QTR	3RD QTR	4TH QTR
Numeration				
Can skip count 2s to 10s the first ten multiples				
Can read and write numbers to 1 million	N/A			
Addition				
Can add 2-digit numbers mentally				
Can add several 4-digit numbers				
Subtraction				
Can subtract 2-digit numbers mentally				
Can subtract 4-digit numbers				
Knows subtraction facts				
Multiplication				
Understands multiplication				
Can multiply 4-digit numbers by 1-digit numbers	N/A			
Knows multiplication facts				
Understands square numbers	N/A	N/A		
Division				
Understands division as inverse of multiplication	N/A	N/A		
Can solve division story problems with remainders	N/A	N/A		
Fractions				
Can find 1/2 and 1/4 of various quantities				
Can show the meaning of 3/4 as three 1/4s	N/A			
Can solve problems, such as 3/4 + __ = 1				
Understands fractions as a type of division	N/A	N/A	N/A	
Calculator				
Can solve multi-step problems				
Can estimate the answer				
Money				
Can make change for amounts less than one dollar				
Problem Solving				
Can solve problems in more than one way				
Geometry				
Can construct an equilateral triangle with drawing tools	N/A			
Understands line symmetry	N/A	N/A	N/A	
Knows terms polygon, hexagon, octagon, right angle, etc.	N/A	N/A	N/A	
Measurement				
Can tell time to the minute				
Can measure in inches	N/A	N/A		
Can measure in centimeters	N/A	N/A		
Can find perimeter	N/A	N/A		
Can find area in square inches or square cm	N/A	N/A		
Can construct and read bar graphs	N/A	N/A	N/A	
Patterns				
Can recognize and continue a simple pattern	N/A	N/A		

How This Program Was Developed

We have been hearing for years that Japanese students do better than U.S. students in math in Japan. The Asian students are ahead by the middle of first grade. And the gap widens every year thereafter.

Many explanations have been given, including less diversity and a longer school year. Japanese students attend school 240 days a year.

A third explanation given is that the Asian public values and supports education more than we do. A first grade teacher has the same status as a university professor. If a student falls behind, the family, not the school, helps the child or hires a tutor. Students often attend after-school classes.

A fourth explanation involves the philosophy of learning. Asians and Europeans believe anyone can learn mathematics or even play the violin. It is not a matter of talent, but of good teaching and hard work.

Although these explanations are valid, I decided to take a careful look at how mathematics is taught in Japanese first grades. Japan has a national curriculum, so there is little variation among teachers.

I found some important differences. One of these is the way the Asians name their numbers. In English we count ten, eleven, twelve, thirteen, and so on, which doesn't give the child a clue about tens and ones. But in Asian languages, one counts by saying ten-1, ten-2, ten-3 for the teens, and 2-ten 1, 2-ten 2, and 2-ten 3 for the twenties.

Still another difference is their criteria for manipulatives. Americans think the more the better. Asians prefer very few, but insist that they be imaginable, that is, visualizable. That is one reason they do not use colored rods. You can imagine the one and the three, but try imagining a brown eight–the quantity eight, not the color. It can't be done without grouping.

Another important difference is the emphasis on non-counting strategies for computation. Japanese children are discouraged from counting; rather they are taught to see quantities in groups of fives and tens.

For example, when an American child wants to know 9 + 4, most likely the child will start with 9 and count up 4. In contrast, the Asian child will think that if he takes 1 from the 4 and puts it with the 9, then he will have 10 and 3, or 13. Unfortunately, very few American first-graders at the end of the year even know that 10 + 3 is 13.

I decided to conduct research using some of these ideas in two similar first grade classrooms. The control group studied math in the traditional workbook-based manner. The other class used the lesson plans I developed. The children used that special number naming for three months.

They also used a special abacus I designed, based on fives and tens. I asked 5-year-old Stan how much is 11 + 6. Then I asked him how he knew. He replied, "I have the abacus in my mind."

The children were working with thousands by the sixth week. They figured out how to add four-place numbers on paper after learning how to do it on the abacus.

Every child in the experimental class, including those enrolled in special education classes, could add numbers like 9 + 4, by changing it to 10 + 3.

I asked the children to explain what the 6 and 2 mean in the number 26. Ninety-three percent of the children in the experimental group explained it correctly while only 50% of third graders did so in another study.

I gave the children some base ten rods (none of them had seen them before) that looked like ones and tens and asked them to make 48. Then I asked them to subtract 14. The children in the control group counted 14 ones, while the experimental class removed 1 ten and 4 ones. This indicated that they saw 14 as 1 ten and 4 ones and not as 14 ones. This view of numbers is vital to understanding algorithms, or procedures, for doing arithmetic.

I asked the experimental class to mentally add 64 + 20, which only 52% of nine-year-olds on the 1986 National test did correctly; 56% of those in the experimental first grade class could do it.

Since children often confuse columns when taught traditionally, I wrote 2304 + 86 = horizontally and asked them to find the sum any way they liked. Fifty-six percent did so correctly, including one child who did it in his head.

This following year I revised the lesson plans and both first grade classes used these methods. I am delighted to report that on a national standardized test, both classes scored at the 98th percentile.

Some General Thoughts on Teaching Mathematics

1. Ninety-five percent of mathematics should be understood; only five percent should be learned by rote.
2. Teaching with understanding depends upon building on what the child already knows. Teaching by rote does not care.
3. The role of the teacher is to encourage thinking by asking questions, not giving answers. Once you give an answer, thinking usually stops.
4. It is easier to understand a new model after you have made one yourself. For example, a child needs to construct graphs before attempting to read ready-made graphs.
5. Good manipulatives cause confusion at first. If the new manipulative makes perfect sense at first sight, it wasn't needed. Trying to understand and relating it to previous knowledge is what leads to greater learning, according to Richard Behr and others.
6. Lauren Resnick says, "Good mathematics learners expect to be able to make sense out of rules they are taught, and they apply some energy and time to the task of making sense. By contrast, those less adept in mathematics try to memorize and apply the rules that are taught, but do not attempt to relate these rules to what they know about mathematics at a more intuitive level."
7. According to Arthur Baroody, "Teaching mathematics is essentially a process of translating mathematics into a form children can comprehend, providing experiences that enable children to discover relationships and construct meanings, and creating opportunities to develop and exercise mathematical reasoning."
8. Mindy Holte puts learning the facts in proper perspective when she says, "In our concern about the memorization of math facts or solving problems, we must not forget that the root of mathematical study is the creation of mental pictures in the imagination and manipulating those images and relationships using the power of reason and logic."
9. The only students who like flash cards are those who don't need them.
10. Mathematics is not a solitary pursuit. According to Richard Skemp, solitary math on paper is like reading music, rather than listening to it; "Mathematics, like music, needs to be expressed in physical actions and human interactions before its symbols can evoke the silent patterns of mathematical ideas (like musical notes), simultaneous relationships (like harmonies) and expositions or proofs (like melodies)."
11. "More than most other school subjects, mathematics offers special opportunities for children to learn the power of thought as distinct from the power of authority. This is a very important lesson to learn, an essential step in the emergence of independent thinking." (A quote from *Everybody Counts*)

12. Putting thoughts into words helps the learning process.
13. The difference between a novice and an expert is that an expert catches errors much more quickly. An expert violinist adjusts pitch so quickly that the audience does not hear it.
14. Europeans and Asians believe learning occurs not because of ability, but primarily because of effort. In the ability model of learning, errors are a sign of failure. In the effort model, errors are natural. In Japanese classrooms, the teachers discuss errors with the whole class.
15. For teaching vocabulary, be sure either the word or the concept is known. For example, if a child is familiar with six-sided figures, we can give him the word, hexagon. Or, if he has heard the word, multiply, we can tell him what it means. It is difficult to learn a new concept and the term simultaneously.
16. Introduce new concepts globally before details. This lets the children know where they are headed.
17. Informal mathematics should precede paper and pencil work. Long before a child learns how to add fractions with unlike denominators, she should be able to add one half and one fourth mentally.
18. Some pairs of concepts are easier to remember if one of them is thought of as dominant. Then the non-dominant concept is simply the other one. For example, if even is dominant over odd; an odd number is one that is not even.
19. Worksheets should also make the child think. Therefore, they should not be a large collection of similar exercises, but should present a variety.
20. In Japan students spend more time on fewer problems. Teachers do not concern themselves with attention spans as is a concern in the U.S.
21. In Japan the goal of the math lesson is that the student has understood a concept, not necessarily has done something (a worksheet).
22. The calendar should show the entire month, so the children can plan ahead. The days passed can be crossed out or the current day circled.
23. A real mathematical problem is one in which the procedures to find the answer or answers are not obvious. It is like a puzzle, needing trial and error. Emphasize the satisfaction of solving problems and the responsibility of not giving away the solution to others.
24. Keep math time enjoyable. A person who dislikes math will avoid it. We store our emotional state along with what we've learned. A child under stress stops learning. If a lesson is too hard, end it and play a game. Try again another day.

RightStart™ Mathematics

There are 13 major characteristics that make this research-based program effective.

1. Refers to quantities of up to 5 as a group; discourages counting individually.
2. Uses fingers and tally sticks to show quantities up to 10; teaches quantities 6 to 10 as 5 plus a quantity, for example 6 = 5 + 1.
3. Avoids counting procedures for finding sums and remainders. Teaches five- and ten-based strategies for the facts that are both visual and visualizable.
4. Employs games, not flash cards, for practice.
5. Once quantities 1 to 10 are known, proceeds to 10 as a unit. Uses the "math way" of naming numbers for several months; for example, "ten-1" for eleven, "ten-2" for twelve, "2-ten" for twenty, and "2-ten 5" for twenty-five.
6. Uses expanded notation (overlapping) place-value cards for recording tens and ones; the ones card is placed on the zero of the tens card. Encourages a child to read numbers starting at the left and not backward by starting at the ones column.
7. Proceeds rapidly to hundreds and thousands using manipulatives and place-value cards. Provides opportunities for trading between ones and tens, tens and hundreds, and hundreds and thousands with manipulatives.
8. Only after the above work, about the fourth month of first grade, introduces the traditional English names for quantities 20 to 99 and then 11 to 19.
9. Teaches mental computation. Investigates informal solutions, often through story problems, before learning procedures.
10. Teaches four-digit addition on the abacus, letting the child discover the paper and pencil algorithm. This occurs in Level B. Four-digit subtraction is mastered in Level C.
11. Introduces fractions with a linear visual model, including all fractions from $\frac{1}{2}$ to $\frac{1}{10}$. "Pies" are not used initially because they cannot show fractions greater than 1. Later, the tenths will become the basis for decimals.
12. Approaches geometry through drawing boards and tools.
13. Teaches short division (where only the answer is written down) for single-digit divisors, before long division. Both are taught in Level E.

Some Pointers

Kindergarten Level A. Most of these lesson plans have two distinct topics, so each lesson has enough activities for two days. Or, if more appropriate for the child, complete two lessons per week.

Transition Lessons. These lessons are designed for children starting Levels C, D, or E who have not been doing RightStart™ Mathematics previously. The lessons need to be completed before the regular manual. The manual tells which lessons to complete before beginning each level .

Counting. Counting needs to be discouraged because it is slow and inaccurate. It also interferes with understanding quantity and learning place-value.

Warm-up. The warm-up is the time for quick review, memory work, and sometimes an introduction to the day's topics. The drawing board, which doubles as a dry erase board, makes an ideal slate for quick responses.

Place value. In order to understand addition algorithms, place-value knowledge is essential. From the very beginning, the children are helped to see quantities grouped in fives and tens. Children can understand place value in first grade and even in kindergarten when it is approached as it is in this program.

Worksheets. The worksheets are designed to give the children a chance to think about and to practice the day's lesson. The children are to do them independently. Some lessons, especially in the early grades, have no worksheet.

Games. Games, not worksheets or flash cards, are used for practice. They can be played as many times as necessary until memorization takes place. Games are as important to math as books are to reading.

Some games are incorporated in this manual. Extra games, found in the book, *Math Card Games,* are suggested in the Review and Practice lessons in Levels C to E. There are games for the child needing extra help, as well as for the advanced child.

Teaching. Establish with the children some indication when you want a quick response and when you want a more thoughtful response. Research shows that the quiet time for thoughtful response should be about three seconds. Avoid talking during this quiet time; resist the temptation to rephrase the question. This quiet time gives the slower child time to think. It also gives a quicker child time to think more deeply.

Encourage the child to develop perseverance. Avoid giving answers too quickly. Children tend to stop thinking once they hear the answer.

Help the children realize that it is their responsibility to ask questions when they do not understand. Do not settle for "I don't get it."

Number of lessons. It is not necessary that each lesson be done in one day. Sometimes two days may be more appropriate. However, do complete each manual in full before going on to the next one.

Visualization. The ability to imagine or visualize is an important skill to develop in mathematics and other subjects as well. Often you are called upon to suggest to the children that they imagine a particular topic.

Questions. I really want to hear how this program is working. Please let me know any improvements and suggestions that you may have.

Joan A. Cotter, Ph.D.

JoanCotter@RightStartMath.com

www.RightStartMath.com

Table of Contents
Level D

Level D–page 2

Lesson 47	Using Check Numbers
Lesson 48	**Review and Practice**
Lesson 49	Finding Check Numbers
Lesson 50	Check Numbers and Multiples of Three
Lesson 51	The "Almost" Subtraction Strategy
Lesson 52	Terry's Subtraction Strategy
Lesson 53	Working With Twos
Lesson 54	**Review and Practice**
Lesson 55	Working With Fives
Lesson 56	Telling Time to the Minute
Lesson 57	Telling Time Practice
Lesson 58	Multiplying With Money
Lesson 59	Multiplying With 1s and 0s
Lesson 60	**Review and Practice**
Lesson 61	Multiplication Problems
Lesson 62	The Multiplication Table
Lesson 63	Working With Threes
Lesson 64 (1 or 2 days)	Representing Thousands
Lesson 65	Reviewing Place Value
Lesson 66	**Review and Practice**
Lesson 67	Working With Fours
Lesson 68	Representing Thousands
Lesson 69	Reading and Writing Large Numbers
Lesson 70 (1 or 2 days)	Working With Large Numbers
Lesson 71	Working With Nines
Lesson 72	**Review and Practice**
Lesson 73	Multiplying and Adding
Lesson 74	Multiplying by a One-Digit Number
Lesson 75	Introducing Area
Lesson 76	Working With Square Inches
Lesson 77	Working With Sixes
Lesson 78	**Review and Practice**
Lesson 79	Working With Centimeters
Lesson 80	Finding Areas
Lesson 81	Area Problems
Lesson 82	Working With Eights
Lesson 83	Multiplying Three Numbers
Lesson 84	**Review and Practice**
Lesson 85	Arrays of Cubes
Lesson 86	Working With Sevens
Lesson 87	Seeing Patterns
Lesson 88	Patterns With Squares
Lesson 89	A Squares Pattern
Lesson 90	**Review and Practice**
Lesson 91	Continuing the Pattern
Lesson 92	The Distributive Law
Lesson 93	Square Inches in a Square Foot
Lesson 94	Multiplying by Two Digits
Lesson 95	The Multiplication Algorithm

Level D–page 3

Lesson 1

The Months of the Year

OBJECTIVES

1. To name, write, and compare the months of the year
2. To write dates several ways
3. To discuss reading the year

MATERIALS

Math journal (found at the end of the worksheets) or the Dry Erase drawing board

A book about calendars or seasons to be read to the child, optional

WARM-UP

Ask the child to recite the months in order and say "Thirty Days has September."

> **Note:** The warm-up activities have several purposes. They provide review and practice of concepts previously learned.

Thirty Days has September
Thirty days has September,
April, June, and November.
The rest have thirty-one to carry,
But only twenty-eight for February,
Except in leap year, that's the time
When February has twenty-nine.

ACTIVITIES

> **Note:** Discussing spelling in math class can help the child integrate the various subjects.

Days in a month. Ask, How many months are in a year? [12] Then ask the child to write the numbers 1 to 12 in a column in her math journal and to write the months in order in another column. You might also use this opportunity to discuss the spelling of the months, all of which are phonetic when pronounced clearly. Also ask her to write the number of days in the month in the third column as shown below.

1	January	31
2	February	28
3	March	31
4	April	30
5	May	31
6	June	30
7	July	31
8	August	31
9	September	30
10	October	31
11	November	30
12	December	31

The number of the month, the month, and the number of days in the month.

How many months have 30 days? [4] What are they? [April, June, September, and November] How many months have 31 days? [7] What are they? [January, March, May, July, August, October, and December] Have we named all the months? [no, not February] Ask her if she can explain about February. [28 days until leap year when it has 29 days]

Writing dates. Ask the child to write today's date in her journal or on the drawing board. Discuss writing the month first, followed by the day, and the year. See the example on the next page.

September 1, 2008	Sep. 1, 2008	9-1-08

Various ways to write the date.

Also ask her to write the date by using abbreviations. Explain that the abbreviation usually is the first three letters of the month. See the second example above.

Next explain that often we use a number for the months. Ask her to look at her list and find the number for September. [9] Demonstrate writing the date as shown above on the right. (Slashes, "/," are sometimes used, but may be confused with a 1.)

For practice in writing dates, ask the child to write the following dates all three ways, or choose other suitable dates:

1. The date for the first day of the next month
2. The date for the last day in September
3. The date for the next day
4. The date for one week from today
5. The date of her last birthday
6. The date for the next New Year's Day

Reading the year. Write 1900 and ask the child for two ways to read the number. [1 thousand 9 hundred, or 19 hundred] Ask how she knows it is 19 hundred. [2 zeroes after the 19] Write the year, for example, 1998, and explain that it really is 19 *hundred* 98, but most of the time we skip the word *hundred* and just say *nineteen ninety-eight.* The year 2000 is pronounced *two thousand* and 2005 is *two thousand five.*

Point to the year and ask the child to read it as a number. [for example, 1998 is 1 thousand 9 hundred ninety-eight]

Tell her that people have used other ways of naming days and years. Ask her what a year is. [the time that the earth takes to go around the sun] Read, if available, a book about calendars or the seasons.

ENRICHMENT Tell the child that in some countries, like Sweden and Germany, calendars begin on Monday, not Sunday. Also in Germany, the numbers go *down,* not across.

Also explain that in other countries, including Europe, and sometimes in the United States, people write the day of the month before the month as shown below.

1 September 2008	1 Sep 2008	1-9-08

The way Europeans write the date.

That makes sense to them because they write in order, from the smallest unit to the largest: day, month, and year.

Lesson 2

Calendar for One Year

OBJECTIVES
1. To use a calendar to gain information
2. To solve calendar problems

MATERIALS
Math journal
A calendar (immediately preceding Worksheet 1-1)

WARM-UP
Ask the child to count by 2s to 20, 5s to 50, and 10s to 100.

Ask the child to recite the months in order and to sing Thirty Days has September."

Thirty Days has September
Thirty days has September,
April, June, and November.
The rest have thirty-one to carry,
But only twenty-eight for February,
Except in leap year, that's the time
When February has twenty-nine.

Ask the child to say the days of the week.

ACTIVITIES

Note: In RightStart Level B, the child made calendars, so calendars are not new.

Introducing the year's calendar. Ask the child to find the current month on the calendar. Give him time to study it. Ask him to circle today's date. Also ask him to find other dates of importance to him, such as his birthday, his family members' birthdays, and holidays.

Ask him what *S M T W T F S* means. [the days of the week] Ask him to write the days of the week. You might want to talk about the spelling. Tell him that in English, the months always start with a capital letter, but that is not true in all languages.

Pronouncing the last syllable as *day,* rather than *dee* helps in spelling. Sunday and Friday are simple. In Monday the first vowel sound is spelled with an *o,* the same as in *month.* In Thursday and Saturday, the "er" sound is spelled with "ur." In Tuesday the vowel sound is made with "ue." Wednesday has two silent letters, the first *d,* and the second *e.*

Ask if any months are the same. [In non-leap years, January and October are identical; also several pairs start on the same day: February & March & November, April & July, and September & December. In leap years, January & July are identical with April also starting the same day.]

In leap years pairs starting on the same day include February & August, March & November, September & December and January & April & July.

Problem 1. What day of the week is New Year's Day of the next year? How do you know? [It is the next day after December 31st.]

Problem 2. Tell the child that Thanksgiving in the United States is the fourth Thursday in November; ask him to find the date.

Then tell him that in Canada, Thanksgiving is the second Monday in October. What is that date?

Problem 3. Tell the child the scouts (or other club) meet on the first Monday of the month. How many times will the club meet in the year? [12] Ask him to list all the dates for the year. Encourage him to use all numbers for the dates.

Problem 4. Tell the child that another club decided to meet on the fourth Tuesday of the month. How many times will the club meet in the year? [12] Ask him to write the dates.

Problem 5. Next tell the child that another club decided to meet on the fifth Wednesday of the month. How many times will the club meet in the year? [It varies.] Ask him to write the dates.

Problem 6. What is the date one week after July 4th [July 11th], What is the date two weeks after July 4th? [July 18th] How could you tell without looking at the calendar? [add 7 or 14]

Problem 7. Tell the child that Memorial Day is the last Monday in May and Labor Day is the first Monday in September. You might tell him that many people think that Memorial Day is the beginning of summer and Labor Day is the end of summer. You could also ask what the holidays mean to him. The problem is: How many days are between Memorial Day and Labor Day? [98 days when Memorial Day falls on May 26 to 31 and 105 days when Memorial Day falls on May 25]

Note: The calendar will be used in the next lesson.

When he has finished, ask him to show his solution and explain how he arrived at that solution. Ask if he could solve it another way. This problem, for example, can be solved by counting the days, or by counting the weeks by 7s, or by adding the number of days in June, July, and August and then adding the extra days.

ENRICHMENT

Problem 8: golden birthdays. Explain that some people celebrate golden birthdays. A golden birthday happens when a person's age is the same as the day of the month as their birthday. For example, a person whose birthday is September 9 has a golden birthday on their 9th birthday.

At what age will a baby born on July 2 have a golden birthday? [age 2] If the baby was born July 2 of this year, when will the baby have a golden birthday? [2 years] What year will a person born February 22, 1990, have a golden birthday? [1990 + 22 = 2012] Encourage the child to solve it mentally. [1990 + 10 = 2000 +12 = 2012] Ask if he has already had his golden birthday. Ask him to find the year of his golden birthday.

For homework, ask him to find the year of the golden birthday of another person.

Lesson 3

Calendars for the Following Years

OBJECTIVES

1. To become aware of similarities and differences of calendars for two consecutive years
2. To solve more calendar problems

MATERIALS

Math journal
AL abacus, optional
2 calendars, current year and the following year (immediately preceding Worksheet 1-1)
Worksheet 1-2, "Calendar Problems"
Worksheet 1-1, "Reviewing Skip Counting"

WARM-UP

Ask the child to count by 3s to 30, 6s to 60, and 9s to 90. If desired, ask him to move the beads on the abacus as he counts. See below. The 6s are every other 3; combine 2 groups of 3s to make the 6s. The 9s are every third 3.

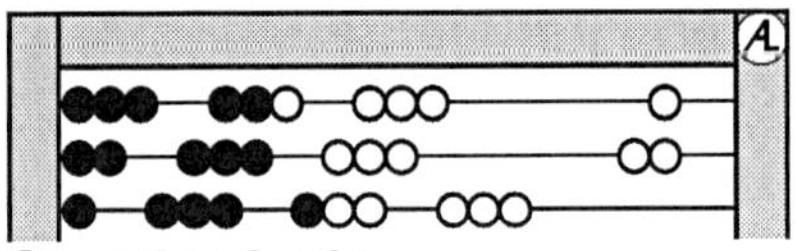

Counting by 3s.

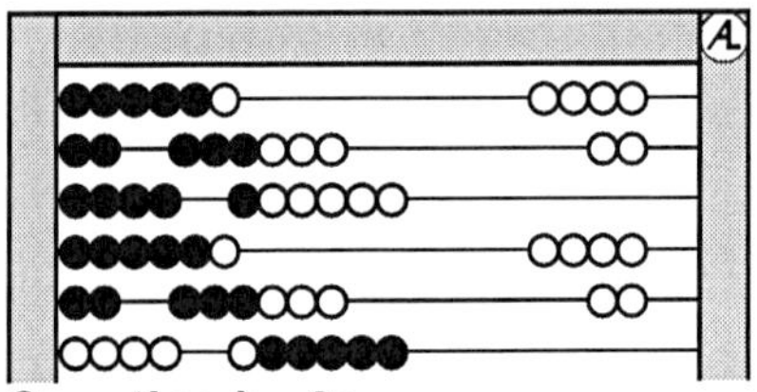

Counting by 6s.

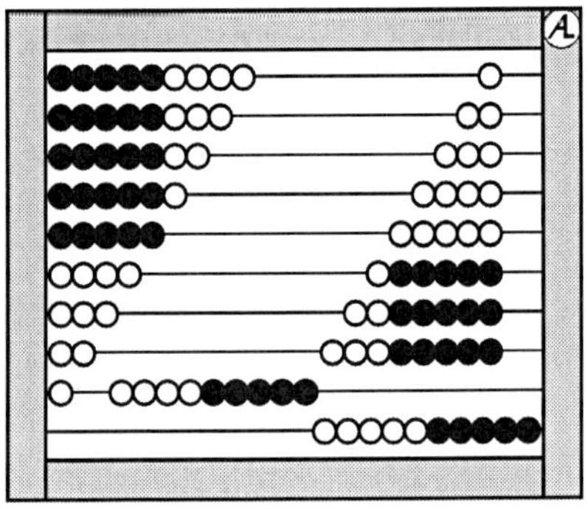

Counting by 9s.

Ask the child to recite "Thirty Days has September" (Lesson 2).

Ask the child to recite the months in order and to say the number of days it has. You might want her to write this information also.

ACTIVITIES

The next year's calendar. Ask the child to find the next year's calendar. How are they the same? [Each month has the same number of days, except possibly February.] How are they different? [The months start on a different day.] Ask him the following or similar questions:

1. Is either year a leap year?
2. Can 2 years in a row be leap years? [No, we have leap years only every 4 years.]
3. What day of the week is New Year's Day in both years?
4. What day of the week is July 4th in both years?
5. Find today's date in the next year. What happens to the day of the week?
6. What days are her birthdays in both years?
7. Ask her to mark the school days, or her vacation days.
8. What day of the week is Thanksgiving this year? Next year?

[Thursday] What is this year's date? Next year's date? How does it change? What would happen in the next several years? Remind them that it must be the fourth Thursday of November.

Problem solving. To solve a problem, ask the child to read it carefully, at least twice. Next she needs to be sure she knows what is being asked. Give her three minutes of uninterrupted time to work on it. Then she can discuss it if she wants. Be sure she can explain her solution.

Problem 1. How many days are left in this year? Find the answer without counting.

Problem 2. How many weeks are in a year? [52 weeks and 1 or 2 days left over] How many weeks are left in this year? If each month had exactly 4 weeks, like February does most of the time, how many months would be in a year? [13, 40 weeks would be 10 months and 12 more weeks would be 3 more months]

Problem 3. How many days from your birthday this year to your birthday the next year? [365, or 366]

Problem 4. If a baby is 3 months old, how many days ago was it born? [around 90] Discuss that *3 months* often means about 3 months.

Note: The child will need this worksheet in future lessons.

Worksheet 1-2. In Level C the children worked extensively with skip counting. This worksheet asks for skip counting in a format that emphasizes the patterns. They are given below.

Note: Children having difficulties with these skip counting patterns should play Multiples Memory (P2) found in *Math Card Games* by Joan A. Cotter.

Multiples of 2

2	4	6	8	10
12	14	16	18	20

Multiples of 4

4	8	12	16	20
24	28	32	36	40

Multiples of 6

6	12	18	24	30
36	42	48	54	60

Multiples of 8

8	16	24	32	40
48	56	64	72	80

Multiples of 9

9	18	27	36	45
90	81	72	63	54

Note: Nines are written with the second line in reverse to show the digits reversing.

Multiples of 3

3	6	9
12	15	18
21	24	27
30		

Multiples of 7

7	14	21
28	35	42
49	56	63
70		

Multiples of 5

5	10
15	20
25	30
35	40
45	50

Lesson 4

Birthday Graphs

OBJECTIVES

1. To construct a bar graph
2. To review an inch

MATERIALS

Paper, at least 36 inches wide and 24 inches long for making a large graph. Draw vertical columns 2 inches wide with 3/4 inch between the columns as shown below. Leave space at the bottom for writing the months of the year and for a title. See below.

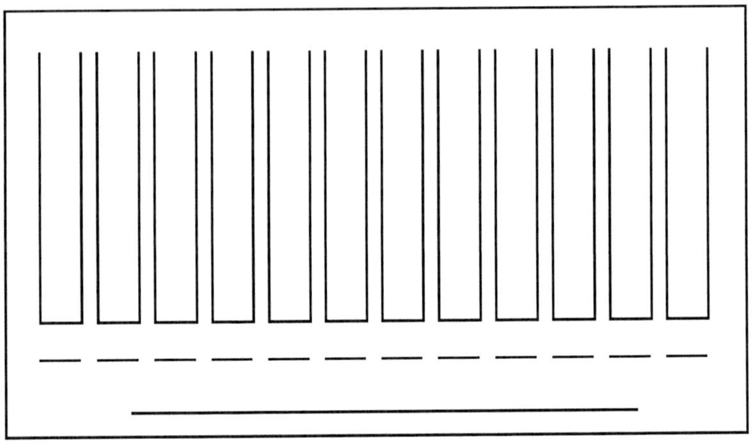

A large piece of paper for graphing the number of birthdays in each month.

Ruler
Crayons

WARM-UP

Say the months in order. Name the months with 31 days. [January, March, May, July, August, October, and December]

Write the following and ask the child to write the answers:

35 + 7 = [42]   73 + 8 = [81]   55 + 10 = [65]   55 + 9 = [64]

Discuss how he found his answers. Discourage counting.

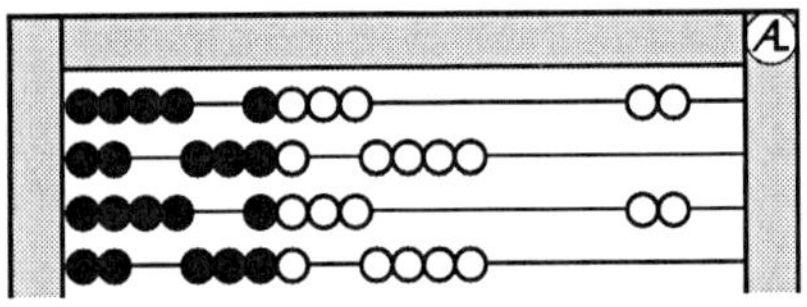

Counting by 4s.

Ask him to count by 4s to 40. See the figure at the left.

ACTIVITIES

One inch on a ruler. Ask the child to draw on a piece of paper a line that he thinks is one inch long. Take out the ruler and discuss where he can find an inch on the ruler. Emphasize it is the space between the long marks near any 2 numbers. Ask if his lines are shorter or longer than an inch.

Note: Emphasizing the space between any two numbers may help the child realize what he is measuring.

Draw a horizontal line about 6 inches long and ask the child to mark it into inches. Note if the spaces between the marks look the same. [They should.]

A horizontal line marked in inches.

Repeat for a vertical line.

Building a graph. Ask the child which month has the greatest number of family and friends' birthdays and the least number of birthdays. Discuss if he could find out by using a graph. This could be a family or homeschool group project, which can be added to over the period of several days.

Show the child the prepared graph paper. Ask him to write the months of the year or their abbreviations below each column as shown below. Help him decide on a good title. Suggest he use one inch of space to show a birthday for a particular month. Then he can color in the rectangle.

Demonstrate with your own birthday month. Measure 1 inch along both sides, draw a short tick mark, and draw the horizontal line. A September birthday is shown below.

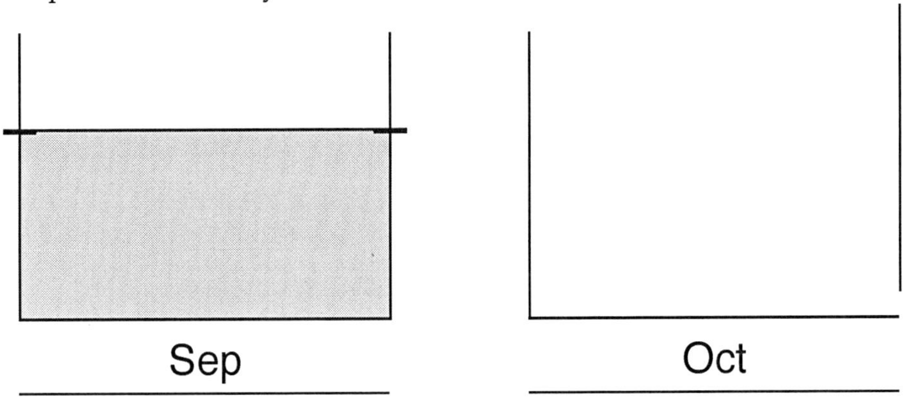

Demonstrating recording a birthday for the month of September.

Ask the child to enter his birthday in the appropriate month. Ask him to enter other birthdays. They could include his family. They could also include his friends. If he does not know the birthday months of some friends, tell him he can enter them the following days. Other family members may also add birthdays.

Individual graphs. Next the child can make an individual graph in his math journal. Starting on a clean page, he needs to draw a horizontal line 2 squares from the bottom and the vertical lines for the columns.

There is space to write only the first letter of the month below the columns. Ask him to record the birthday months of his family and friends. He might want to write the initials in the squares. See the figure below.

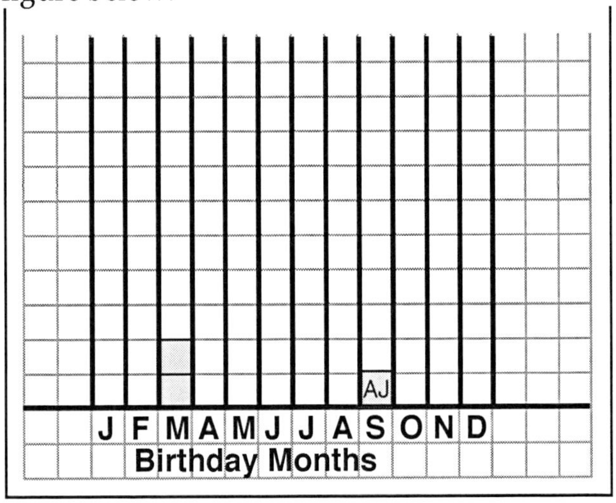

A birthday month graph done in a math journal.

Lesson 5

Even Numbers for Sums

OBJECTIVES
1. To review even and odd numbers
2. To learn the term *multiple*

MATERIALS Worksheet 1, "Reviewing Skip Counting," done previously
1-inch tiles

WARM-UP Ask the child to recite the days of the week backward.

Give the child time to enter more birthdays on the large graph started in Lesson 4.

Ask her to practice skip counting by 3s, 6s, and 9s.

ACTIVITIES ***Reviewing even numbers.*** Ask the following questions.

1. What is an even number and how do you know? [a number with a partner, a multiple of 2]

2. Is 14 an even number? [yes] Ask the child to prove it to you using the tiles. See the figure below.

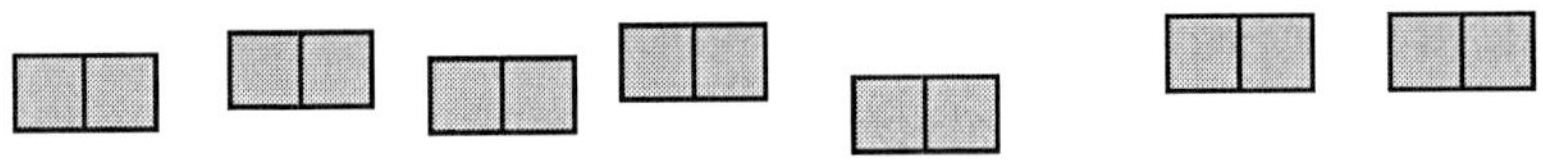

3. What do you call a number that is not even? [odd]

> **Note:** Avoid teaching the rote rule that a number is even or odd depending upon the ones digit. The child needs to understand it first.

4. How many letters are in the word *even;* is it even or odd? [4, even]

5. How many letters are in the word *odd;* is it even or odd? [3, odd]

6. Is 13 an even number? [no] Ask the child to prove it to you. See the figure below.

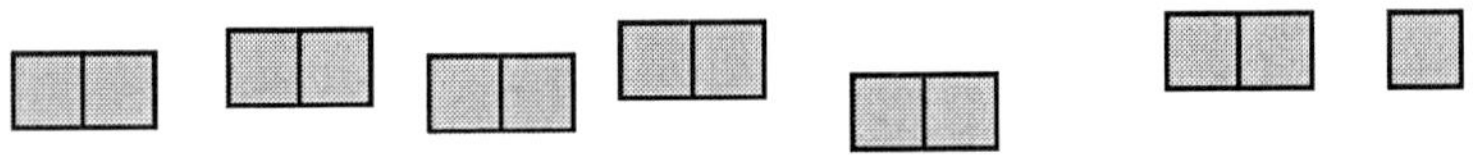

7. What kind of a number do you get when you add 2 even numbers, such as 8 and 4? [even]

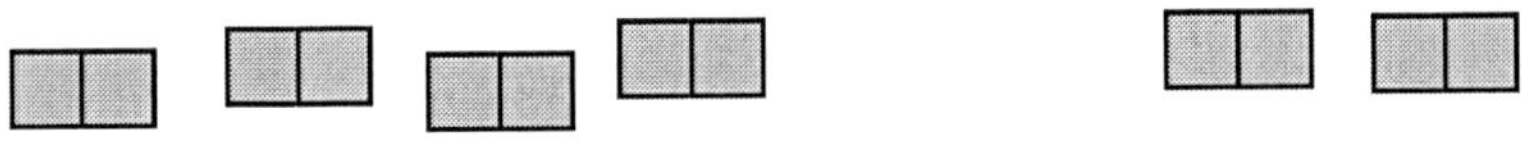

8. What kind of a number do you get when you add 2 odd numbers, such as 7 and 5? [even, the 2 extras form a group of 2.]

9. Is 42 even or odd and how do you know? [even, 10 and 10 and 10 and 10 and 2 are all even] Ask for a demonstration.

10. Is 63 even or odd and how do you know? [odd, the tens are all even, but 3 is odd.]

Multiples. Explain that 2, 4, 6, and 8 are called *multiples* of 2 and that 10, 20, 30, and 40 are multiples of 10.

Worksheet. Ask the child to circle all the even numbers on their multiples (skip counting) worksheet. See the solutions below. What patterns do you see? [Multiples of even numbers are always even. Multiples of odd numbers alternate.]

2	4	6	8	10
12	14	16	18	20

4	8	12	16	20
24	28	32	36	40

6	12	18	24	30
36	42	48	54	60

8	16	24	32	40
48	56	64	72	80

9	18	27	36	45
90	81	72	63	54

3	6	9
12	15	18
21	24	27
30		

7	14	21
28	35	42
49	56	63
70		

5	10
15	20
25	30
35	40
45	50

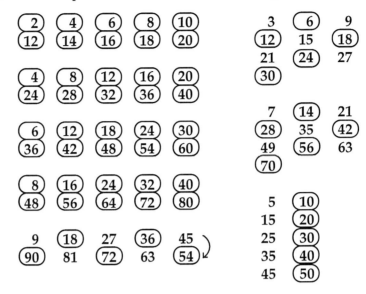

Sums with even numbers. Write

8 =

and ask the child to finish the equation using only **even** numbers. When she has found one solution, write "8 = " again and ask if she can find another solution. Note for this problem 8 = 6 + 2 may be assumed to be the same as 8 = 2 + 6. Continue until she finds all 4 solutions. You might need to remind her that she can use more than 2 numbers on the right.

Note: The child has used equations, such as 8 = 4 + 4, in earlier levels.

Note: The term *equation* should be used, not "number sentence." The word *sentence* is a grammatical, not mathematical, term according to the dictionary.

8 = 6 + 2
8 = 4 + 4
8 = 4 + 2 + 2
8 = 2 + 2 + 2 + 2

Now ask the child to write all the sums for 12 = in her journal using only even numbers. The solutions are below. Stress some kind of order to keep track of her work.

12 = 10 + 2
12 = 8 + 4
12 = 8 + 2 + 2
12 = 6 + 6
12 = 6 + 4 + 2
12 = 6 + 2 + 2 + 2
12 = 4 + 4 + 4
12 = 4 + 4 + 2 + 2
12 = 4 + 2 + 2 + 2 + 2
12 = 2 + 2 + 2 + 2 + 2 + 2

Homework. Ask the child to write in her journal all the ways to make sums for 10 = using even numbers. To help her remember the assignment, ask her to write "10 = " six times in her journal. Solutions are listed below.

10 = 8 + 2
10 = 6 + 4
10 = 6 + 2 + 2
10 = 4 + 4 + 2
10 = 4 + 2 + 2 + 2
10 = 2 + 2 + 2 + 2 + 2

Lesson 6

Reviewing Addition Strategies and Facts

OBJECTIVES
1. To review addition strategies
2. To review the addition combinations

MATERIALS Worksheet 2, "Reviewing Addition"

WARM-UP Ask the child to recite the months in order and to write *January, February,* and *March,* spelling them correctly.

Write the following and ask the child to write the answers:

21 + 9 = [30]   28 + 5 = [33]   16 + 7 = [23]   87 + 9 = [96]

Discuss how he found his answers. Discourage counting.

ACTIVITIES ***Reviewing strategies.*** Write

7 + 5 =

and ask the child how he could figure it out without counting if he did not know. Encourage several responses. One possibility is to think that 3 is needed with 7 to make 10; when 3 is taken from the 5, 2 remains, giving 10 and 2, or 12. Then the 3 combines with the 7 to make 10 and 2, or 12. Another way is to think that 7 is 5 and 2; that 5 and the other 5 combine to make 10; so 10 and 2 is 12.

Note: These strategies also work when adding two-digit numbers mentally.

Then write

4 + 5 =

and ask for strategies. One strategy is to realize he knows adding low fives from looking at his hands. One hand and 4 more is 9. Another strategy involves knowing 5 + 5 = 10, then 4 + 5 must be 1 less, or 9.

Next ask him how he could find

8 + 9 =

One way is to take 1 from the 8 and combine it with the 9, getting 7 and 10, or 17. Another way is take 2 from the 9 and giving it to the 8, leaving 10 and 7, or 17. A third way is to employ doubles; if one knows 8 + 8 = 16, then 8 + 9 must be 1 more, or 17. A fourth way is think that 8 is 5 and 3 and 9 is 5 and 4. The two 5s make 10 and the leftover numbers 3 and 4 give 7, so the answer is 10 and 7, or 17.

Ask the child to name addition combinations that are hard for him. Discuss several strategies that could help.

Timed tests. Timed tests where the child must quit after so many minutes often causes the child unnecessary stress. A child not finishing the page feels like a failure. Many adults blame these tests as one cause of their math anxiety. Another serious problem is that a child still insecure with strategies will regress to counting to find the sums, further delaying his mastery of the facts.

Note: Research shows that a child under stress stops learning.

A better way to introduce timing is to ask the child to time himself and write down his time. This way the child finishes the page and works to better his time, similar to running a race.

Worksheet. Explain to the child that this worksheet reviews almost all of the basic addition facts. (Some of the simpler facts with a number + 1 and 1 + a number are missing.)

Before the child writes on the sheet, tell him he will have about 10 minutes to practice it orally. What kinds of things do people practice? [sports, music] Why do people practice? [to get better or faster] Practice must be done correctly. For math facts that means using strategies, not counting.

Explain the following procedure–it might help to demonstrate it. Look at the problem and enter the problem on the calculator, but do not press the equals key. Then softly say the sum and press the equals key to see if it is correct. If his oral answer does not agree with the calculator, discuss what strategy he used.

After the practice, explain that he is to start writing on his paper when you say, Start. Tell them to raise his hand when he has finished and you will write the time, which he is to copy onto his paper.

Note: A good time for doing these is about 5 minutes.

After he has finished, read off the answers, given below, and ask him to mark the wrong answers, perhaps with a colored pen.

31 + 7 = **38**	98 + 5 = **103**	63 + 2 = **65**	15 + 2 = **17**
39 + 2 = **41**	73 + 8 = **81**	56 + 3 = **59**	33 + 9 = **42**
94 + 2 = **96**	46 + 8 = **54**	28 + 1 = **29**	23 + 6 = **29**
38 + 9 = **47**	52 + 9 = **61**	29 + 7 = **36**	81 + 8 = **89**
48 + 7 = **55**	76 + 4 = **80**	98 + 3 = **101**	88 + 8 = **96**
89 + 9 = **98**	47 + 9 = **56**	15 + 5 = **20**	14 + 5 = **19**
34 + 8 = **42**	65 + 9 = **74**	36 + 2 = **38**	73 + 4 = **77**
59 + 4 = **63**	74 + 3 = **77**	62 + 6 = **68**	66 + 6 = **72**
99 + 5 = **104**	29 + 8 = **37**	41 + 9 = **50**	27 + 7 = **34**
95 + 6 = **101**	22 + 5 = **27**	62 + 3 = **65**	17 + 2 = **19**
87 + 8 = **95**	73 + 7 = **80**	59 + 1 = **60**	61 + 6 = **67**
38 + 2 = **40**	82 + 8 = **90**	24 + 6 = **30**	17 + 5 = **22**
57 + 4 = **61**	47 + 3 = **50**	53 + 5 = **58**	32 + 7 = **39**
91 + 5 = **96**	45 + 4 = **49**	54 + 4 = **58**	82 + 2 = **84**
84 + 7 = **91**	24 + 9 = **33**	48 + 6 = **54**	99 + 3 = **102**
27 + 6 = **33**	73 + 3 = **76**	86 + 5 = **91**	55 + 7 = **62**
16 + 9 = **25**	15 + 3 = **18**	55 + 8 = **63**	82 + 4 = **86**
41 + 4 = **45**	59 + 6 = **65**	58 + 4 = **62**	76 + 7 = **83**

Homework. Ask him to think of 2 or 3 facts that are hard for him and to write them down. For homework he is to think of strategies that could help him remember them and explain them in his journal.

REMEDIAL A child having difficulties with the facts needs to understand he must work extra hard to catch up. See the Level C manual for practice sheets.

Lesson 7

Working With Sums

OBJECTIVE 1. To use addition facts to build sums

MATERIALS 1 set of 10 slips of paper with the following digits: 1, 2, 5, 5, 6, 6, 7, 8, 9, and 0 (Appendix pg. 1)
Math journal

WARM-UP Write the following. Remind the child to add the 10 first and then the 5 when adding 15. For example, 42 + 15 = 42 + 10 [52] + 5 [57].

42 + 10 = [52]	42 + 15 = [57]	38 + 10 = [48]	38 + 15 = [53]
25 + 15 = [40]	40 + 15 = [55]	55 + 15 = [70]	97 + 15 = [112]

Then ask the child to write the answers.

What happens when an even number is added to an even number? [even] even plus odd? [odd] odd plus odd? [even]

ACTIVITIES ***Reviewing addition.*** Tell the child to find the number of days in a whole year. Ask her to show the addition. Ask about the carrying.

```
  1
 31
 28
 31
 30
 31
 30
 31
 31
 30
 31
 30
+31
365
```

Adding the days in the months to find the number of days in a year.

Problem. Write the following problem for the child to solve.

In Spring City there is a factory with shifts 1, 2, and 3; 1308 persons work the shifts. In the first shift there are 459 workers and in the second shift there are 487 workers. How many workers are in the third shift? [362 workers]

Ask the child to solve it by herself. Then ask her to share her work with you. Tell her that she should explain what she did and listen very carefully to your questions or suggestions. Play devil's advocate and ask her questions about carrying, or how the subtraction was done. Tell her it is all right to change her work, but she must agree and understand.

```
  459     1308
+ 487    - 946
  946      362
```

Solving the number of workers in the third shift problem.

Note: A child may subtract starting at the left, like division. Do not attempt to teach a new algorithm to a child who has one that works.

Rally behind her if needed. Also ask if she could solve the problem in a different way.

Building sums. Explain to the child that someone had a deck of digit cards and took the 10 top digits. The ones she took were the following. Take out the numbers.

Note: Just as we use letters to make words, the digits 0 to 9 are the symbols we use to make numbers.

1	2	5	5	6	6	7	8	9	0

The 10 cards with digits.

Explain that she is to use the digits on the slips of paper to make a column of numbers and their sum. A number is not used to show the carries. <u>What is a sum?</u> [the answer you get when you add]

Give her some examples, such as the following.

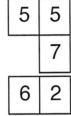

5	5
	7
6	2

5	2	1
	6	6
5	8	7

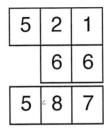

Two solutions using several digits.

Also give her the following example with an error; ask if it is a solution. Commend her for finding the error.

Note: This type of problem solving involves guessing and checking–and persevering.

		9
	2	5
6	8	0
7	1	5

A WRONG solution using 7 digits.

Note: Ask the child to write the date before starting new work in her journal or on worksheets.

Ask her to write her solutions in her journal. Ask her to use as many of the digits as possible for each sum. It is possible to use all ten digits. (Zeroes may not be used alone or preceding a number.) Allow time for the child to show some of her solutions.

Homework. If desired, ask the child to draw a line under today's work and to find several more for homework.

Lesson 8

Corners Game

OBJECTIVES 1. To review the addition facts totaling 5, 10, and 15
2. To mentally add 5, 10, 15, and 20

MATERIALS Math journal for scoring
Corner Cards

WARM-UP Ask the child to recite the months in order.

Write the following and ask the child to write the answers:

175 + 5 [180]   185 + 10 [195]   150 + 15 [165]   105 + 15 [120]

Discuss how he found his answers. Stress adding the tens first when they add the 15.

ACTIVITIES ***Corners Game.*** Ask the child to play the Corners Game (instructions below,) but everyone starts with 100 points. Tell him you will play until no one can play or all the cards are played. Remind him to do his own scoring, which is done mentally.

Note: The child has played the Corners Game in first and Level C with several variations.

Tell him you will play until 10 minutes before the end of the lesson time. At that time he is to find the total when all the players' scores are added together. <u>What would the individual scores be if you had started out at 0?</u> [100 less]

Be sure the child adds cards only to the LAST card or to a Corner. Assess if the child needs help with mental adding or the facts.

CORNERS GAME

Objectives To mentally add 5, 10, and 15

Number of players 2 to 4

Cards The set of Corners cards; each of the 50 cards has four colored numbers. No two cards are alike.

Deal The stack of cards is placed face down on the table. Each player draws four cards initially and draws another card each time after playing a card. Players' cards are laid out face up in full view of all players.

Object of the game To join the cards to make the highest possible score

Play The player with the card having the lowest green number starts by placing that card in the center of the table. In the event there is more than one card with the lowest green number, of those cards, the card with the lowest blue number starts.

The player to the left of the one who started takes a turn, playing one card while observing the following rules:

1.The *colors* of the numbers where two cards join must be the same. (One number will be upside down.)

2. The *sum* of two adjoining numbers must total 5, 10, 15, or 20, or the two numbers may be the same. However, only those sums that are 5, 10, 15, or 20 will add to the score. That is, joining 1 with 1, 2 with 2, 3 with 3,

4 with 4, 6 with 6, 7 with 7, 8 with 8, or 9 with 9 is legal, but does not give points for scoring. See Figs. 1 and 2 for an example.

3. Play is to the *last* card played or to a *corner*. A corner is a space where a card could fit by joining edges with two or more cards. See Fig. 3. *All* edges must match according to the rules above. Beginners usually ignore the corners.

Players continue by taking turns playing one card. A player must play if possible. The game is over when the all the cards are played or no one can play.

Scoring Players do their own scoring, which is done mentally with only the final result of each turn written down. See scoring in the figures below. (A person playing to a Corner may write down the results for each side joined.)

Sample game Shown below is a progression of a sample game between two players. Four type styles are used to designate the four colors.

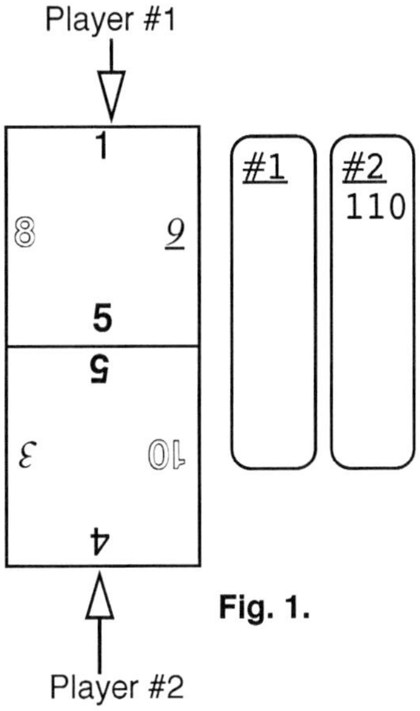

Fig. 1.

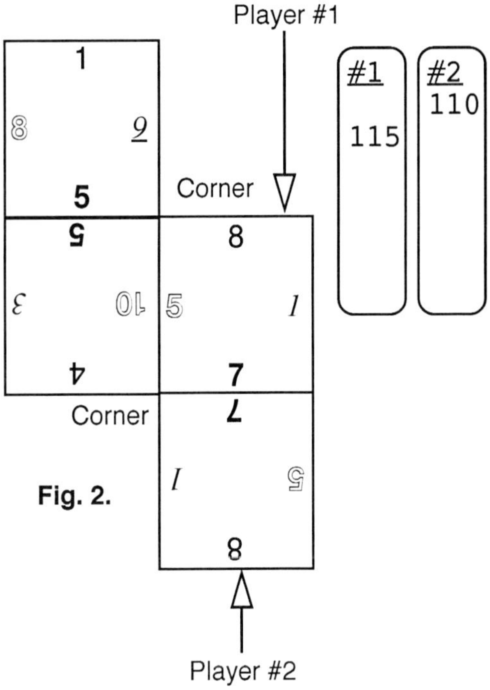

Fig. 2.

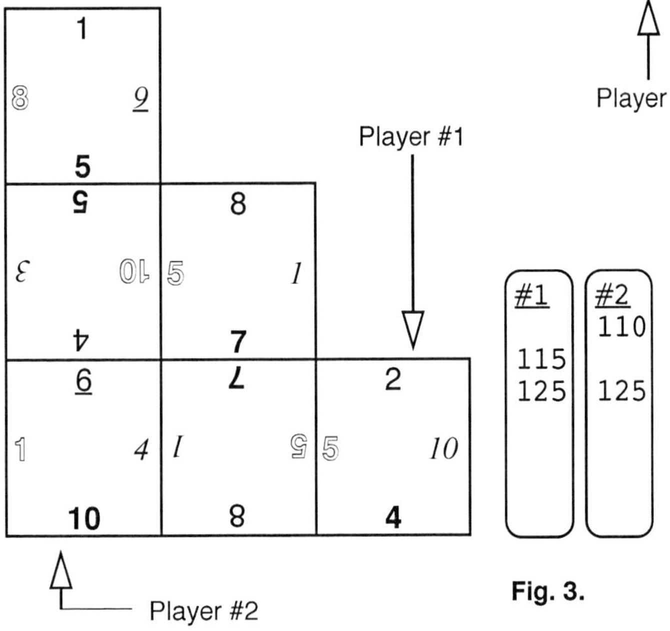

Fig. 3.

Lesson 9

Addition Practice

OBJECTIVE 1. To review adding multidigit numbers

MATERIALS Worksheet 3, "Addition Practice"

WARM-UP Ask the child what is 8 + 8. [16] What is 16 + 8? [24] Continue with 24 + 8 [32] to 72 + 8. [80] Then ask her to say the multiples of 8 (skip counting by 8s) and ask them to write them in her journal. What even and odd pattern do you see? [all even] Read only the ones; what is the pattern? [counting by 2s backward] Read only the tens; what is the pattern? [adding 1 each time in the row]

8	16	24	32	40
48	56	64	72	80

ACTIVITIES ***Reviewing tens, hundreds, thousands.*** Write

700

and ask the child to read it. [7 hundred] Ask how she knows it is 7 **hundred.** [2 zeroes after the 7] Annex another zero and ask the same questions. [7 thousand, because of the 3 zeroes]

> **Note:** When a 0 is attached to the end of a number, the term is *annex.* If a zero were added, the number wouldn't change.

Ask what is 6 tens plus 6 tens. [12 tens] Ask the child to write the problem. Then ask for another name for 12 tens. [1 hundred twenty] (If she says 120 initially, ask how many tens it is. [12 tens])

> **Note:** To continue to develop their mathematical abilities, the child MUST understand the procedures she uses.

$$\begin{array}{r} 60 \\ +\ 60 \\ \hline 120 \end{array}$$

Adding several numbers. Write the following numbers and ask the child to copy them and add.

$$\begin{array}{r} 326 \\ 489 \\ +\ 216 \\ \hline \end{array}$$

> **Note:** It is appropriate to use the word *carry,* according to the dictionary. However, "rename" is not in the dictionary and "regroup" often has a different meaning.

[1031] Ask her to show the process. Explain that you are going to ask lots of questions. For example, after she adds the ones [21], ask how she knows where to write the 2 and the 1. Ask why we can't put 21 in the ones place. Ask why we carry the 2 over to the tens. [We are trading 20 ones for 2 tens because 10 ones is 1 ten.]

$$\begin{array}{r} {\scriptstyle 2} \\ 326 \\ 489 \\ +\ 216 \\ \hline 1 \end{array}$$

Ask the child to add the next column. Ask what the 2 and 8 are; ask if they are billions, bananas, or more ones. [tens] After they are added, ask why only the 3 is written with the tens. [13 tens is 1 hundred and 3 tens] Exclaim that the last time she put a 2 on top and this time a 1 is put on top.

```
   12
  326
  489
+ 216
   31
```

Ask her to add the next column. Ask what we are adding this time. [hundreds] When the sum of 10 hundreds is reached, ask what that is. [1 thousand] It is not necessary to write the 1 in the thousands place. After she writes the 1 and the 0, ask, Why do we need a 0; isn't zero nothing? [It means no hundreds. Without the 0, it would be 1 hundred, not 1 thousand.]

```
   12
  326
  489
+ 216
 1031
```

Now tell the child that Kevin (avoid the name of a friend or family member) added it a different way. Ask her to watch and to see if it is right. Explain that you are going to first add the hundreds, 3 hundred plus 4 hundred plus 2 hundred is 9 hundred. Write 9 hundred as shown below.

```
  326       326       326       326
  489       489       489       489
+ 216     + 216     + 216     + 216
  900       900       900       900
            110       110       110
                       21        21
                               1031
```

Next add the tens in the same way, 2 tens plus 8 tens plus 1 ten equals 11 tens. Write it on the next line as shown above. Continue with the ones, writing 21 on the third line. Lastly draw a line and add all quantities as shown in the fourth column above.

Ask the child for her opinion, It is correct?

Worksheet. Tell the child on the top half of page she is to explain how to add the numbers shown to a another child. She must explain all the steps, everything she does and why. The lower half is practice. The answers are shown below.

1760			
1416	**8420**	**9841**	**9802**
1460	**5720**	**1633**	**9857**

Lesson 10

Adding Time

OBJECTIVES

1. To review telling time
2. To understand the *carrying* process in adding time

MATERIALS

A mechanical clock with geared hands that are easily moved
Worksheet 4, "Working With Clocks"

WARM-UP

Ask the child to say the multiples of 4 to 80, but with alternating voices. Say 4 with a soft voice, then say 8 with a louder voice, say 12 with a soft voice, and so on. What is special about what was said with a louder voice? [counted by 8s]

Ask the child to write the multiples of 4 to 80 and circle the multiples of 8. Also ask him to write the multiples of 8. See below.

4	(8)	12	(16)	20
(24)	28	(32)	36	(40)
44	(48)	52	(56)	60
(64)	68	(72)	76	(80)

8	16	24	32	40
48	56	64	72	80

ACTIVITIES

Note: The terms *hour* hand and *minute* hand are more descriptive than "big" and "little" hands.

Reviewing an hour. How long is an hour? [answers may relate to favorite TV programs] How many minutes are in an hour? [60 min.] Show the child the mechanical clock. How are the minutes shown on most clocks? [the *spaces* between the marks of any 2 adjacent numbers] If desired, ask him to count the spaces. How many minutes go by while the minute hand moves from one number to the next? [5 min.] Then set the minute hand at 12, and move it around to every number while he counts the minutes. [5, 10, 15, . . . 60]

Reviewing telling time. Why does the clock have 2 hands? [The hour hand tells the number of hours after midnight or after noon. The minute hand tells the number of minutes after the hour.] Where are the numbers for the hour hand? [on the clock] Where are the numbers for the minute hand? [invisible on most clocks] How we can find the number of minutes? [counting by 5s]

Set the clock at 10:15 and ask the child for the time. [10:15] How do we know? [The hour hand is after 10, but not at 11. The minute hand is three 5s, or 15 minutes, after 12.] Ask the child to set the clock for 11:15 and 12:15. Then reset the clock to 10:15.

One hour later. What will the time be 1 hour from now? [11:15] How many minutes between 10:15 and 11:15? [60 min.] Count around to show the 60 minutes. Ask what the time will be in 1 more hour. [12:15] Ask the child to set the clock one hour later. [1:15] Why couldn't we say 13:15? [People in Europe and in the military do say 13:15 for 1:15 p.m.]

Half and quarter of an hour. How many minutes are in a half hour? [30] Set the clock for 1:30 and ask the child what is special about the time. [The hour is half over.] Where is the hour hand? [halfway between the 1 and 2] What are 2 ways to say the time? [one thirty, half past one]

What does the word *quarter* mean? [one-fourth, or half of a half] How many minutes are in a quarter of an hour? [15 minutes] Set the clock for 1:45. What is special about this time? [a quarter before the hour] Where is the hour hand? [a little before the 2] What are 2 ways to say the time? [one forty-five, quarter to two, (13;45)]

Repeat for 2:15. [The hour hand is a little after the 2. The time is two fifteen, quarter after two.]

For practice, set the clock to the following times and ask the child to give the time two different ways.

2:30 [two thirty, half past two]

2:45 [two forty-five, quarter to three]

3:15 [three fifteen, quarter after three]

3:30 [three thirty, half past three]

3:45 [three forty-five, quarter to four]

Ask him to solve the following problems on the worksheet. Ask for various ways of solving them.

Problem 1. Jamie gets on the school bus at 8:05. The ride is a half hour long. What time does Jamie arrive at school? [8:35]

This can be solved by adding 30 to 5. It can also be solved by counting on a clock.

Problem 2. Jamie gets on the bus after school at 3:35. What time does Jamie arrive home? [4:05] The ride takes a half hour.

Adding 30 to the 35 results in 65 minutes. Since there are only 60 minutes in an hour, 1 hour is carried to the hours and the extra 5 is the minutes.

Problem 3. Jay and his family are going to visit his grandmother, who lives an hour and a half away. If they leave at 1:45, what time will they arrive? [3:15]

Problem 4. The next day they leave Jay's grandmother's house at 6:30. What time will they be home? [8:00]

Problem 5. A cuckoo clock just chimed at 12:15. It chimes every quarter hour. Name the next four times it will chime. [12:30, 12:45, 1:00, 1:15]

Problem 6. Joleen finished eating her dinner at 6:30. She went to bed ninety minutes later. What time did she go to bed? [8:00]

Worksheet. Ask the child to complete the worksheet. The answers are given below.

5:15, a quarter after 5

	3:45	1:20	6:05	9:55
1 hour later	**4:45**	**2:20**	**7:05**	**10:55**
30 minutes later	**4:15**	**1:50**	**6:35**	**10:25**
5 minutes later	**3:50**	**1:25**	**6:10**	**10:00**

Lesson 11

Finding Perimeter in Inches

OBJECTIVES
1. To find the perimeter of a large rectangle
2. To measure and work with halves of an inch

MATERIALS
Math journal
12" rulers (preferably where the zero is not at the edge)
Prepare a rectangle at least 36" by 36", for example, a table, a carpeted area, or even a large sheet of paper
Worksheet 5, "Perimeter in Inches"

WARM-UP
Ask the child to double various numbers, for example: 4 [8], 7 [14], 9 [18], 6 [12], 12 [24], 25 [50], 26 [52], 49 [98], and ½. [1]

Then ask her to write the multiples of 4 and then to double each number. What happens? [multiples of 8]

4	8	12	16	20		8	16	24	32	40
24	28	32	36	40		48	56	64	72	80

ACTIVITIES
Reviewing perimeter. Discuss what perimeter is. [distance, or length, around a shape] Write the word perimeter and circle the part saying "meter." Do you know what it means? [measure, as in a water meter or an electric meter] Next circle the part that spells, "peri." What does it means? [around] So, what is perimeter? [measuring around] Seeing *rim* in perimeter is also a good reminder.

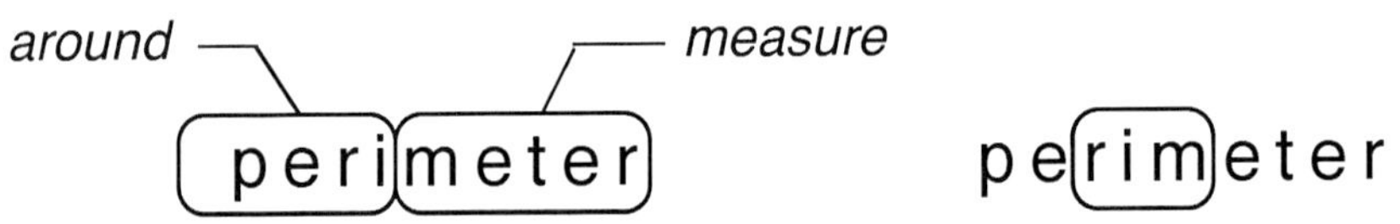

The derivation of the word *perimeter.*

Worksheet Problem 1. Take out the worksheet and ruler. Ask the child to do the first problem. Ask her to explain her work. Review the abbreviation for inches. [in or in.]

> **Note:** Asking the child to mark the inches helps her realize what she is measuring. Once she understands, there is no need to mark the inches.

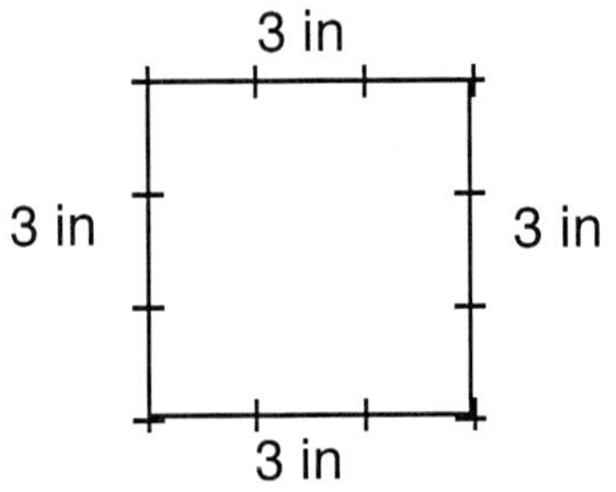

P = 3 + 3 + 3 + 3
P = 12 in

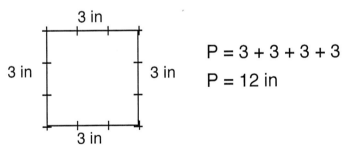

The solution for Problem 1.

> **Note:** Although the abbreviation for inches can be written as "in.", the periods after most measurement abbreviations are omitted, as in cm and m, for example.

Emphasize writing the equations

P = 3 + 3 + 3 + 3
P = 12 in

Also stress she is to do what makes sense to her.

You might ask her for other names of that figure. [quadrilateral, parallelogram, rhombus, and square]

Worksheet Problem 2. Ask the child to measure the top side of the rectangle for problem 2. [5 ½"] Discuss that the long side measures 5 inches and half of another inch. <u>How could you write it?</u> If she writes a slanted line, explain that slanted fraction lines are used in cooking and on roadway signs. But in mathematics the fraction line is usually written horizontally as shown below.

$$5\frac{1}{2}$$

Ask her to write the measurements and to find the perimeter. See the figure below.

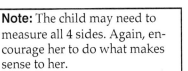

Note: The child may need to measure all 4 sides. Again, encourage her to do what makes sense to her.

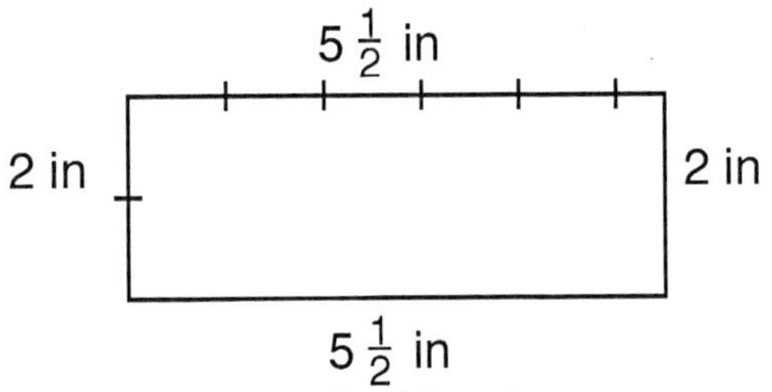

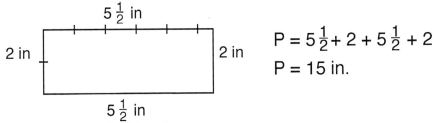

$$P = 5\frac{1}{2} + 2 + 5\frac{1}{2} + 2$$
$$P = 15 \text{ in.}$$

The solution for Problem 2.

Ask her how she added the numbers to find the perimeter. Emphasize that the 2 halves make another inch.

Finding the perimeter of the large rectangle. Show the child the large rectangle. Tell her that she is to find the perimeter. Discuss using 2 rulers to measure. One method is shown below, where the second ruler overlaps at the mark for 10 inches.

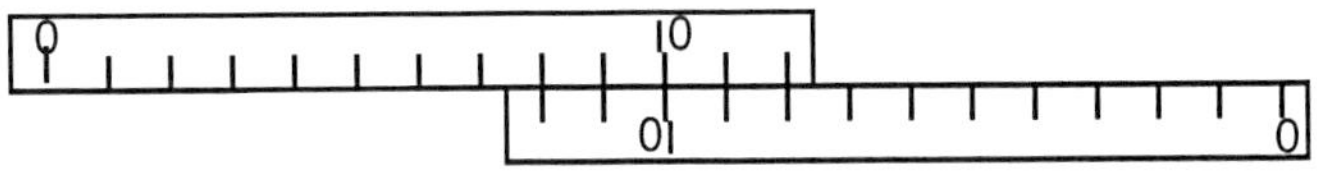

One way using 12-inch rulers to measure lengths longer than 10 inches.

Ask the child to draw a picture of the rectangle in her journal. Then ask her to measure the sides of the rectangle in inches.

There are several ways she could do the addition. She could simply add the 4 numbers. She also might add 2 numbers and add the third to that sum and so on. An advanced child might add the 2 unlike numbers and double the sum. Ask the child to share her work with you.

Worksheet Problem 3. Assign the third problem. First ask the child to take a guess. The solution is below.

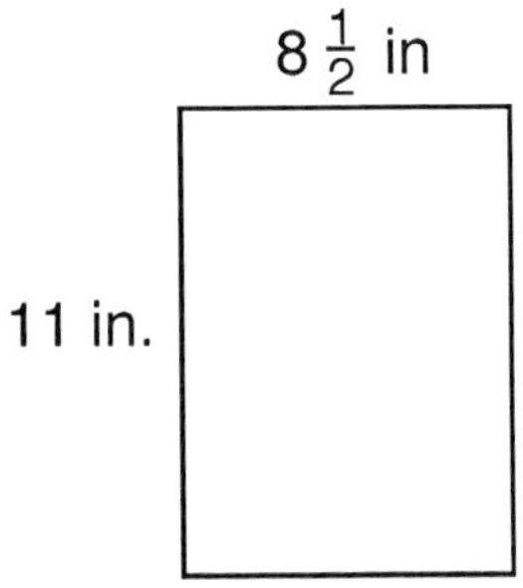

$$P = 8\frac{1}{2} + 11 + 8\frac{1}{2} + 11$$
$$P = 39 \text{ in}$$

The solution for Problem 3.

Lesson 12

Review

OBJECTIVES 1. To review and practice

MATERIALS Worksheet 6-A and 6-B, "Review" (2 versions)
Calculators
Corners cards

ACTIVITIES ***Review worksheet.*** Give the child a review sheet.

Note: Two review sheets are provided. They can used as a pretest and posttest.

The oral problems to be read to the child are as follows:

54 + 4 =   78 + 8 =   49 + 9 =

Review Sheet 6-A.

1. Write only the answers to the oral questions. **58** **86** **58**

4. Write only the answers. 85 + 25 = **110** 63 + 58 = **121** 125 + 15 = **140**

7. Draw the hands.

quarter after 1

Draw the hands.

12:05

Write the time two ways.

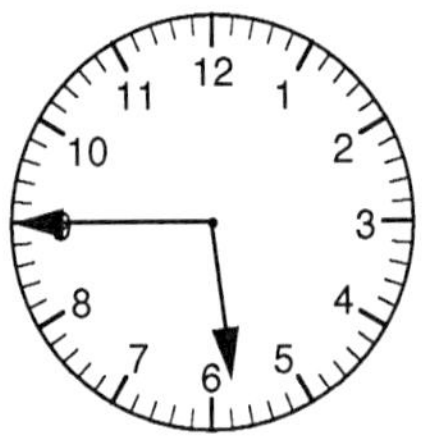

5:45

quarter to 6

11. Write the first three months of the year. How many days are in those months?

January	**31**	**There are 90 days,**
February	**28**	**but in leap year**
March	**+31**	**there are 91 days.**
	90	

12. Add 467 + 28 + 3165.

		1	2	
		4	6	7
			2	8
+	3	1	6	5
	3	6	6	0

13. Write the multiples of 3.
Circle the even numbers.

3 **(6)** **9**
(12) **15** **(18)**
21 **(24)** **27**
(30)

The oral problems to be read to the child are as follows:

63 + 8 =   57 + 10 =   28 + 4 =

Review Sheet 6-B.

1. Write only the answers to the oral questions. **71** **67** **32**

4. Write only the answers. 96 + 34 = **130** 47 + 74 = **121** 136 + 14 = **150**

7. Draw the hands.

quarter to 3

Draw the hands.

7:05

Write the time two ways.

10:15

quarter after 10

11. Write the last three months of the year. How many days are in those months?

October	**31**
November	**30**
December	**+31**
	92

There are 92 days.

12. Add 74 + 2388 + 19.

		1	2	
			7	4
	2	3	8	8
		+	1	9
	2	4	8	1

13. Write the multiples of 5.
Circle the even numbers.

5 **(10)**
15 **(20)**
25 **(30)**
35 **(40)**
45 **(50)**

Multiples practice. Ask the child if they remember how to find the multiples on a calculator. For example, to see the multiples of 2 on a 4-function Casio:

Press 2 [+] [+]. Then [=] [=] [=].

For other calculators, press 2 [+]. Then [=] [=] [=].

Ask the child to first practice saying the multiples 2 to 9 by themselves. Then ask them to practice with their partners. One child says the multiples while the other child checks with the calculator.

Note: It is very important that children know skip counting before attempting to learn the multiplication combinations

Lesson 13

Finding Perimeter in Feet and Inches

OBJECTIVES

1. To review that 1 foot equals 12 inches
2. To measure in feet and inches
3. To find perimeter in feet and inches

MATERIALS

12-inch rulers
2 sheets of construction paper 9 inches by 12 inches
Worksheet 7, "Perimeter in Feet and Inches"

WARM-UP

Ask the child to recite all doubles. [1 + 1 = 2, 2 + 2 = 4, . . . 9 + 9 = 18] What was special about the sums? [all evens, the multiples of 2]

Ask the following orally. Remind the child to add them in the order; for example, 78 + 20 [98] + 9. [107] They can write the answers on their slates.

78 + 29 = [107]   33 + 28 = [61]   55 + 39 = [94]   49 + 36 = [85]

ACTIVITIES

Reviewing a foot. Ask the child to show with her hands how long she thinks 12 inches is. Ask her to measure the distance.

What is the special name for 12 inches? [foot] Tell her at one time it was the length of the king's foot. Is that a good way to measure? [No, king's feet aren't all the same.] What is its abbreviation? [ft or ft.]

Perimeter in feet and inches. Tell the child that today she will use feet and inches to find perimeter. Explain that when using feet and inches, the inches cannot be greater, or more, than 11.

Note: Since the word *greater* may not be in the child's vocabulary, use it often in context.

Show the child 2 sheets of construction paper joined the long way and 2 sheets joined the short way. Which do you think has the *greater* perimeter? See the figures below.

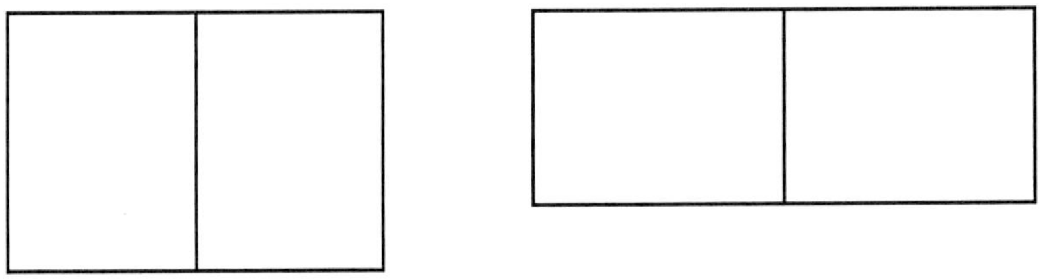

Two adjacent 9 by 12 inch sheets of paper for use in finding perimeter.

Take out the worksheet. Ask the child to lay 2 papers side by side as in the left figure above. Ask her to measure the short edge [1 ft] and to write it on her worksheet.

Next ask her to measure the longer edge. [1 ft 6 in] She can use 2 rulers for measuring. Discuss where to place the second ruler, not at the edge of the ruler, but at the marking for 12 inches. See the figure below.

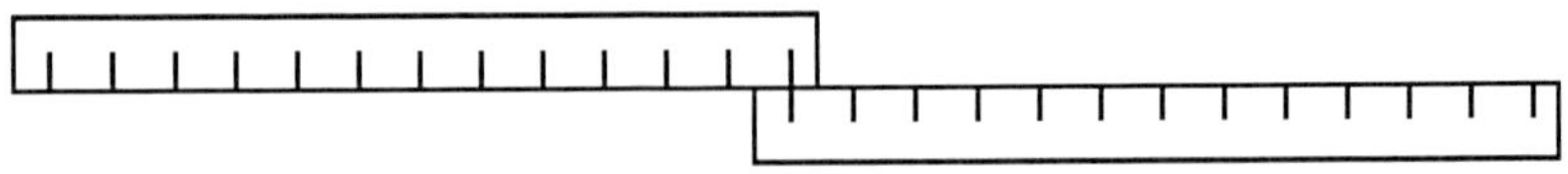

One way using 12-inch rulers to measure lengths over 1 foot.

> **Note:** It is not necessary to write the measurements on all 4 sides. Let the child decide.

Ask her to write it on her worksheet. [1 ft 6 in] Ask her to complete writing the measurements and to calculate the perimeter on her worksheet. See the figure on the previous page.

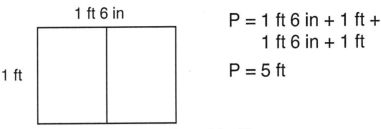

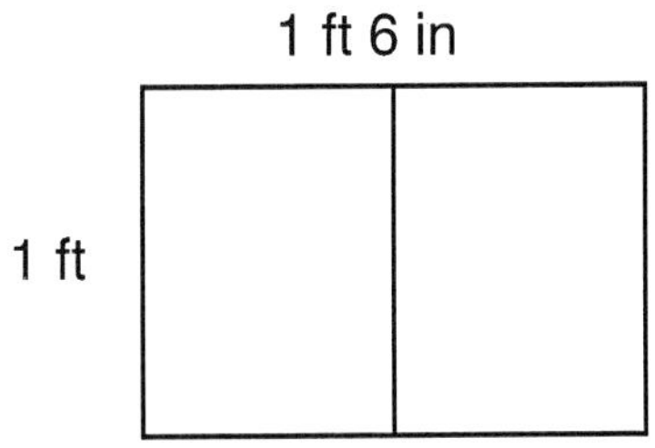

P = 1 ft 6 in + 1 ft +
1 ft 6 in + 1 ft

P = 5 ft

Finding the perimeter around two 9 by 12 inch sheets of paper laid side by side.

Ask the child to explain how she added.

Next ask the child to lay the sheets of paper end to end and to measure and find the perimeter. See the figure below.

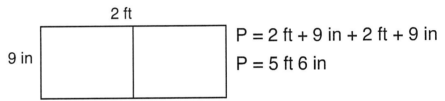

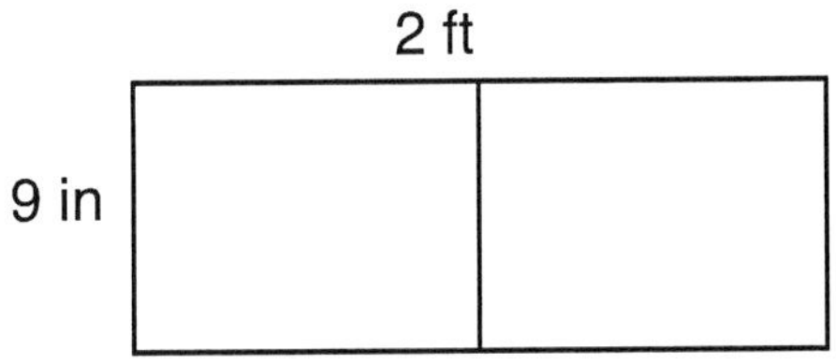

P = 2 ft + 9 in + 2 ft + 9 in

P = 5 ft 6 in

Finding the perimeter around two 9 by 12 inch sheets of paper laid end to end.

This addition is a bit more complicated than the previous problem. Ask the child to explain how she added. If she gives 4 ft 18 in, ask her if she can find an answer with less than 12 inches.

Measuring heights. Ask the child to measure you. Then have her write your height in both inches and in feet and inches.

Worksheet problem 3. Ask the child to read and solve problem 3. Ask her to show her solution. The solution is shown below.

P = 569 + 687 + 569 + 687

P = 2512 ft

It was less than a mile.

```
  3 3
  569
  687
  569
+ 687
 2512
```

Homework. Assign problem 4. The solution is shown below.

> **Note:** The comparison can be done mentally.

P = 596 + 658 + 596 + 658

P = 2508 ft

Marc walked farther, 4 ft farther.

```
  3 2
  596
  658
  596
+ 658
 2508
```

Lesson 14

Finding Halves and Fourths

OBJECTIVES
1. To review the fraction ¼
2. To divide a string into halves and fourths

MATERIALS
12-inch ruler
A piece of string or yarn 10 inches long

WARM-UP
Ask the child to write the following problem in his journal and add: 5426 + 3659. [9085]

Note: Giving the addition problems orally gives the children practice in translating spoken numbers into written symbols.

Ask the child to say the multiples of 2 to 40, but with alternating voice volumes. Then ask what was special about the numbers said with the second voice volume. [counted by 4s]

Also ask him to write the multiples of 4.

4	8	12	16	20
24	28	32	36	40

Then ask him to look at his multiples to answer the following: How much is one 4? [4], two 4s? [8], three 4s? [12], five 4s? [20], ten 4s? [40] nine 4s? [36], and six 4s? [24]

ACTIVITIES
Reviewing quarters and fourths. Draw a long rectangle and tell the child that is a whole; write a 1 in it as shown below. Draw an identical rectangle directly below the rectangle and divide it into halves as shown below. What do we call each part? [one-half] Ask the child to write the fraction in each part.

Note: This linear model of fractions is superior to circles because comparisons are simpler and fractions over 1 can easily be shown.

The child has used these in earlier levels.

1			
$\frac{1}{2}$		$\frac{1}{2}$	
$\frac{1}{4}$	$\frac{1}{4}$	$\frac{1}{4}$	$\frac{1}{4}$

Fraction strips showing whole, half, and fourths.

What do we get when we divide each half into 2 equal parts? [a quarter, or one-fourth] Ask for both names. Draw another rectangle and divide it into fourths and ask the child to write the fractions. See the figure above.

Ask the following questions for the child to respond.

1. How many halves do you need to make a whole? [2]
2. How many fourths do you need to make a whole? [4]
3. How many quarters do you need to make a whole? [4]
4. How many quarters are in a half? [2]
5. What is the same as 2 quarters? [a half]
6. Which is greater (or more), 2 halves or a whole? [same]
7. What is the same as 2 halves? [a whole]
8. What is a half of one-half? [quarter or a fourth]

Ten-inch string. Show the child the 10-inch piece of string. How long do you think it is? Then ask him to measure it. [10 in]

Now fold the string in half and ask how long it is. [5 in] Ask him to explain the answer. [perhaps, because 5 + 5 = 10] Ask him to measure it. [5 in] What fraction is the folded piece compared to the whole piece? [1/2] Ask him to draw the string and write the measurements. See the figures below.

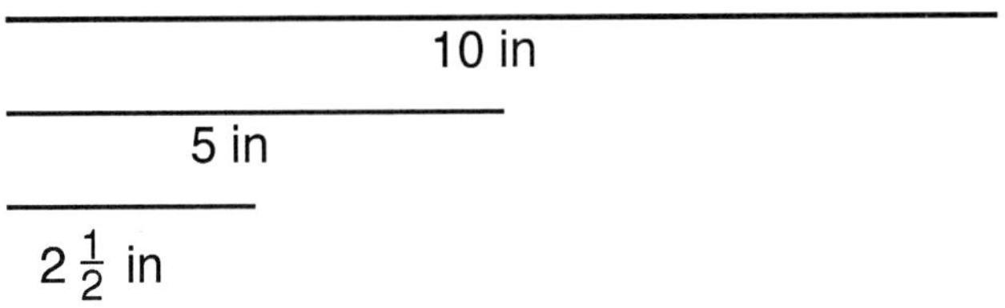

A piece of string 10 in long folded in half twice.

Fold it once more. Now what is its length? [$2\frac{1}{2}$ in] Ask him to explain how he knew. [half of 4 is 2 and half of 1 is $\frac{1}{2}$] What fraction is the folded piece compared to the whole piece? [$\frac{1}{4}$] Write the length and measurement as shown above.

Nine-inch string. Unfold the string and explain that you are going to cut off 1 inch. Cut off 1 inch and ask how long the string is now. [9 in] Repeat the steps above. Fold the string and ask how much half is. [$4\frac{1}{2}$ in] Ask how much one-fourth is. [$2\frac{1}{4}$ in, because half of 4 is 2 and half of $\frac{1}{2}$ is $\frac{1}{4}$] See the figures below.

Note: Learning how to read rulers to $\frac{1}{4}$ of an inch will be taught later.

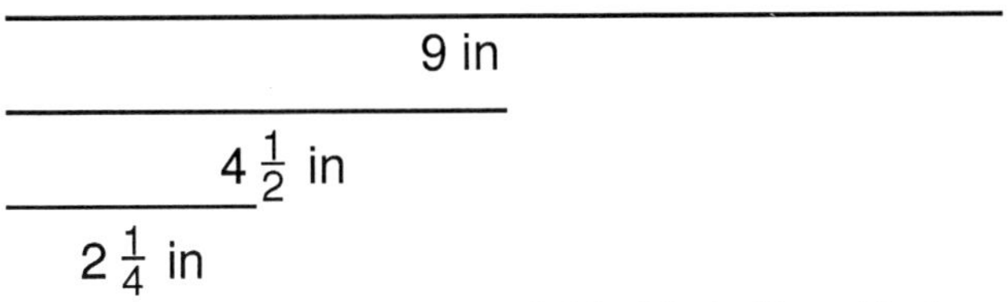

A piece of string 9 in long folded in half and in fourths.

Written work. Tell the child that he is to do the same thing in his journal with 2 lines, one 6 inches and one 5 inches long. Explain he is to think of each square in his journal as 1 inch. Ask him to draw a half and a fourth of the lines and to write the measurements.

The solutions are shown below.

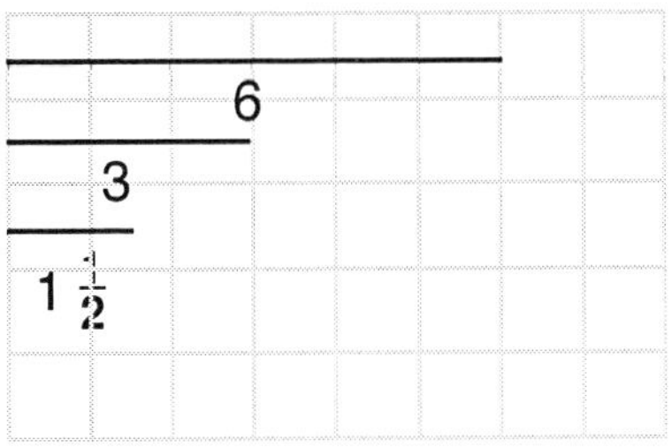

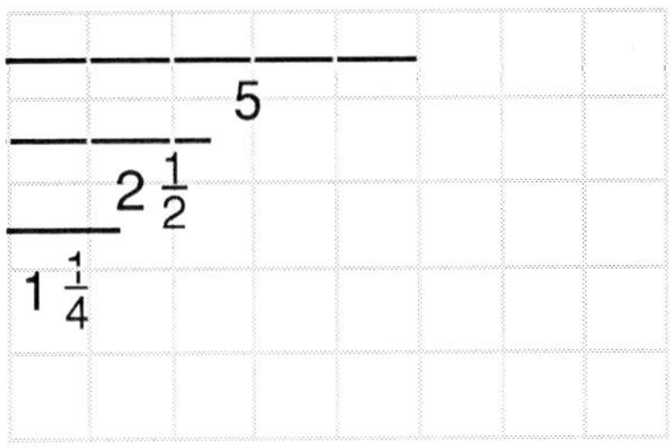

Lesson 15

Adding Halves and Fourths

OBJECTIVES
1. To divide rectangles into fourths 3 different ways
2. To add simple mixed fractions with halves and fourths

MATERIALS
4 paper rectangles, 8½ by 5 inches (2 will fit on a sheet of paper)
Ruler
Worksheet 8, "Perimeter With Fractions"

WARM-UP
Ask the child to write the following problem in her journal and add: 6259 + 2462. [8721]

Also ask her to write the multiples of 4. After they are written, ask her to say them.

4	8	12	16	20
24	28	32	36	40

Then ask her to look at her multiples to answer the following: How much is onc 4? [4], two 4s? [8], three 4s? [12], five 4s? [20], ten 4s? [40] seven 4s? [28], and nine 4s? [36]

ACTIVITIES
Rectangles into halves. Give 3 paper rectangles to the child. Ask her to fold them into halves 3 different ways. The solutions are shown below.

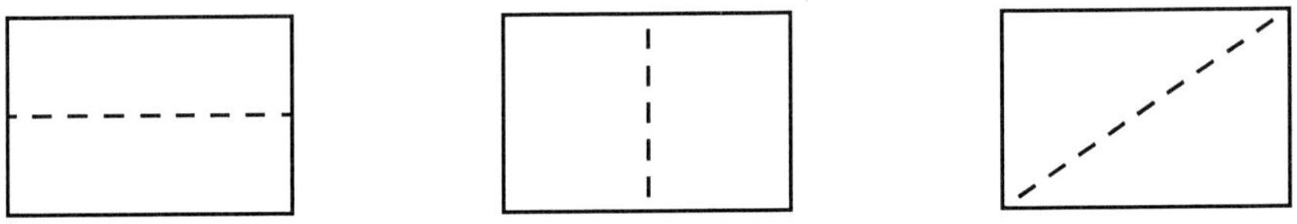

Three different ways to divide a rectangle into halves.

Perimeters of the rectangles. Which of the smaller rectangles do you think has the larger perimeter? Ask her to sketch the largest rectangle and to write the measurements. [8½ by 5] Then she is to sketch the two "half rectangles." Next she figures out the measurements from the large rectangle, writes them down, and calculates the perimeters. The drawings and equations are below.

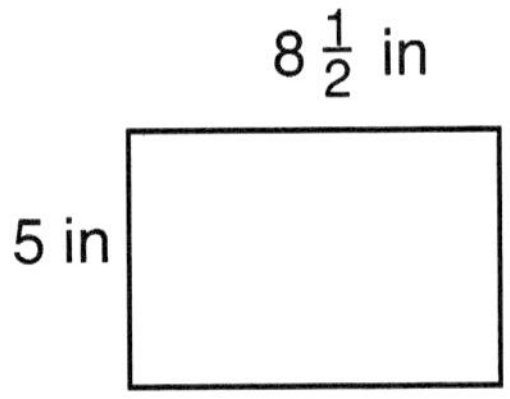

The measurements of the whole rectangle.

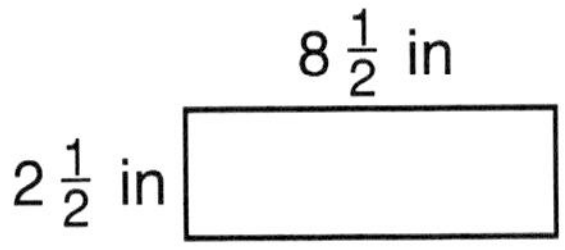

$$P = 8\frac{1}{2} + 2\frac{1}{2} + 8\frac{1}{2} + 2\frac{1}{2} = 22 \text{ in}$$

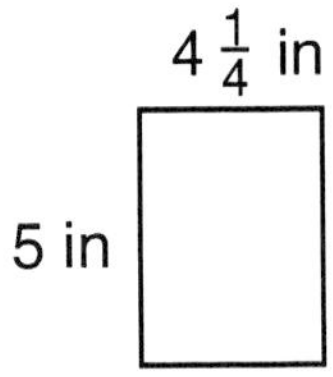

$$P = 4\frac{1}{4} + 5 + 4\frac{1}{4} + 5 = 18\frac{1}{2} \text{ in}$$

Note: Learning how to read rulers to ¼ of an inch will be taught later.

Ask the child to explain how she added the numbers. The simplest way of adding the numbers is to first add the whole numbers and then to add the fractions.

In the first example, the numbers total 20 and 2 halves make a one, giving a total of 20 + 1 + 1, or 22. In the second example, the whole numbers total 18 and the 2 one-fourths equal ½.

Discuss which perimeter is greater.

Rectangles into fourths. Now ask the child to divide each of the paper rectangles into fourths. Take out the fourth paper rectangle and ask her to find a fourth way. The solutions are shown below.

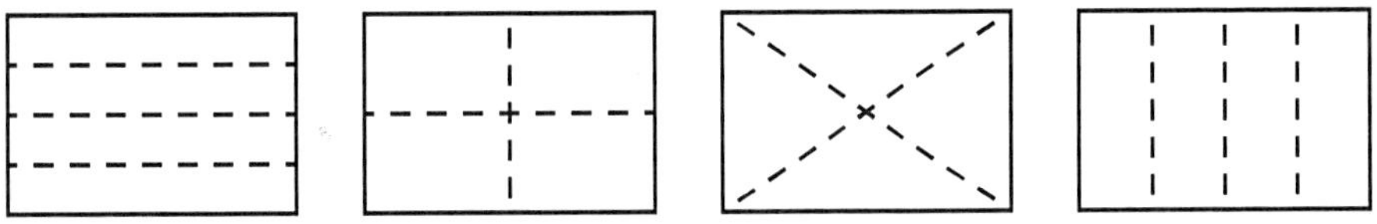

Four different ways to divide a rectangle into fourths.

Worksheet. Ask the child to do the first problem. Ask her to refer to the folded rectangles for ideas; explain, however, that she cannot use the folded rectangle with diagonals.

Tell her to write the measurements for the fourths near the sides. Ask her to explain how she found the size of the fourth and calculated the perimeters. The measurements and solutions are shown below.

When she has finished, you might discuss a pattern she sees. [The closer the rectangle is to a square, the less the perimeter.]

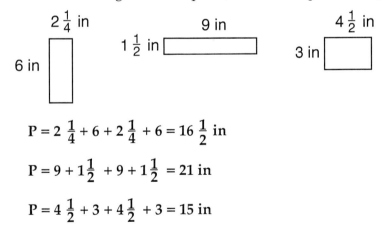

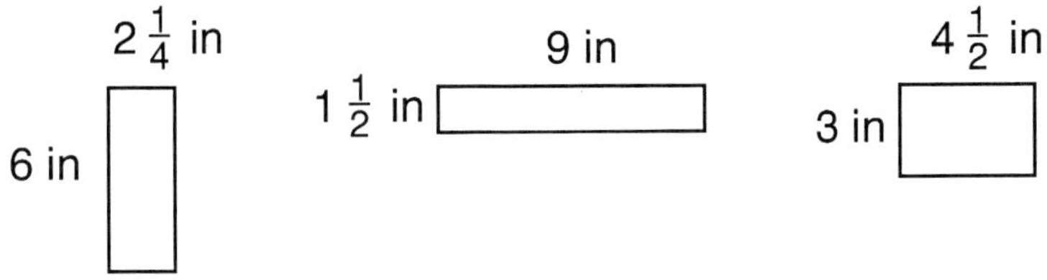

$$P = 2\tfrac{1}{4} + 6 + 2\tfrac{1}{4} + 6 = 16\tfrac{1}{2}\text{ in}$$

$$P = 9 + 1\tfrac{1}{2} + 9 + 1\tfrac{1}{2} = 21\text{ in}$$

$$P = 4\tfrac{1}{2} + 3 + 4\tfrac{1}{2} + 3 = 15\text{ in}$$

Homework. Assign problem 4. The solution is below.

$$P = 2\tfrac{1}{2} + 1\tfrac{1}{4} + 2\tfrac{1}{2} + 1\tfrac{1}{4} = 7\tfrac{1}{2}\text{ in}$$

Lesson 16

Quarters of an Hour

OBJECTIVES
1. To review a quarter of an hour
2. To work informally with quarters of an hour
3. To review the abbreviations for minute (min) and hour (hr)

MATERIALS
A geared clock
Worksheet 9, "Quarters of an Hour"

WARM-UP
Ask the child to add the following problem in his journal: 539 + 4787. [5326]

Also ask the child to say the multiples of 4.

4	8	12	16	20
24	28	32	36	40

Then ask him to write the multiples. How much is one 4? [4], six 4s? [24], two 4s? [8], seven 4s? [28], three 4s? [12], eight 4s? [32], four 4s? [16], nine 4s? [36], five 4s? [20], and ten 4s? [40]

ACTIVITIES
Reviewing quarters and fourths. Draw the same fraction strips with 1, halves, and fourths as shown below. Point to the last row; What are the 2 names for that fraction? [quarter, 1⁄4] Write the fractions at the left.

<table>
<tr><td colspan="4">1</td></tr>
<tr><td colspan="2">$\frac{1}{2}$</td><td colspan="2">$\frac{1}{2}$</td></tr>
<tr><td>$\frac{1}{4}$</td><td>$\frac{1}{4}$</td><td>$\frac{1}{4}$</td><td>$\frac{1}{4}$</td></tr>
</table>

Fractions strips.

Quarter of an hour. How many minutes are in an hour? [60] Write 60 min in the top rectangle as shown below. Remind him that the abbreviation for minute is *min* and the abbreviation for hour is *hr*. How many minutes are in half an hour? [30] Write it next in the half strips. How many minutes are in a quarter of an hour? [15] Write it next in the quarter strips. See the figure below.

<table>
<tr><td colspan="4">1 60 min</td></tr>
<tr><td colspan="2">$\frac{1}{2}$ 30 min</td><td colspan="2">$\frac{1}{2}$ 30 min</td></tr>
<tr><td>$\frac{1}{4}$ 15 min</td><td>$\frac{1}{4}$ 15 min</td><td>$\frac{1}{4}$ 15 min</td><td>$\frac{1}{4}$ 15 min</td></tr>
</table>

Relating fractions and the clock.

Ask the following questions for the child to answer.

1. How many half hours make a whole hour? [2]
2. How many minutes are in a half hour? [30 min]

3. How long is 2 half hours? [60 min, 1 hr]
4. How long is a quarter of an hour? [15 min]
5. How long is 2 quarters of an hour? [30 min, half hour]
6. How long is 3 quarters of an hour? [45 min]
7. Which is longer, 1 hour or 3 quarters of a hour? [1hr]
8. How long is 4 quarters of an hour? [60 min, or 1 hr]
9. How long is 3 halves of an hour? [90 min, or 1 hr and 30 min]

Note: A child needs to learn the two ways time is used: the time of day and elapsed time. For example, 1:20 is a particular time, but 1 hr 20 min is an amount of time.

Set the hands of the clock at 6: 00 and ask what the time would be 1 quarter of an hour later. [6:15] Ask what time 2 quarters of an hour later would be (from 6:00). [6:30] And ask what time 3 quarters of an hour later would be. [6:45]

Problem 1. Sammy can swim a mile in three quarters of an hour. If Sammy starts at 8:15, when will Sammy finish swimming the mile? [9:00] Ask for an explanation on how he solved the problem.

Problem 2. Fast walkers can walk a mile in a quarter of an hour. How long will it take them to walk 5 miles? [1 hr 15 min, or 1 hr and a quarter]

The following problems all assume walking a mile in a quarter of an hour.

The Anderson family starts their 5-mile hike at 9:30. When will they finish? [10:45]

The Garcia family starts their 5-mile hike at 9:45. When will they finish? [11:00]

The Smith family decides to walk 3 miles, leaving at 9:45. When will they finish? [10:30]

The Vu family leaves home at 11:00 and walks 3 miles. Then they spend a half hour eating their picnic lunch before walking back 3 miles. What time will they arrive home? [1:00] There are several ways this problem can be done. Work with the child until he has found several solutions.

Worksheet. Give the child time to work on the worksheet. Remind him to explain how he solved the problems. The answers are as follows:

1. 2½ hrs

2. 10½ hrs

3. They will be 15 min (a quarter of an hour) late.

4.

1st Qtr	2nd Qtr	3rd Qtr	4th Qtr
January	**April**	**July**	**October**
February	**May**	**August**	**November**
March	**June**	**September**	**December**

5. The third and fourth quarters have the greatest number of days, 92.

Lesson 17

Fractions of a Dollar

OBJECTIVES
1. To review the parts of a dollar
2. To relate the value of coins to fractions

MATERIALS
U.S. coins
Worksheet 10, "Fractions of a Dollar"

WARM-UP
Ask the child to write the following problem in her journal and add: 6427 + 848. [7275]

> **Note:** The child needs only look at the position to find the product. For example, 4 × 8 is the eighth position, or 32.

Also ask the child to say the multiples of 4.

4	8	12	16	20
24	28	32	36	40

Then ask her to write the multiples. How much is one 4? [4], six 4s? [24], two 4s? [8], seven 4s? [28], three 4s? [12], eight 4s? [32], four 4s? [16], nine 4s? [36], five 4s? [20], and ten 4s? [40]

ACTIVITIES
Parts of a dollar. Draw the fraction strips with 1, halves, fourths, and tenths as shown below. Write the fractions at the left as shown below.

1									
$\frac{1}{2}$					$\frac{1}{2}$				
$\frac{1}{4}$			$\frac{1}{4}$			$\frac{1}{4}$			$\frac{1}{4}$
$\frac{1}{10}$	$\frac{1}{10}$	$\frac{1}{10}$	$\frac{1}{10}$	$\frac{1}{10}$	$\frac{1}{10}$	$\frac{1}{10}$	$\frac{1}{10}$	$\frac{1}{10}$	$\frac{1}{10}$

Fractions strips.

How many cents are in a dollar? [100] Tell her that the word *cent* means hundred. Can you think of any other words that start like *cent* and means hundred? [*Century* and *centennial* refer to one hundred years.] Write 100¢ in the top rectangle as shown below.

1 100¢									
$\frac{1}{2}$ 50¢					$\frac{1}{2}$ 50¢				
$\frac{1}{4}$ 25¢			$\frac{1}{4}$ 25¢			$\frac{1}{4}$ 25¢			$\frac{1}{4}$ 25¢
$\frac{1}{10}$ 10¢	$\frac{1}{10}$ 10¢	$\frac{1}{10}$ 10¢	$\frac{1}{10}$ 10¢	$\frac{1}{10}$ 10¢	$\frac{1}{10}$ 10¢	$\frac{1}{10}$ 10¢	$\frac{1}{10}$ 10¢	$\frac{1}{10}$ 10¢	$\frac{1}{10}$ 10¢

Relating fractions and a dollar.

How much is half of a dollar, or half of 100? [50¢] Show her a half dollar, if available; what it is called? [half dollar] Explain they are hardly ever used any more. Ask her to write 50¢ next to the 1/2 fractions, as shown above.

How much is one-fourth of a dollar, or half of 50? [25¢] Show her a quarter and ask what it is called. [quarter] Ask her to write 25¢ next to the 1/4 fractions, as shown above.

How much is one-tenth of a dollar? This is a more difficult question; it might help to rephrase it as: If 100 is broken into 10 parts, how much is in each part? [10¢] Show her the dime and ask for its name. [dime] Write 10¢ next to the 1/10 fractions.

Note: Do not be concerned if the child does not follow the discussion on the fractions below 1⁄10.

What do you think is the name for the fraction that is half of one-tenth? [one-twentieth] What coin is worth half of a dime? [nickel] How much it is worth? [5¢] Show her the nickel and ask if she remembers why it is larger than the dime. [The dime used to be made from silver, which is very expensive so a smaller size was chosen.] Explain that it is hard to make fractions that small, so you will not write them.

Finally show a penny: What it is worth? [1¢] What fraction of a dollar it is? [1/100]

Worksheet. Give the child Worksheet 10. The solutions are as follows.

<table>
<tr><td colspan="10">100¢
dollar</td></tr>
<tr><td colspan="5">**50¢**
half dollar</td><td colspan="5">**50¢**
half dollar</td></tr>
<tr><td colspan="2">**25¢**
quarter</td><td colspan="3">**25¢**
quarter</td><td colspan="2">**25¢**
quarter</td><td colspan="3">**25¢**
quarter</td></tr>
<tr><td>**10¢**
dime</td><td>**10¢**
dime</td><td>**10¢**
dime</td><td>**10¢**
dime</td><td>**10¢**
dime</td><td>**10¢**
dime</td><td>**10¢**
dime</td><td>**10¢**
dime</td><td>**10¢**
dime</td><td>**10¢**
dime</td></tr>
</table>

1. How many quarters equal a dollar? **[4]**
2. How many quarters are equal to 2 dollars? **[8]**
3. How many dimes are equal to a dollar? **[10]**
4. How many pennies in a dollar and a half? **[150]**
5. Which is greater, 3 quarters or 1 dollar? **[1 dollar]**
6. How much does 2 quarters equal? **[50¢]**
7. How many quarters are in a dollar and a half? **[6]**
8. How many dimes are equal to half a dollar? **[5]**
9. Which is greater, 6 quarters or 3 half dollars? **[same]**
10. If half a pizza costs a dollar and a half, what does a whole pizza cost? **[three dollars]**

Lesson 18

Review

OBJECTIVES 1. To review and practice
2. To enhance skills through playing games

MATERIALS Worksheet 11-A and 11-B, "Review" (2 versions)
Cards for playing games

ACTIVITIES ***Review worksheet.*** Give the child a review sheet.

The oral problems to be read to the child are as follows:

73 + 7 =   68 + 4 =   27 + 10 =

Review Sheet 11-A.

1. Write only the answers to the oral questions. 80 72 37

4. Write only the answers. 75 + 19 = 94 47 + 26 = 73 81 + 24 = 105

7. Write the fractions in the rectangles.

1			
$\frac{1}{2}$		$\frac{1}{2}$	
$\frac{1}{4}$	$\frac{1}{4}$	$\frac{1}{4}$	$\frac{1}{4}$

13. Show $3\frac{1}{2}$ inches on the ruler. Circle or shade it.

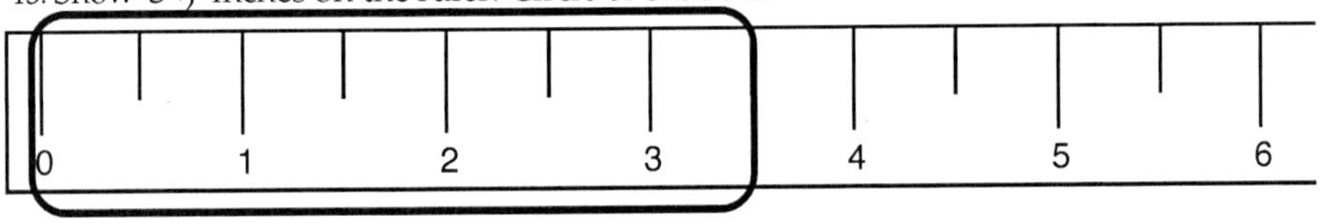

14. How many minutes in a quarter of an hour? 15 min

15. How many minutes in three quarters of an hour? 45 min

16. How much money is a quarter of a dollar? 25¢

17. How much money is three quarters of a dollar? 75¢

18. How much is half of $4\frac{1}{2}$ inches? $2\frac{1}{4}$ in

19. Write the first 15 multiples of 2.

2	4	6	8	10
12	14	16	18	20
22	24	26	28	30

36. Add 637 + 6456.

			1	
		6	3	7
+	6	4	5	6
	7	0	9	3

35. What patterns do you see in the multiples of 2?
All even numbers. The ones are the same in each column.

The oral problems to be read to the child are as follows:

86 + 5 =   47 + 7 =   58 + 10 =

Review Sheet 11-B.

1. Write only the answers to the oral questions. **91** **54** **68**

4. Write only the answers. 16 + 76 = **92**   53 + 38 = **91**   87 + 24 = **111**

7. Write the fractions in the rectangles.

1			
$\frac{1}{2}$		$\frac{1}{2}$	
$\frac{1}{4}$	$\frac{1}{4}$	$\frac{1}{4}$	$\frac{1}{4}$

13. Show $4\frac{1}{2}$ inches on the ruler. Circle or shade it.

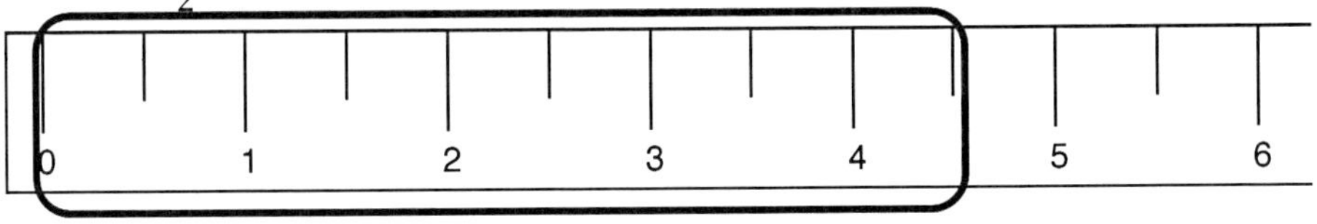

14. How much money is a quarter of a dollar? **25¢**

15. How much money is four quarters of a dollar? **100¢ or $1**

16. How many minutes in a quarter of an hour? **15 min**

17. How many minutes in four quarters of an hour? **60 min or 1 hr**

18. How much is half of $6\frac{1}{2}$ inches? **$3\frac{1}{4}$ in**

19. Write the first 15 multiples of 2.

2	4	6	8	10
12	14	16	18	20
22	24	26	28	30

36. Add 729 + 8576.

		1	1	
		7	2	9
+	8	5	7	6
	9	3	0	5

35. What patterns do you see in the multiples of 2?

All even numbers. The ones are the same in each column.

Games. Spend the remaining time playing games.

SKIP COUNTING. Multiples Memory (P2), Mystery Multiplication Card (P4), and Missing Multiplication Cards (P5), *(Math Card Games).*

MULTIPLICATION. Multiplication Memory (*Math Card Games,* P10).

MENTAL ADDING. Corners variations *(Math Card Games,* A10-12).

Lesson 19

Adding Money as Fractions

OBJECTIVE 1. To work with money as fractions

MATERIALS Dimes and pennies, optional
Worksheet 12, "Adding Money as Fractions"

WARM-UP Ask the child to say and then write the multiples of 2.

2	4	6	8	10
12	14	16	18	20

How much is six 2s? [12] What is another way of saying it? [2 times 6] What is 2 × 5? [10] What is 2 × 3? [6] 2 × 7? [14] What is 2 × 9? [18] What is 2 × 1? [2]

ACTIVITIES ***Adding dimes.*** Draw the following fraction chart with 2 sets of 10 subdivisions. Mark the halfway point with a heavy line.

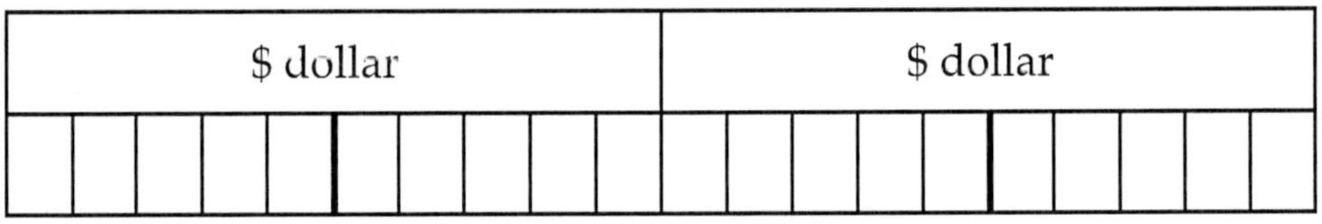

Showing 2 dollars divided into tenths.

What does this chart show? [2 dollars, each divided into tenths] What do we call one-tenth of a dollar? [dime] Write *d*'s for dime.

Ask the child to show where thirty cents is on the chart. Demonstrate crosshatching by drawing parallel lines from the lower left to the upper right as shown below.

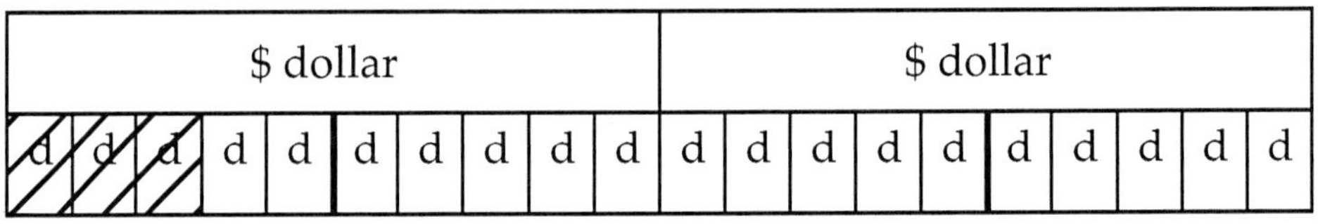

Showing 30¢.

Ask her to write it. [30¢ or $0.30] If necessary, ask: Is there another way to write it?

Where is sixty more cents? Demonstrate it by crosshatching from the upper left to the lower right as shown below.

Note: It is traditional to write a zero before the decimal point for dollar amounts under one dollar.

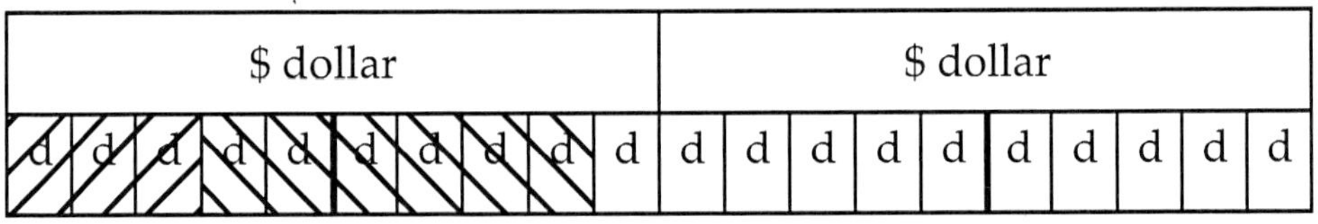

Showing 30¢ + 60¢.

Ask her to write the equation both ways. [30¢ + 60¢ = 90¢, $0.30 + $0.60 = $0.90]

Now ask her to add forty more cents and to write the sums. See the figure on the next page. [$1.30] Explain that we usually do not use a cent sign for amounts over one dollar.

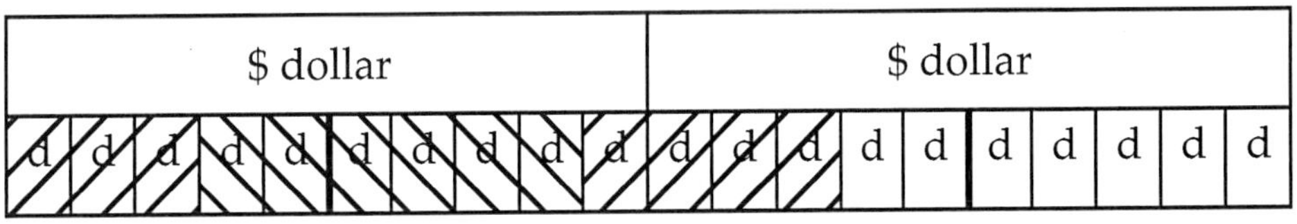

Showing 30¢ + 60¢.

Adding pennies. If we break the dime into tenths, what do we have? [pennies] Use *p* for pennies. Draw the chart and ask a child to show adding 7 pennies and 8 pennies.

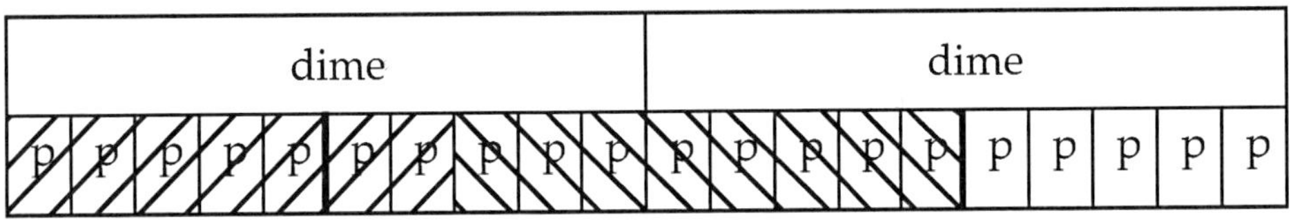

Showing 7¢ + 8¢.

Ask the child to write the equations. [7¢ + 8¢ = 15¢, $0.07 + $0.08 = $0.15]

Worksheet. Discuss the first few problems after the child has completed them. Then have her finish the page. The answers are given below.

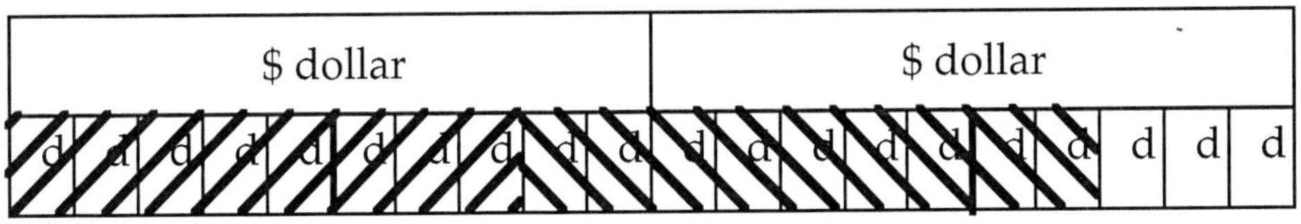

1. **$0.80 + $0.90 = $1.70**

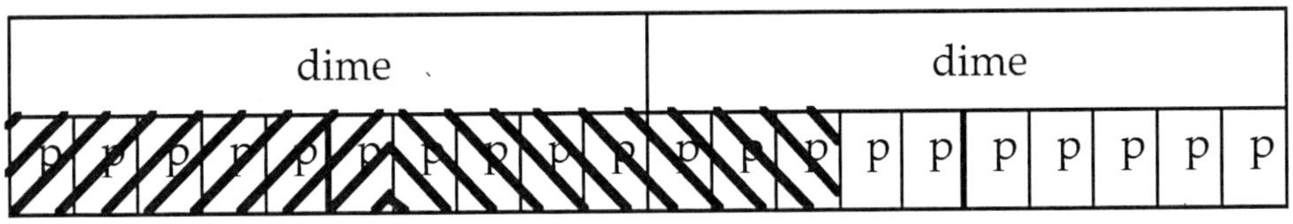

2. **6¢ + 7¢ = 13¢**

$0.06 + $0.07 = $0.13

3.	0	dollars	9	dimes	6	pennies	**$0.96**
4.	0	dollars	13	dimes	5	pennies	**$1.35**
5.	1	dollar	4	dimes	16	pennies	**$1.56**
6.	0	dollars	8	dimes	20	pennies	**$1.00**
7.	2	dollars	12	dimes	20	pennies	**$3.40**
8.	1	dollar	20	dimes	20	pennies	**$3.20**
9.	1	dollar	0	dimes	200	pennies	**$3.00**
10.	0	dollars	100	dimes	0	pennies	**$10.00**
11.	3	dollars	90	dimes	100	pennies	**$13.00**
12.	0	dollars	5	dimes	150	pennies	**$2.00**

Lesson 20

Making Change Different Ways

OBJECTIVES 1. To understand money as fractions
2. To review adding money

MATERIALS Replicas of dimes, nickels, and pennies
Worksheet 13, "Making Change Different Ways"

WARM-UP Ask the child to say and then write the multiples of 4.

4	8	12	16	20
24	28	32	36	40

What is two 4s? [8] five 4s? [20] ten 4s? [40] and nine 4s? [36]

ACTIVITIES ***Expanding the dimes fraction.*** Draw the chart from the previous lesson.

dime										dime									
p	p	p	p	p	p	p	p	p	p	p	p	p	p	p	p	p	p	p	p

Dimes and pennies shown as fractions.

Where could be put nickels in this chart? [between the dime and pennies as shown below]

dime										dime									
nickel					nickel					nickel					nickel				
p	p	p	p	p	p	p	p	p	p	p	p	p	p	p	p	p	p	p	p

Dimes, nickels, and pennies shown as fractions.

Note: Some children may be helped by the abacus as shown below.

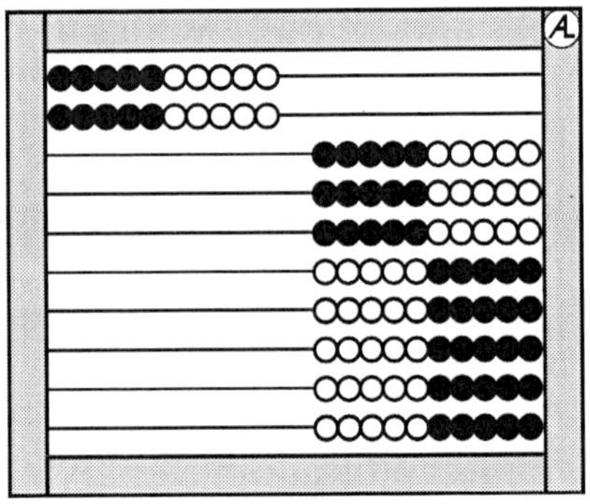

Ways to make 20¢ problem. Tell the child he is to find all the ways he can make 20¢, using dimes, nickels, and pennies. Tell him he can use the coins or the chart, whatever he wants. Stress his work must be organized in some way.

The 9 solutions shown here are organized from greater to less in each denomination. The reverse is also good organizing.

2 dimes	0 nickels	0 pennies
1 dime	2 nickels	0 pennies
1 dime	1 nickel	5 pennies
1 dime	0 nickels	10 pennies
0 dimes	4 nickels	0 pennies
0 dimes	3 nickels	5 pennies
0 dimes	2 nickels	10 pennies
0 dimes	1 nickel	15 pennies
0 dimes	0 nickels	20 pennies

When he has finished, ask how many he found. [9] If needed, give him a few more minutes to find more solutions or remove duplicates.

Encourage him to show some of his solutions on the chart. A few examples are shown below.

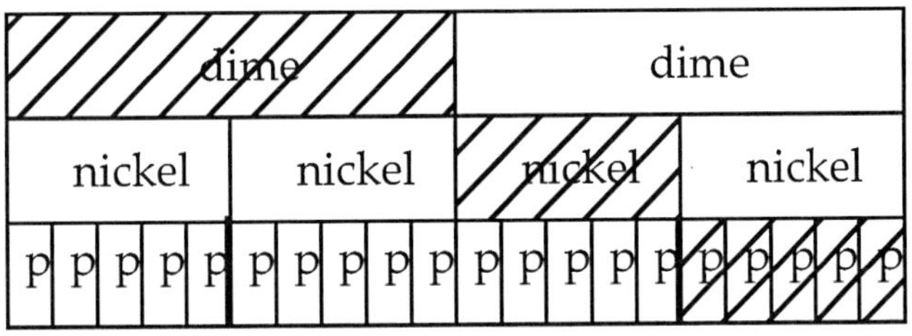

Making 20¢ with 1 dime, 1 nickel, and 5 pennies.

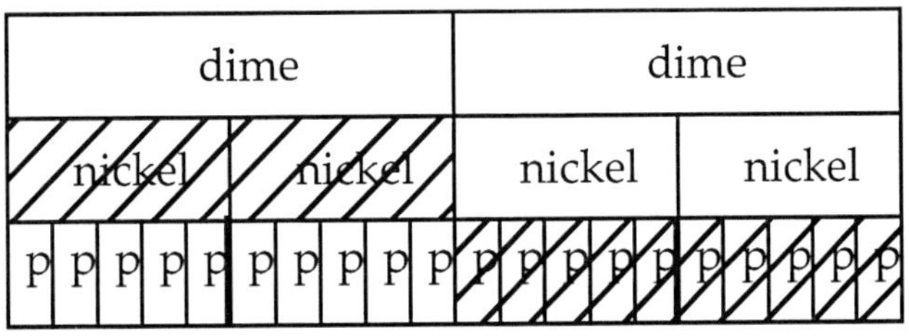

Making 20¢ with 2 nickels, and 10 pennies.

Worksheet. Ask: <u>What is different about the chart on the worksheet?</u> [It includes a quarter. It has 5 nickels rather than 4, and 25, not 20, pennies. The row with dimes looks strange.] Discuss these differences.

Ask the child to read the instructions; ask him to explain what he is to do. The solutions are shown below.

1 quarter	0 dimes	0 nickels	0 pennies
0 quarter	2 dimes	1 nickel	0 pennies
0 quarter	2 dimes	0 nickels	5 pennies
0 quarter	1 dime	3 nickels	0 pennies
0 quarter	1 dime	2 nickels	5 pennies
0 quarter	1 dime	1 nickel	10 pennies
0 quarter	1 dime	0 nickels	15 pennies
0 quarter	0 dimes	5 nickels	0 pennies
0 quarter	0 dimes	4 nickels	5 pennies
0 quarter	0 dimes	3 nickels	10 pennies
0 quarter	0 dimes	2 nickels	15 pennies
0 quarter	0 dimes	1 nickel	20 pennies
0 quarter	0 dimes	0 nickels	25 pennies

Lesson 21

Gallons and Quarts

OBJECTIVES

1. To learn the fractions of a gallon
2. To become familiar with gallons and quarts, relating them to fractions
3. To review writing the greater than and less than symbols, (> <)

MATERIALS

1 gallon container, filled with water
2 half-gallon containers
At least 2 quart containers
A funnel, optional
Worksheet 14, "Gallons and Quarts"

WARM-UP

Ask the child to write the following problem in her journal and add: 6427 + 848. [7275]

Ask the child to say and then write the multiples of 6.

6	12	18	24	30
36	42	48	54	60

Ask them to look at the multiples and state the 6s facts: $6 \times 1 = 6$; $6 \times 2 = 12$, and so forth.

ACTIVITIES

Note: You might want to do at least the first part of this lesson outdoors.

Gallons and quarts. Show the child the gallon container and ask what it is called. [gallon container] How much water does it holds? [gallon] How much milk would fit? [gallon] How much juice would fit? [gallon]

Note: These questions are asked to help the child realize that the word *gallon* is a quantity, or amount of liquid.

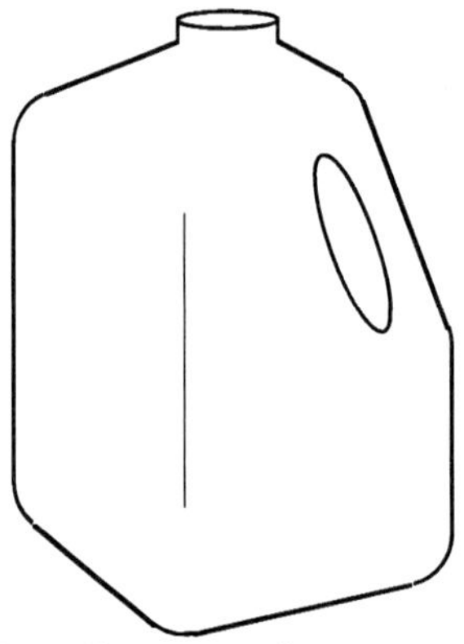

A gallon container.

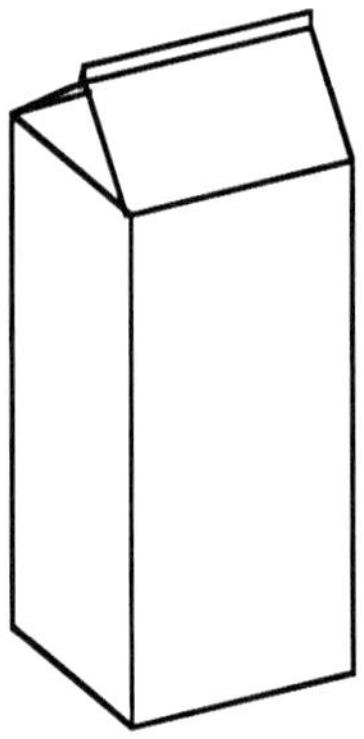

A half-gallon container.

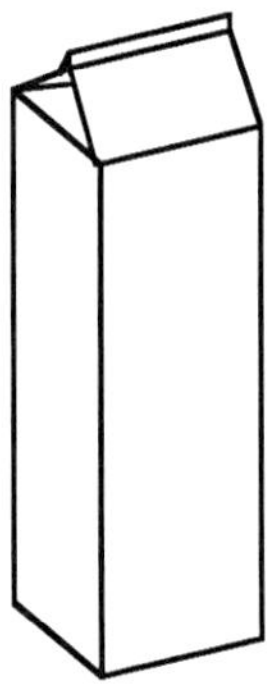

A quart container.

Next show her the half-gallon container. What is it called? [half-gallon container] How much water from the gallon will fit in the half-gallon container? [one half] Fill the half-gallon container and show the level of the water remaining in the gallon jug. Ask how many half-gallon containers are needed to hold the water in the gallon container. [2] Pour it into the second half-gallon container.

Now show her the quart containers and ask how many she thinks will be needed to empty the half-gallon. [2] What do we call half of a half-gallon? [quart] The child might think to say a "quarter" gallon, but stress that the word is *quart.* How many quart containers are needed to hold the water in the half-gallon? [2] How many quarts equal a gallon? [4]

Draw the usual fraction strips (the figure below without the words). Point to the longest strip and ask: <u>What word do we write here?</u> [gallon] Write the word *gallon* as shown. Then ask for the words in the remaining strips. [half-gallon, quart] See below.

1 gallon			
$\frac{1}{2}$ half-gallon		$\frac{1}{2}$ half-gallon	
$\frac{1}{4}$ quart	$\frac{1}{4}$ quart	$\frac{1}{4}$ quart	$\frac{1}{4}$ quart

Relating gallons and quarts to fractions.

To summarize ask the following:

1. <u>How many half-gallons in a gallon?</u> [2]
2. <u>How many quarts in a gallon?</u> [4]
3. <u>What is one fourth of a gallon?</u> [quart]
4. <u>What is one half of a half-gallon?</u> [quart]

Reviewing the > and < symbols. Write

gallon quart

Point to the word *gallon* and ask if it is greater than or less than (point to the word) *quart*. [greater] <u>How do we write the *greater than* symbol?</u>

If necessary, show her how to make it by placing 2 dots by the larger quantity, one at the top and one at the bottom, and one dot in the middle nearer the smaller quantity as shown below. Then connect the dots, starting at the larger number. The same procedure applies for the *less than* symbol. See the figure below. Repeat for quart and half-gallon. <u>At which number do we start when we draw the symbols, the larger or smaller?</u> [larger in both cases]

Note: The child used this way to write these in earlier levels.

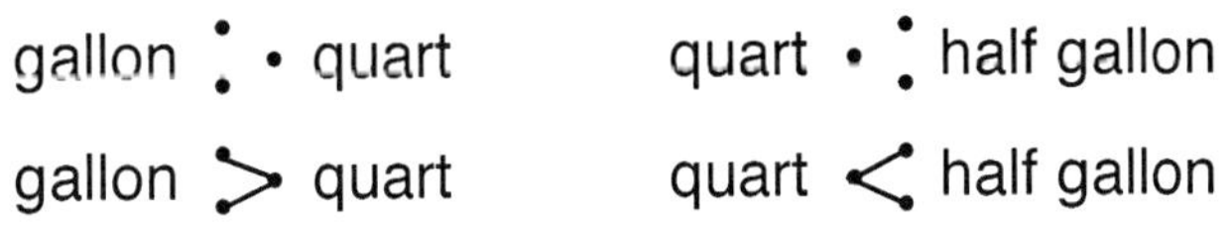

Using the dot-to-dot method to write > and < symbols.

Worksheet. Ask the child to do the chart and problems a-f. Discuss them. The others may be assigned for homework. The solutions are as follows:

a. 4 quarts (=) 1 gallon
b. 1 half-gallon (>) 1 quart
c. 1 gallon (>) 3 quarts
d. 2 quarts (<) 2 half-gallons
e. ¼ gallon (=) 1 quart
f. 6 quarts (<) 2 gallons
g. 1 half-gallon (=) 2 quarts
h. ½ gallon (>) 1 quart
i. 10 quarts (<) 10 gallons
j. 3 gallons (=) 12 quarts
k. **64 ounces**
l. **64 ounces**
m. **128 ounces**
n. **128 ounces**
i. **96 ounces**

Lesson 22

Gallon Problems

OBJECTIVE 1. To use gallons, half-gallons, and quarts to solve problems

MATERIALS 1 gallon container, filled with water
1 plastic gallon ice cream pail
A funnel, optional
Worksheet 15, "Gallon Problems"

Note: Try to get a gallon pail, not a 5-quart pail.

WARM-UP Write the following problem horizontally and ask the child to add it in his journal:

$37.54 + $213.87 [$251.41]

Use the following for writing the multiples of 3. Write 3. What is 3 + 3? [6] Write 6. What is 6 + 3? [9] Write 9. What is 9 + 3? Write 12 on the second line. Continue to 30. What is special about these numbers? [multiples of 3] Then ask him to write the 3s multiples.

Note: The child needs only look at the position to find the product. For example, 3 × 8 is the eighth position, or 24.

3	6	9
12	15	18
21	24	27
30		

Ask the child to look at his multiples and to answer the following:

What is 3 times 3? [9] What is 3 times 6? [18] What is 3 times 9? [27] What is 3 times 4? [12] What is 3 times 5? [15] What is 3 times 10? [30] What is 3 times 1? [3]

ACTIVITIES ***Problem 1.*** Take out the worksheet. Ask the child to read the first problem and to explain what it means. Then ask him to solve it. The problem is:

1. Ms Black wants to buy a gallon of milk. Which is the cheapest way for her to buy it– gallon, half-gallons, or quarts? Use the chart below for the prices. Explain your answer.

gallon	$2.98
half gallon	$1.54
quart	$0.78

Ask him to explain his solution.

The cost of buying milk by the gallon is **$2.98**.

The cost of buying milk by the half-gallon is **$3.08**. [$1.54 + $1.54]

The cost of buying milk by the quart is **$3.12**. [$0.78 + $0.78+ $0.78+ $0.78]

Why do you think milk usually costs more when you buy it in smaller sizes? [more work putting it on the shelves, more containers to fill, more material to make the containers]

Ice cream pail. Show the child the pail and ask: What is it? [ice cream pail] How much do you think it holds? [one gallon] Do you think the water in the gallon container will fit in the pail? Why? [They are both gallons.]

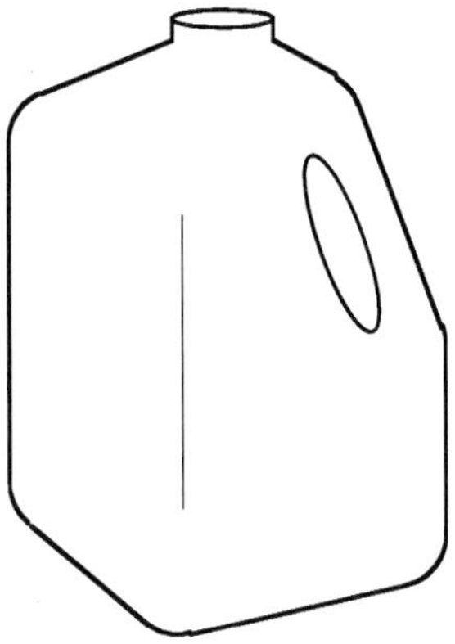

A gallon container.

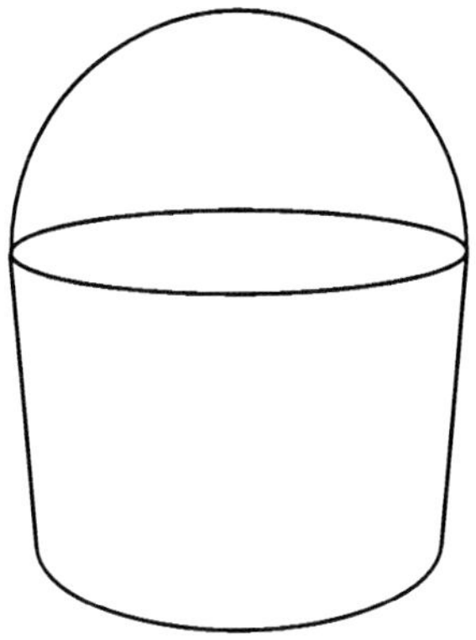

A gallon ice cream pail.

> **Note:** It might seem surprising that ice cream is sold like a liquid–which it is: a frozen liquid.

Pour the water from the gallon container into the pail.

We can buy ice cream in gallon pails: how else can we buy it? [half-gallons and quarts]

Problem 2. Ask the child to read the second problem on the worksheet:

2. Mr. Black wants to buy exactly 2 gallons of ice cream. He wants to buy 3 different flavors–vanilla, chocolate, and strawberry. He can buy it in gallon, half-gallon, and quart containers. What are all the different ways he can buy it?

Ask for one solution. If he does not suggest 2 gallons, ask him for his opinion on that solution. If necessary ask: How could he get the 3 flavors? (Actually, that is the only way to get 2 gallons that will not work.) Remind the child to explain his work, using good organization.

The solutions are below.

1 gallon	**2 half-gallons**	**0 quarts**
1 gallon	**1 half-gallon**	**2 quarts**
1 gallon	**0 half-gallon**	**4 quarts**
0 gallons	**4 half-gallons**	**0 quarts**
0 gallons	**3 half-gallons**	**2 quarts**
0 gallons	**2 half-gallons**	**4 quarts**
0 gallons	**1 half-gallon**	**6 quarts**
0 gallons	**0 half-gallons**	**8 quarts**

Discuss which way he should buy the ice cream.

Lesson 23

Musical Notes

OBJECTIVES 1. To review the term *ellipse*
2. To understand the value of musical notes as related to fractions

MATERIALS A circular object, such as a paper plate
Worksheet 16, "Musical Notes"

WARM-UP Read the following to the child and ask her to write it in her journal and add: $2.08 + $19.94. [$22.02]

Ask the child to write the multiples of 6.

6	12	18	24	30
36	42	48	54	60

What is 6 × 5? [30] 6 × 3? [18] 6 × 7? [42] 6 × 9? [54] and 6 × 1? [6]

ACTIVITIES ***Ellipses.*** Show the child the paper plate held vertically at their eye level. What shape do you see? [circle] Ask her to watch as you slowly turn it horizontally. See the figures below. Tell her the name of a circle that is "squashed" is *ellipse.* Ask her to find other ellipses in the room.

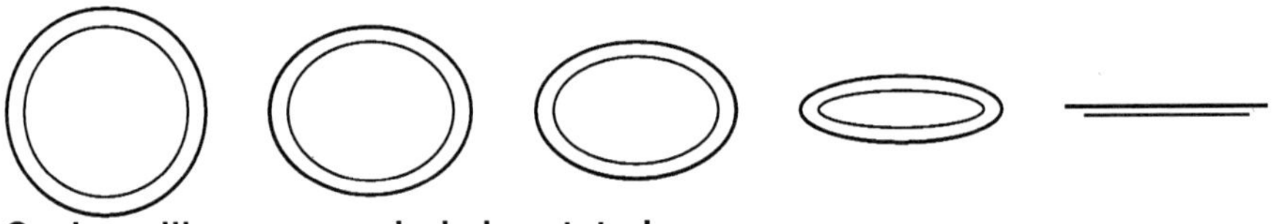

Seeing ellipses as a circle is rotated.

Ways to write time. Introduce the topic of written music by relating it to time. What is time? [maybe, how long it takes for something to happen] What words do we use to talk about time? [hours, minutes, days, weeks, months, years] How do we show time on paper? [circles, numbers, hour hand, minute hand, calendars]

> **Note:** This discussion can help the child understand that writing things can be very different.

Tell the child there is another kind of time we talk and write about—in music. What are some words we use? [beat, rhythm, claps] Ask her to sing and clap to the beat of "Twinkle, Twinkle, Little Star" (Twinkle, twinkle, little star; How I wonder what you are), or some other simple melody. Do it a second time, but ask her, How many beats do you think *How* and *I* get? [1] You may need to say that a beat is the same as a clap. Sing it a third time, asking her to find 2 words that each get 2 beats. [star, are]

What kinds of things do we need to write down on paper when we write music? [number of beats, pitch, and so on] Explain that today we are going to talk about writing beats, or musical time.

Musical notes. Draw the usual fraction strips, including eighths (the figure on the next page without the notes). Point to the various strips and ask for their names. [1 or whole, half, fourth or quarter, eighth] Stress that in music we use the term *quarter*, not fourth.

Draw the whole note and tell her it is called a *whole note*; write the word. What is the shape of the note? [ellipse] Next draw the half notes and ask what she thinks its name is. [half note] Continue with quarter notes and eighth notes.

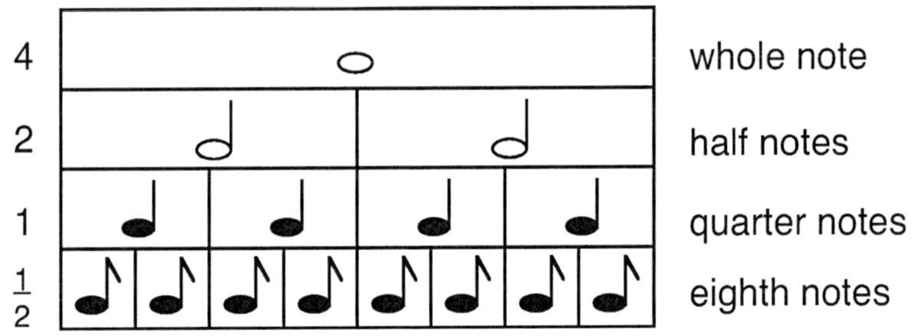

Relating fractions and musical notes.

Tell her for the music today, a whole note gets 4 beats. Write a 4 near the chart as shown above. How many beats does the half note get? [2] Repeat for a quarter note [1] and an eighth note? [one half]

Beats per measure. Explain that music is organized in measures. Draw a rectangle and say we will call it a measure. See the figure below. Write the time signature of 4/4 and say that the top number tells how many beats are in a measure.

Draw a whole note and ask: How many beats does it get? [4] How many beats does the measure need? [4] Can any more notes fit in the measure? [no] See the figure above on the right.

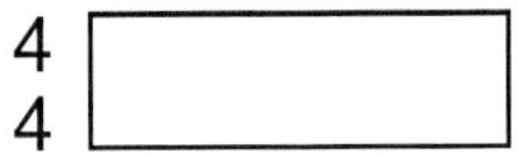

An empty measure.

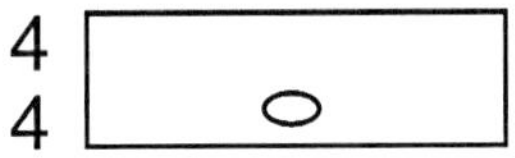

A complete measure.

Draw another measure with a half note and 1 quarter note. Ask how many beats it has in the measure. [3] Is it enough? [No, it needs 1 more beat.] Ask the child to draw it in. See below.

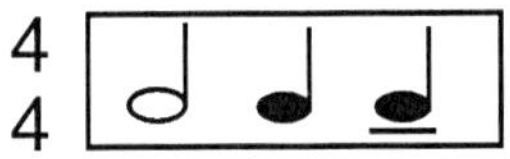

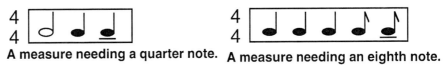

A measure needing a quarter note.

A measure needing an eighth note.

Repeat for a measure with 3 quarter notes and an eighth note. [1 more eighth note is needed. See the figure above.

Worksheet. Ask the child to complete the worksheet; the measures are shown below.

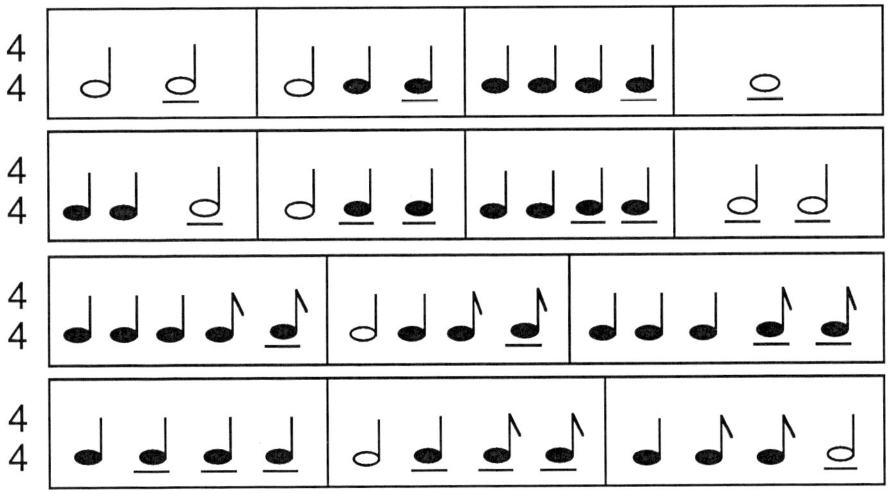

Lesson 24

Review

OBJECTIVES 1 To review and practice
2. To enhance skills through playing games

MATERIALS Worksheet 17-A and 17-B, "Review" (2 versions)
Cards for playing games

ACTIVITIES ***Review worksheet.*** Give the child a review sheet.

The oral problems to be read to the child are as follows:

27 + 7 = 38 + 8 = 27 + 5 =

Review Sheet 17-A.

1. Write only the answers to the oral questions. **34** **46** **32**

4. Write only the answers. 64 + 36 = **100** 58 + 39 = **97** 76 + 28 = **104**

7. How many minutes are in two quarters of an hour? **30 min**

8. What do we call one fourth of a gallon? **quart**

9. What do we call one fourth of a dollar? **quarter**

10. What do we call one fourth of a whole note in music? **quarter note**

11. How much money is three halves of a dollar? **$1.50**

12. How much is half of $8\frac{1}{2}$ inches? **$4\frac{1}{4}$ in**

13. Jill is 3 feet 9 inches tall. Jack is 6 inches taller than Jill. How tall is Jack? (Remember that 1 foot equals 12 inches.)

4 ft 3 in

14. Chris has 9 quarters and 4 dimes. How much money does Chris have altogether?

9 quarters = $2.25
4 dimes = .40
$2.65

15. Write the multiples of 6.

6 12 18 24 30
36 42 48 54 60

25. Add $17.89 + $9.37.

		1	1		
$	1	7	.8	9	
	+	9	.3	7	
$	2	7	.2	6	

The oral problems to be read to the child are as follows:

86 + 5 =   47 + 7 =   58 + 10 =

Review Sheet 17-B.

1. Write only the answers to the oral questions. **91** **54** **68**

4. Write only the answers. 27 + 73 = **100**   67 + 58 = **125**   38 + 84 = **122**

7. How many minutes are in three quarters of an hour? **45 min**

8. What do we call one fourth of a gallon? **quart**

9. What do we call one fourth of a dollar? **quarter**

10. What do we call one fourth of a whole note in music? **quarter note**

11. How much money is two halves of a dollar? **$1.00**

12. How much is half of $10\frac{1}{2}$ inches? **$5\frac{1}{4}$ in**

13. Ray is 3 feet 10 inches tall. Rachel is 4 inches taller than Ray. How tall is Rachel? (Remember that 1 foot equals 12 inches.)

4 ft 2 in

14. Jordan has 7 quarters and 6 nickels. How much money does Jordan have altogether?

7 quarters = $1.75

6 nickels = .30

$2.05

15. Write the multiples of 6.

6 12 18 24 30

36 42 48 54 60

25. Add $9.65 + $16.75.

		1	1	
	$	9.	6	5
+	1	6.	7	5
$	2	6.	4	0

Games. Spend the remaining time playing games.

SKIP-COUNTING. Multiples Memory (12), Mystery Multiplication Card (P4), and Missing Multiplication Cards (P5) *(Math Card Games)*.

MULTIPLICATION. Multiplication Memory *(Math Card Games,* P10).

MENTAL ADDING. Corners variations *(Math Card Games,* A10-12).

Lesson 25

Degrees in a Circle

OBJECTIVES

1. To review the terms *parallel, perpendicular,* and *right angle*
2. To introduce the concept of rotation
3. To learn the meaning of 90°, 180°, 270°, and 360°

MATERIALS

Cards with the words, North, East, South, and West, attached to the appropriate walls in the room or on the floor
2 rulers or similar objects for showing angles
Worksheet 18, "Right Angles and Degrees"

WARM-UP

Ask the child to write the following problem in her journal and add: $68.06 + $43.99. [$112.05]

Ask the child to say the multiples of 9 while you write them.

Note: The 9s are written with the second row reversed to emphasize the pattern.

9	18	27	36	45
90	81	72	63	54

What patterns do you see? [The ones decrease by 1 as the tens increase by 1. The digits equal 9. The digits reverse.] Ask her to write the 9s multiples.

ACTIVITIES

Making turns. Ask the child to stand and face straight ahead. Then ask her to turn all the way around. Tell her that was 1 revolution. Now ask her to turn half way around and then turn another half. What happened when you turned half way? [She faced the back.]

Now ask her to turn to the right a quarter of the way around. Ask her to turn another quarter, and another quarter, and 1 more quarter. Ask how many quarters she needed to make a whole turn? [4] How many quarters made a half turn? [2]

Reviewing parallel and perpendicular. Hold the two rulers in a parallel position and ask: What do we call these lines? [parallel] Hold them in the other positions shown below and ask: Are these lines parallel?

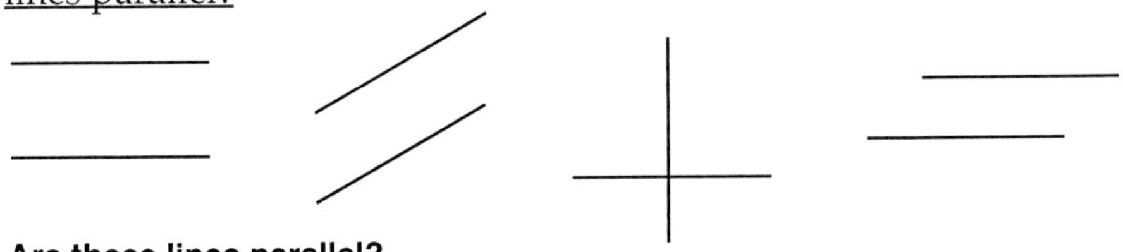

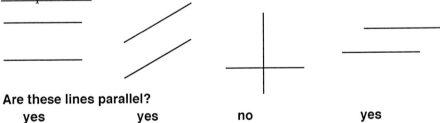

Are these lines parallel?
yes **yes** **no** **yes**

Repeat for perpendicular lines. [perpendicular] Hold them in the other positions shown and ask: Are these lines perpendicular?

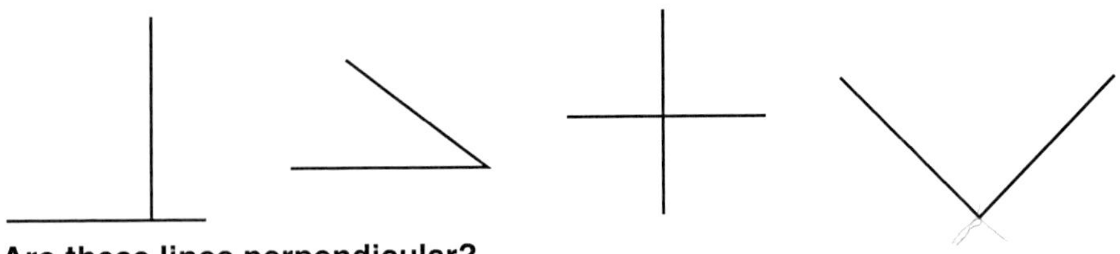

Are these lines perpendicular?
yes **no** **yes** **yes**

> **Note:** The children were introduced to the term, "right angle," in Level C.

Right angles. Hold the rulers perpendicular as shown below on the left and tell the child that it has another name, *right angle.* Tell her it is called right angle because when one line is horizontal, the other seems to stand up right. It does not mean left or right. Show the rulers in the positions shown below and say that these also are right angles.

Right angles.

Ask the child to find right angles in the room.

> **Note:** Research has shown that dynamic angles, rather than static angles, are easier for a child to understand.

Measuring angles. Hold the rulers horizontally (3 o'clock position for the child). Slowly move a ruler vertically; ask her to tell you when you reach a right angle. See below. Tell her that we measure how far something turns in *degrees.* It has nothing to do with degrees in temperature. Stress that a quarter turn is 90°.

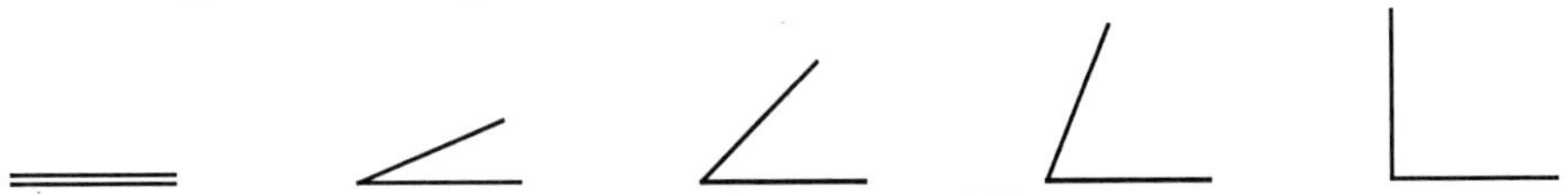

Rotating a ruler through various degrees to 90°.

Move one ruler through 180° as shown below and ask her how many degrees that is. [twice 90, or 180] Move it to three quarters and ask how many degrees. [270] Continue with a full circle and ask how many degrees. [360] Write the numbers and ask: What pattern are the numbers 90, 180, 270, and 360? [multiples of 90]

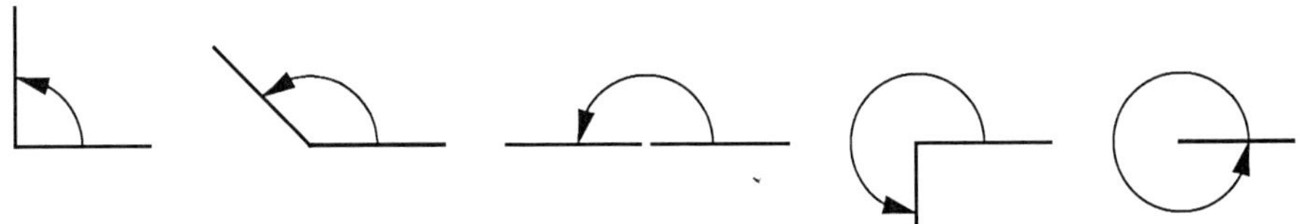

Rotating to 90°. **180°** **270°** **360°**

Repeat the activity, asking the child to follow along with her rulers.

Ask the child to face north. Ask her to turn 90° to the right. What direction are you facing now? [east] Turn 180°. What direction are you facing now? [west] Turn 90° left. What direction are you facing now? [south] Turn 180° left. What direction are you facing now? [north]

> **Note:** It is important that a child reads to learn. Hence, the symbol for degree, °, is explained on the worksheet.

Worksheet. The child can complete the worksheet on her own. The solutions follow:

parallel	**perpendicular**		**90**	
all but the second				
E, F, H, L, T, and possible I				
180		**180**		
90°	**90°**	**90°**	**90°**	
South	**North**	**South**	**East**	**East**

Lesson 26

Skip Counting Patterns

OBJECTIVES 1. To study the skip counting patterns
2. To learn the term *digit*

MATERIALS Worksheet 19, "Skip Counting Patterns"

WARM-UP Read following problem and ask the child to write it in his journal and add: $39.59 + $3.87. [$43.46]

Write

$$26 + 26 = \_\_$$

Ask the child to solve it in his head. How did you do it? Ignore any counting strategies, or ask if it could be done without counting. Some methods include: knowing 25 + 25 = 50, so must be 52; adding 26 + 20, giving 46, then adding 6 to get 52; and taking 4 from second 26, giving 30 and 22.

ACTIVITIES ***Worksheet patterns.*** Ask the child to fill in the blanks for the skip counting patterns on the worksheet. See below.

Note: These formats were emphasized in earlier levels.

2	4	6	8	10
12	14	16	18	20

4	8	12	16	20
24	28	32	36	40

6	12	18	24	30
36	42	48	54	60

8	16	24	32	40
48	56	64	72	80

3	6	9
12	15	18
21	24	27
30		

7	14	21
28	35	42
49	56	63
70		

5	10
15	20
25	30
35	40
45	50

9	18	27	36	45
90	81	72	63	54

Discuss the patterns. What is the last number in every pattern? [2s end in 2 tens, 3s end in 3 tens, and so forth.]

TWOS. What patterns do you see? [They are the even numbers. The second row is 10 plus the first row. The ones repeat in the second row.]

FOURS. What patterns do you see? [The second row is 20 more than the first row. The multiples are every other even number. The ones repeat in the second row.]

SIXES. What patterns do you see? [The first row is the even 3s. Second row is 30 more than the first row. The ones repeat in the second row.]

EIGHTS. What patterns do you see? [In each row the ones are the even numbers backward. The second row is 40 more than the first row, also every other 4. The ones repeat in the second row.]

THREES. What patterns do you see? [Consider the tens: the first row is below 10; the second row is in the teens; the third row is in the twenties. Next consider the ones: they are the numbers 0 to 9 starting at the lower left with 0 (30) and continuing up the column and over to the bottom of the next column. The numbers alternate between even and odd.]

SEVENS. What patterns do you see? [Within each row the tens increase by 1 (0, 1, 2; 2, 3, 4; and 4, 5, 6. The ones are the numbers 0 to 9 starting at the upper right (21) and continuing down the column and over to the next column on the left. The numbers alternate between even and odd.]

FIVES. What patterns do you see? [The ones are either 5 or 0, 5 in first column and 0 in the second column. Also the tens go up by 1 in both columns.]

NINES. What patterns do you see? [The sum of the digits in all cases is 9. The ones go down by 1 while the tens go up by 1. Also the second row has the digits of the first row reversed, as indicated by the arrow. The numbers alternate even and odd.]

Writing about patterns. Ask the child to write a sentence in his journal for each set of multiples, giving a way to help him learn the multiples.

Digits. Explain that a *digit* is a number from 0 to 9 that we use to make all our numbers. What are the digits are in the number 81? [8 and 1] What are the digits in 365? [3, 6, and 5] What are the digits in 1000? [1 and three 0s]

Worksheet. For homework give him the worksheet. The solutions are given below:

2, 4, 6, 8

none

8

6

7

3, 7, 9

9

Ask the child to practice the multiples outside of lesson time with another family member.

Lesson 27

Multiplying with Multiples

OBJECTIVES 1. To review the meaning of multiples
2. To use the multiples tables to multiply

MATERIALS 40 tiles
Worksheet 20, "Multiplying with Multiples"

WARM-UP Write 39 + 39 = __

Ask the child to solve it mentally. Then ask how she did it. Ignore any counting strategies. Some methods include: 40 + 40 = 80, so must be 78; adding 39 + 30, giving 69, then adding 9 to get 78; and taking 1 from second 39, giving 40 and 38.

Read following problem and ask the child to write it in her journal and add: $73.61 + $48.29. [$121.90]

ACTIVITIES ***Meaning of multiples.*** Place 4 tiles in a row as shown in the left figure below. How many groups of 4 are there? [1] How many there are altogether? [4]

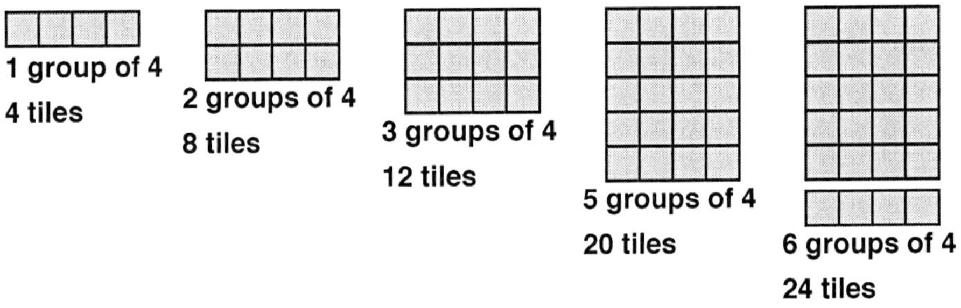

Repeat with a second row of 4s as shown above. How many groups of 4 are there? [2] How many are there altogether? [8]

Note: This activity will also help the slower child, or one new to the program, to understand multiples.

Lay out another row of 4 and ask the same questions. Continue to 10; leave a gap after 5 as shown above.

Now show 1 row—cover the others—and ask how many tiles the child sees. [4] Uncover another row and ask again how many. [8] Continue uncovering rows and repeating the questions. [12, 16, 20, 24, 28, 32, 36, 40] What pattern are you saying? [multiples of 4]

Ask her to write the multiples of 4 as shown below.

4	8	(12)	16	20
24	28	32	36	40

4 taken 3 times (4 × 3)

4	8	12	16	20
24	(28)	32	36	40

4 taken 7 times (4 × 7)

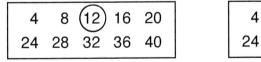

Ask her to point to and circle the number that tells how much 4 taken 3 times is. [third number in the first row] How much is 4 taken 7 times? [second number in the second row] How much is 4 taken 9 times? [fourth number in the second row] How much is 4 taken 1 time? [first number in the first row]

Reading a multiples table. Ask the child to write the multiples of 6 and 8 as shown below.

Writing equations. Then circle the number 24 on the 6 table as shown below and ask for the equation. [$6 \times 4 = 24$] Circle other numbers and ask for the equations, for example, the 48 on the 8 table. [$8 \times 6 = 48$]

6	12	18	24	30
36	42	48	54	60

6 taken 4 times (6 × 4)

8	16	24	32	40
48	56	64	72	80

8 taken 6 times (8 × 6)

Note: The child worked with multiplication equations in Level C.

Ask the child to read the equation. [6 taken 4 times equals 24, or 6 times 4 equals 24] Stress that 6 times 4 means 6 taken 4 times or 4 groups of 6.

Finding products with 3s, 7s, and 9s. Ask the child to write the multiples of 3, 7, and 9 as shown below.

3	6	9
12	15	18
21	24	27
30		

The multiples of 3 with 3 × 4 and 3 × 7 circled.

7	14	21
28	35	42
49	56	63
70		

The multiples of 7 with 7 × 3 and 7 × 5 circled.

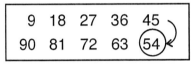

9	18	27	36	45
90	81	72	63	54

The multiples of 9 with 9 × 6 circled.

Then write

$$7 \times 3 =$$

and ask her to read and find the answer on the table. Repeat for 7×5, 3×4, and 3×7. The answers are circled above.

Next write $9 \times 6 =$ and ask where the answer is. Emphasis that the second row of the 9s is written backward. Repeat with a few other examples.

Worksheet. Give the child the worksheet. Answers are below.

12	**24**
12	**24**
32	**36**
32	**16**
21	**14**
21	**49**
42	**28**
63	**35**

$2 \times 8 =$ **16**	$8 \times 9 =$ **72**	$6 \times 4 =$ **24**
$3 \times 6 =$ **18**	$3 \times 1 =$ **3**	$8 \times 6 =$ **48**
$8 \times 4 =$ **32**	$9 \times 4 =$ **36**	$3 \times 8 =$ **24**
$9 \times 3 =$ **27**	$6 \times 6 =$ **36**	$9 \times 8 =$ **72**
		8 × 3 = 24
		8 × 7 = 56

Lesson 28

Adding the Same Number

OBJECTIVES 1. To connect adding the same number with skip counting
2. To use the skip counting patterns to add the same number

MATERIALS Worksheet 21, "Adding the Same Number"

WARM-UP Write $54 + 32 = \_\_$

to be solved mentally. Ask how the child did it. Some methods include: changing it to 56 + 30, giving 86; adding 54 + 30, giving 84, then adding 2 to get 86; and adding 50 + 30 to get 80 and 4 + 2, to get 86.

Read following problem and ask the child to write it in his journal and add: \$85.36 + \$17.49. [\$102.85]

ACTIVITIES ***Adding 3s and 4s.*** Ask the child to add the 3s and the 4s on the worksheet and to describe the pattern. The sums are shown below.

3	3	3	3	3	3	3	3	3
+3	3	3	3	3	3	3	3	3
6	+3	3	3	3	3	3	3	3
	9	+3	3	3	3	3	3	3
		12	+3	3	3	3	3	3
			15	+3	3	3	3	3
				18	+3	3	3	3
					21	+3	3	3
						24	+3	3
							27	+3
								30

4	4	4	4	4	4	4	4	4
+4	4	4	4	4	4	4	4	4
8	+4	4	4	4	4	4	4	4
	12	+4	4	4	4	4	4	4
		16	+4	4	4	4	4	4
			20	+4	4	4	4	4
				24	+4	4	4	4
					28	+4	4	4
						32	+4	4
							36	+4
								40

Then ask how he did it and what patterns he noticed. [skip counting patterns for 3 and 4] Ask him to read the sums and to correct any mistakes. Then ask the following:

1. What is two 3s [6] and two 4s [8]? How much more is 8 over 6? [2]

2. What is three 3s [9] and three 4s [12]? What is the difference between 9 and 12? [3]

3. What is five 3s [15] and five 4s [20]? What is the difference between 15 and 20? [5]

4. What is ten 3s [30] and ten 4s [40]? What is the difference between 30 and 40? [10]

Note: Finding generalizations such as these is one component of algebra.

What pattern do you notice? [The differences in the sums is the same as the number of times the numbers are added.]

Problem 1. Ask the child to solve this and the next problem in his journal. Juanita bought 7 dozen eggs. How many eggs did she buy? [84] Then ask him to share the solution with you.

```
   1
  12
  12
  12
  12
  12
  12
+ 12
  84
```

Ask how he found the sum of seven 2s. [possibly by counting by 2s or thinking that seven 2s is 14] Be sure both methods are discussed.

Problem 2. Juan is buying a treat for each member in his family. The treat costs 46¢. There are 6 members in his family. What will the treats cost? [$2.76]

```
   3            3
  46¢       $0.46
  46¢        0.46
  46¢        0.46
  46¢        0.46
  46¢        0.46
+ 46¢      + 0.46
$2.76       $2.76
```

> **Note:** Either method is correct, but tell the child he cannot use both a decimal point (dot) and a cent sign.
>
> The dollar sign needs to be written only once at the top of the column and for the answer.

Tell the child that he can use the multiples he wrote in the previous lesson to find six 6s and six 4s, if he does not know yet. Ask him to show the solution. Discuss why a 3 is carried. [36 has 3 tens]

Worksheet problem. Ask the child to read the problem about Chris and to solve it. [2920 days (actually 2922, if leap years are considered)]

```
   43
   365
   365
   365
   365
   365
   365
   365
 + 365
  2920
```

Ask him to show the work; discuss all aspects of the problem.

Worksheet. The child can complete the computations on his own. Suggest he use his multiples tables.

The solutions are given below.

576 200 3276 2835 1998 970

Lesson 29

Continuing Geometric Patterns

OBJECTIVE 1. To understand a pattern and to continue it

MATERIALS 1-inch tiles
Worksheet 22, "Continuing Geometric Patterns"

WARM-UP Write 98 + 19 = __

to be solved mentally. Ask how the child did it. Some methods include: changing it to 100 + 17, giving 117; adding 98 + 10, giving 108, then adding 9 to get 117; and adding 90 + 10 to get 100 and 8 + 9, to get 17, for a total of 117.

Read following problem and ask the child to write it in her journal and add: 5689 + 5689. [11378] Ask her to read the answer. [11 thousand 3 hundred seventy-eight]

ACTIVITIES ***Geometric patterns.*** 1. Lay out the following pattern with tiles and ask the child to describe the pattern she sees. [adding 2 to the previous figure]

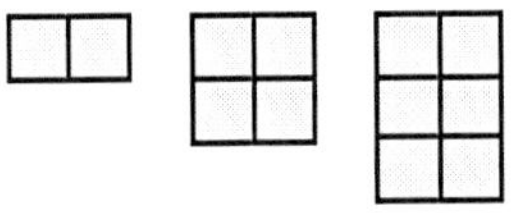

What is the pattern? [adding 2]

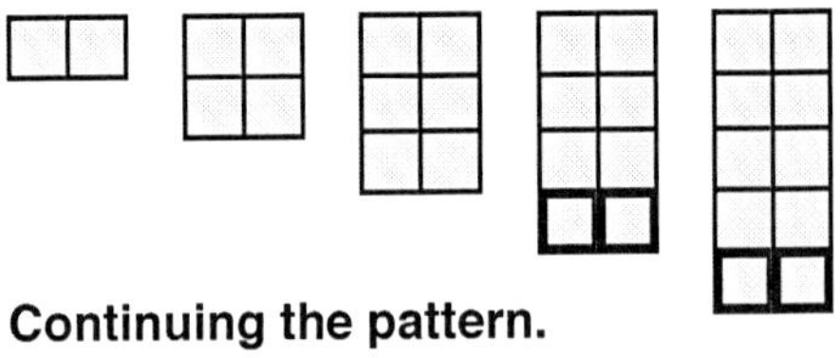

Continuing the pattern.

Then ask the child to continue the pattern 2 more times as shown above on the right. How many tiles are in each figure? [2, 4, 6, 8, 10]

2. Lay out the following pattern. Ask her to describe the pattern she sees. [It starts with a row of 9 tiles; each row is 2 less than the previous row.]

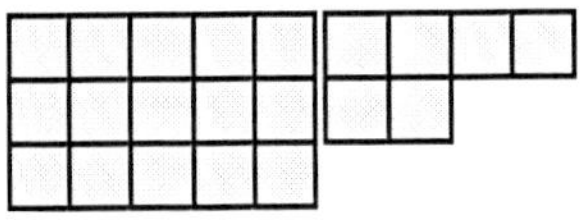

What is the pattern? [subtracting 2]

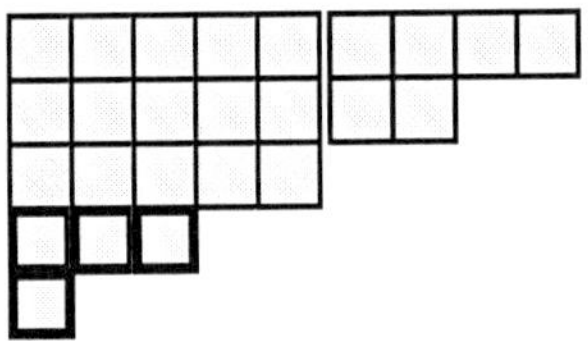

Continuing the pattern.

Note: The child had an introduction to negative numbers in Level C.

Ask the child to continue the pattern. See the figure above on the right. Ask her to say the number of tiles in each row. [9, 7, 5, 3, 1] Could you continue the pattern with tiles? [no] Could you continue with numbers? [yes, negative 1, negative 3, negative 5]

3. Next lay out the squares pattern shown below. Ask for a description of the pattern. [sides increasing by 1 each term: 1 × 1 square, 2 × 2 square, and 3 × 3 square]

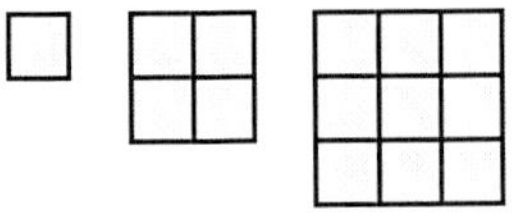

What is the pattern? [constructing squares]

Ask the child to continue it in her journal. See below. Then ask how many tiles are in each square? [1, 4, 9, 16, 25] Ask her to say the multiplication equation for each term. [$1 \times 1 = 1$, $2 \times 2 = 4$, $3 \times 3 = 9$, $4 \times 4 = 16$, $5 \times 5 = 25$]

A second way to see this pattern is to notice that an odd number is added each time. How many squares did we add going from the first term to the second? [3] How many squares did we add going from the second term to the third? [5] How many squares did we add going from the third term to the fourth? [7] What is the pattern? [adding consecutive odd numbers] See the figure below.

Note: The square pattern is one of the important patterns in mathematics.

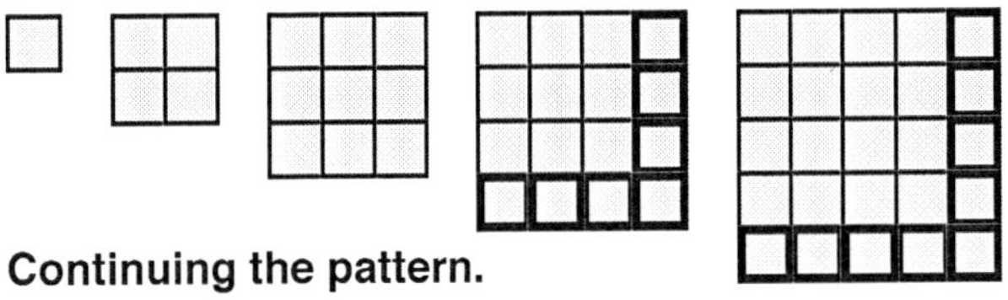

Continuing the pattern.

4. Draw 3 rectangles as shown below. Divide the second rectangle into halves and the third rectangle into thirds. What comes next? [a rectangle divided into fourths]

Note: This pattern will come up again with numeric patterns.

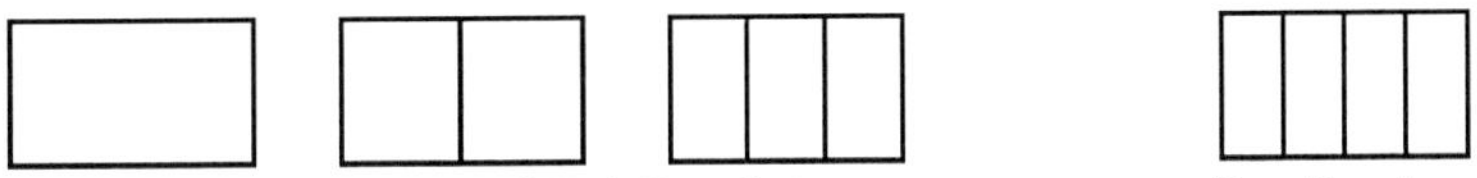

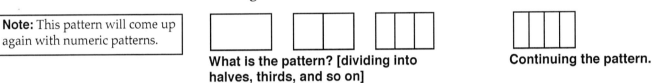

What is the pattern? [dividing into halves, thirds, and so on]

Continuing the pattern.

Worksheet. The worksheet has similar patterns to continue. The solutions are shown below.

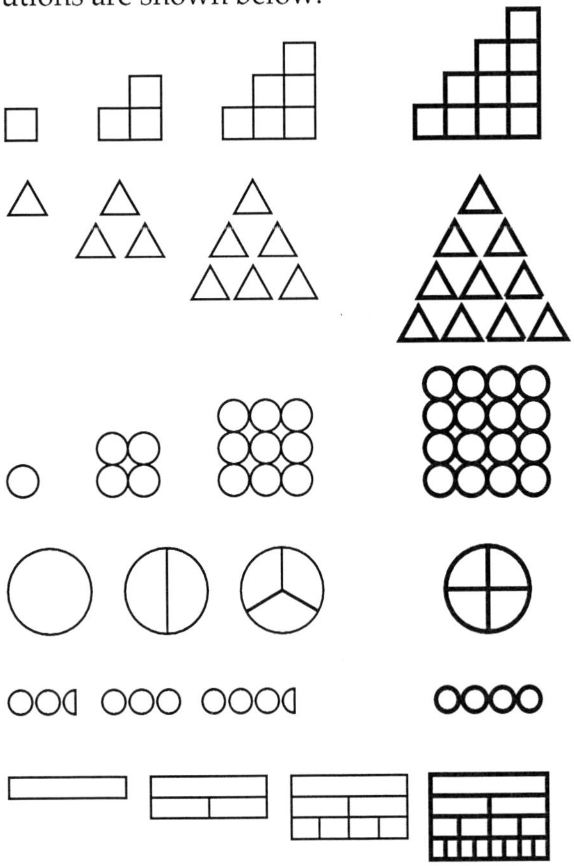

Lesson 30

Review

OBJECTIVE 1. To review and practice

MATERIALS Worksheet 23-A and 23-B, "Review" (2 versions)
Cards for playing games

ACTIVITIES ***Review worksheet.*** Give the child a review sheet.

The oral problems to be read are as follows:

39 + 10 = 81 + 15 = 47 + 15 =

Review Sheet 23-A.

1. Write only the answers to the oral questions. **49** **96** **62**

4. Write only the answers. 56 + 14 = **70** 39 + 29 = **68** 67 + 98 = **165**

7. Write the multiples of 8.

8 **16** **24** **32** **40**

48 **56** **64** **72** **80**

17. Jan and Jay are buying trees to plant. Each tree costs $3.89. How much will eight trees cost?

		7		7
	$	3	.8	9
		3	.8	9
		3	.8	9
		3	.8	9
		3	.8	9
		3	.8	9
		3	.8	9
	+	3	.8	9
$	3	1	.1	2

18. Draw $7\frac{1}{2}$ cookies.

19. Draw a right angle. How many degrees does it have?

90°

21. A quart holds 32 ounces. How many ounces are in a half gallon? How many ounces in a gallon?

64 oz half gallon

128 oz in gallon

23. Jan and Jay plant five trees. Each tree takes them a quarter of an hour. If they start at 9:15, what time will they finish?

10:30

24. How many minutes is 5 quarters of an hour?

60 + 15 = 75 min

25. How much money is 5 quarters of a dollar?

$1.25

The oral problems to be read to the child are as follows:

86 + 5 = 47 + 7 = 58 + 10 =

Review Sheet 23-B.

1. Write only the answers to the oral questions. **91** **54** **68**

4. Write only the answers. 73 + 17 = **90** 47 + 49 = **96** 78 + 99 = **177**

7. Write the multiples of 9.

9 18 27 36 45

54 63 72 81 90

17. Jordan is buying lilac bushes to plant. Each bush costs $4.79. How much will nine bushes cost?

			7	8
	$	4	.7	9
		4	.7	9
		4	.7	9
		4	.7	9
		4	.7	9
		4	.7	9
		4	.7	9
		4	.7	9
	+	4	.7	9
$	4	3	.1	1

18. Draw $6\frac{1}{2}$ cookies.

19. Draw a right angle. How many degrees does it have?

90°

21. A quart holds 4 cups. How many cups are in a half gallon? How many cups in a gallon?

8 cups half gallon

16 cups in gallon

23. Jordan is planting 6 lilac bushes on Monday. Each takes a quarter of an hour. If Jordan starts at 9:30, what time will Jordan finish?

11:00

24. How many minutes is 3 quarters of an hour?

45 min

25. How much money is 6 quarters of a dollar?

$1.50

Games. Spend the remaining time playing games.

SKIP-COUNTING. Multiples Memory (12), Mystery Multiplication Card (P4), and Missing Multiplication Cards (P5) *(Math Card Games).*

MULTIPLICATION. Multiplication Memory *(Math Card Games,* P10).

MENTAL ADDING. Corners variations *(Math Card Games,* A10-12).

Lesson 31

Continuing Numeric Patterns

OBJECTIVE 1. To discover a numeric pattern and to continue it

MATERIALS 1-inch tiles
Worksheet 24, "Continuing Numeric Patterns"

WARM-UP Write

$$99 + 24 = \_\_$$

to be solved mentally. Ask how she did it. Here the simplest and most obvious is to take 1 from the 24 to make 100 + 23, or 123.

Read the following problem and ask the child to write it in her journal and add, 79,016 + 8809. [87,825] Ask her to read the answer. [87 thousand 8 hundred twenty-five]

What is 10 doubled? [20] What is 42 doubled? [84] What is 12½ doubled? [25] What is 19½ doubled? [39] What is 4 ¼ doubled? [8½] What is a quarter hour doubled? [half hour]

ACTIVITIES ***Numeric patterns.*** 1. Write the following pattern:

4 6 8 __ __

and ask what 2 terms come next. To help the child understand the procedure, ask how she could get from 4 to 6. [add 2] To show this, write a "+ 2" in the space between the 4 and 6 as shown. Does the "+ 2" also work for going from 6 to 8? [yes] What is the rule for continuing the pattern? [adding 2] What are the next two terms? [10, 12]

4 **+2** 6 **+2** 8 10 12

2. Write the next pattern:

90 80 70 __ __

and ask for the next two terms. [60, 50] What operation do we do to get to the next number? [subtract 10] Ask the child to write it and the next two terms. See below.

90 **– 10** 80 **– 10** 70 **60** **50**

3. For the third example, use:

2 4 8 __ __

What operation do we do to get to the next number? [+ 2, × 2] Does adding 2 work for going from 4 to 8? [no] What other operation could we try? [multiply by 2, or double] Does it work from 4 to 8? [yes] What are the next two terms? See below.

2 **× 2** 4 **× 2** 8 **16** **32**

4. The next example is very important in mathematics, but harder to identify. Write

1 $^{+3}$ 4 $^{+5}$ 9 __ __

The pattern can be continued with + 7 and + 9 and is given below.

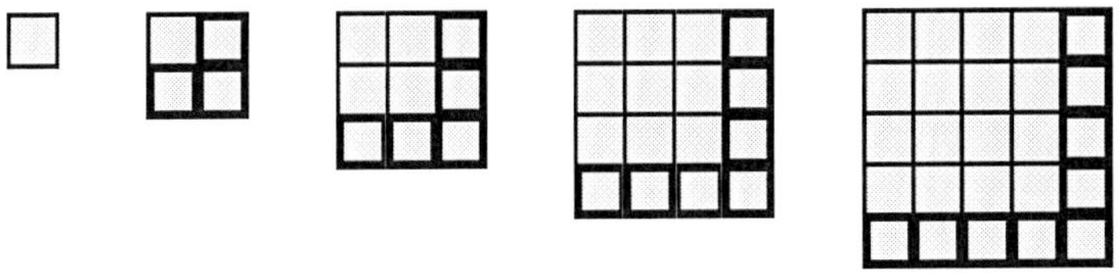

The square pattern formed by adding consecutive odd numbers.

Note: Discovering the pattern involves guessing and then checking to see if it works for the next number.

1 4 9 <u>**16**</u> <u>**25**</u>

<u>Does this remind you of a pattern you did with the tiles?</u> Lay out the pattern below.

Summarize by asking: <u>How can we find the pattern?</u> [by guessing and checking. Sometimes writing little numbers to show the change helps.]

Worksheet. The worksheet has 18 numeric patterns to be continued. The solutions are shown below.

35	37	39	**41**	**43**
102	101	100	**99**	**98**
9	18	27	**36**	**45**
16	24	32	**40**	**48**
35	30	25	**20**	**15**
18	24	30	**36**	**42**
3	6	9	**12**	**15**
3	6	12	**24**	**48**
4	40	400	**4000**	**40,000**
80	40	20	**10**	**5**
1	4	9	**16**	**25**
0¢	25¢	50¢	**75¢**	**$1.00**
$0.05	$0.10	$0.15	**$0.20**	**$0.25**
$0.01	$0.10	$1.00	**$10.00**	**$100.00**
$1.00	$1.50	$2.00	**$2.50**	**$3.00**
1	1½	2	**2½**	**3**
0	¼	½	**¾**	**1**
5	4½	4	**3½**	**3**
½	⅓	¼	**⅕**	**⅙**

Lesson 32

Subtracting by Going Up

OBJECTIVES

1. To explore the relationship between adding and subtracting
2. To practice the *going up* strategy for subtraction

MATERIALS Worksheet 25, "Subtracting by Going Up"

WARM-UP Write 45 + 39 = __

> **Note:** Skip this lesson and the next 3 lessons if the child has done the transition lessons or does not need the review.

to be solved mentally. Ask how the child did it. Some methods include: changing it to 44 + 40, giving 84; adding 45 + 30, giving 75, then adding 9 to get 84; and adding 40 + 30 to get 70 and 5 + 9, to get 14, for a total of 84.

Read following problem and ask the child to write it in his journal and add: $7011.06 + $209.04. [$7220.10]

What is needed with 9 to make 11? [2] What is needed with 9 to make 17? [8] What is needed with 9 to make 15? [6] What is needed with 9 to make 16? [7] What is needed with 9 to make 13? [4]

ACTIVITIES ***Comparing addition and subtraction.*** Ask the child what is the opposite of north. [south] Tell him that words that are opposite are called *antonyms*. Ask for the antonym of east and [west] south. [north]

> **Note:** The term *take away* is not a good way to describe subtraction. It emphasizes only one aspect of subtraction.
>
> Indeed, that phrase is not found in the dictionary.

Ask for the antonyms of addition [subtraction], plus [minus], and subtract. [add] Ask the child to describe subtraction. [breaking into parts, comparing, opposite of adding] Also ask him to pose several problems to be solved by subtraction.

Problem 1. Terry read 8 pages and just finished page 29. What page did Terry start on? [21] Ask the child to write the numbers in the part-whole circle set shown.

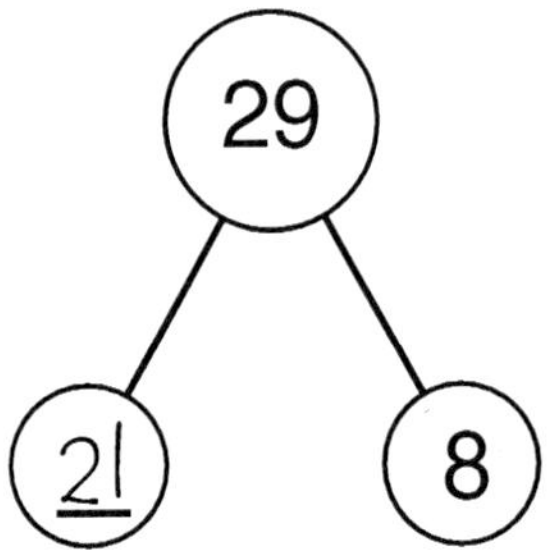

A part-whole circle set. The whole is written in the large circle and the parts in the smaller circles.

Ask him to write the equation. Then ask him to write it a different way. The addition and subtraction equations are below.

21 + 8 = 29

29 – 8 = 21

Problem 2. Ask the child to solve the following problem and to write the equation. Shawn wants to own 100 cards. He already has 95 cards. How many more does he need? [5, 95 + 5 = 100 or 100 – 95 = 5]

To solve this problem most people would start at 95 and *go up* 5 more to reach 100. Note that the child may write the equation *after* he solves the problem.

> **Note:** This is similar to the procedure for making change in money transactions.

Practice. Write

29 + __ = 33

and ask how the child could solve it. Stress the solution of finding what makes 30, that is 1, and then adding that to 3 to reach 33. [4] Also ask him to write the subtraction equation. [33 – 29 = 4]

Repeat for

26 + __ = 42

Here first 4 needed to reach 30, then 10 is needed to reach 40, and 2 more to reach 42, giving 16 as the answer. Another way is first to add 10 to reach 36 and then add the 4 to reach 40 and the 2 to reach 42.

> **Note:** The *going up* procedure is easier for many subtraction problems.

Subtracting by going up. Give the child this problem. There are 85 blocks kept in Gerry's toy box. On a certain day, 77 were there. How many were missing? [8] Ask him to write the subtraction equation. [85 – 77 = 8] Ask how he found his answer.

Write and ask the child to solve the following by going up:

64 – 39 = __ [25]

Ask how he did it. One way is to add 1 to reach 40, then 20 to reach 60, and finally 4 to reach 64, making 25. A second way is to add 20 to reach 59, then 5 more to reach 64.

Ask the child to solve the following:

72 – 59 = __ [13]

Again ask how he did it.

Worksheet. The first part of Worksheet 25 asks the child to solve missing addend problems and to rewrite them as subtraction problems. The last part are subtractions, to be done by going up.

6 + **4** = 10	**10 – 6 = 4**
8 + **12** = 20	**20 – 8 = 12**
27 + **4** = 31	**31 – 27 = 4**
46 + **7** = 53	**53 – 46 = 7**
39 + **9** = 48	**48 – 39 = 9**
8 + 9 = 17	**17 – 9 = 8**
7 + 17 = 24	**24 – 17 = 7**
18 + 12 = 30	**30 – 12 = 18**
9 + 26 = 35	**35 – 26 = 9**
11 + 51 = 62	**62 – 51 = 11**

39 – 33 = **6**	50 – 32 = **18**
31 – 28 = **3**	62 – 49 = **13**
13 – 8 = **5**	79 – 62 = **17**
74 – 71 = **3**	100 – 82 = **18**
51 – 31 = **20**	101 – 92 = **9**
105 – 96 = **9**	41 – 23 = **18**
75 – 50 = **25**	56 – 39 = **17**

Lesson 33

Subtracting by Going Down

OBJECTIVES 1. To practice the *going down* strategy for subtraction
2. To discuss when trading is necessary

MATERIALS A ream of paper, if available
Corner Cards (see Lesson 8)

WARM-UP Ask the child to say the multiples of 5. Then ask her to write them and to say the tables. [$5 \times 1 = 5$ and so forth]

Ask the child to write 55 six times in her journal and add. [330] Repeat for 5555 eight times. [44,440]

Ask, What how much is 39 minus 37? [2] How much is 20 minus 17? [3] How much is 50 minus 46? [4] How much is 82 minus 78? [4] How much is 90 minus 85? [5]

ACTIVITIES ***Problem 1.*** Show the child the ream of paper, if available, and explain that is how paper is packaged for copy machines (and some computer printers). Show her a stack of 20 sheets and ask how many sheets of paper she thinks are in the ream. [500]

How many sheets are left if someone takes out 1 sheet from the ream? [499] Ask for an explanation and ask her to write the equation.

$$500 - 1 = 499$$

How many are left if someone takes 10 more sheets? [489]

$$499 - 10 = 489$$

How many are left if someone takes 100 more sheets? [389]

$$489 - 100 = 389$$

How many sheets are taken from the ream? [111] Ask for at least two solutions. Either add the 1, 10, and 100 or start with 389 and go up to 500 or do the subtraction 500 – 389.

Problem 2. Give the child this problem. Sarah is selling apples. She started with 64 apples and sold 3. How many apples does she have left? [61] Ask her to write the problem in the first column below.

$$\begin{array}{r} 64 \\ -3 \\ \hline 61 \end{array} \qquad \begin{array}{r} 61 \\ -3 \\ \hline 58 \end{array} \qquad \begin{array}{r} 58 \\ -3 \\ \hline 55 \end{array}$$

Note: This line of reasoning is being pursued because a child may not be sure when to trade.

Sarah sells 3 more; now how many are left? [58] Again ask for the numbers to be written in a column as shown above. Repeat the question once more. [55]

Now ask the child to notice that the tens in the first and third examples stayed the same in the answers—64, 61 and 58, 55. But the in second example, the 6 tens turned to 5 tens—61, 58. Why is that so? [There are enough ones in the first and third examples, but in the second example a ten must be traded to get enough ones.

Ask how she can tell if she needs to trade. [The number being subtracted is greater, or more, than the top number.]

Now write the following examples and ask which will need a trade. [first, fourth, fifth]

43	55	87	36	15
– 5	– 3	– 6	– 7	– 8

Ask the child to write the answers.

43	55	87	36	15
– 5	– 3	– 6	– 7	– 8
38	52	81	29	7

Subtracting 5s and 10s. Tell the child she will be playing Subtraction Corners. She will be subtracting her scores mentally. Practice with her as follows. Start with 150. When 15 is subtracted, first subtract the 10 then the 5.

	150
subtract 10	140
subtract 5	135
subtract 10	125
subtract 15	110
subtract 15	95

Subtraction Corners. The rules for playing Subtraction Corners are the same as regular Corners. However, the scoring is different. Players start with 150 points and the player with the lowest score is the winner. You might ask, Can you think of other games where the person with the lowest score is the winner? [golf, track]

Scoring can be done in the math journal with only the results written. The computations are done mentally as above. Be sure she does her own scoring.

Lesson 34

Reviewing Subtraction Strategies

OBJECTIVES 1. To review the subtraction facts
2. To review subtraction strategies for minuends ≤ 10

MATERIALS An AL abacus, if available
Worksheet 26, "Reviewing Subtraction Facts $\leq$ 10"

WARM-UP Ask the child to say the multiples of 7. Then ask them to write them and to say the tables. [$7 \times 1 = 7$ and so forth]

7	14	21
28	35	42
49	56	63
70		

Ask him to write 706 seven times in his journal and add. [4942]

Say the following while he writes only the answers:

37 – 2 = [35]   91 – 2 = [89]   20 – 4 = [16]

ACTIVITIES ***Subtracting from 10 or less.*** Take out the worksheet and ask the child to do the facts in the first rectangle. Then ask him to circle any problems that were hard for him. Next ask him to correct his answers while you read the answers given below.

8 – 6 = **2**	8 – 5 = **3**	10 – 2 = **8**	9 – 7 = **2**	8 – 7 = **1**
3 – 2 = **1**	6 – 2 = **4**	10 – 8 = **2**	8 – 2 = **6**	10 – 9 = **1**
9 – 4 = **5**	9 – 6 = **3**	6 – 1 = **5**	9 – 8 = **1**	9 – 2 = **7**
6 – 4 = **2**	2 – 1 = **1**	8 – 4 = **4**	9 – 3 = **6**	6 – 5 = **1**
7 – 3 = **4**	7 – 6 = **1**	5 – 4 = **1**	8 – 3 = **5**	8 – 1 = **7**
9 – 5 = **4**	10 – 7 = **3**	9 – 1 = **8**	5 – 3 = **2**	10 – 1 = **9**
10 – 3 = **7**	7 – 2 = **5**	10 – 4 = **6**	3 – 1 = **2**	4 – 2 = **2**
10 – 5 = **5**	5 – 2 = **3**	7 – 5 = **2**	6 – 3 = **3**	4 – 3 = **1**
4 – 1 = **3**	7 – 1 = **6**	10 – 6 = 4	5 – 1 = **4**	7 – 4 = **3**

Strategies for 10 or less. Ask the child to think of as many subtraction strategies as he can. Remind him that counting is not a strategy. Some strategies are listed below.

<u>Subtracting 1.</u> Subtracting 1 from a number is the previous number. Examples: 4 – 1 = 3, 9 – 1 = 8.

<u>Subtracting 2.</u> Subtracting 2 from an even number is the previous even number. Subtracting 2 from an odd number is the previous odd number.

<u>Subtracting 5 (from numbers 5 to 10).</u> Subtracting 5 from a number is obvious on fingers or the abacus; it is almost a definition with sums up to 10.

<u>Subtracting from 10.</u> These were learned first with the Go to the Dump game. They are easy to see on the abacus.

<u>Going Up.</u> To subtract 9 – 6, start with 6 and think how much is needed to get to 9. See the figure on the next page on the left.

If the number being subtracted is less than 5, first find how much is needed to go to 5 and then add the amount over 5. For example in 7 – 4, first compute the amount to go to 5 (1) and then add the 1 with 2, the amount 7 is over 5. See the figure below on the right.

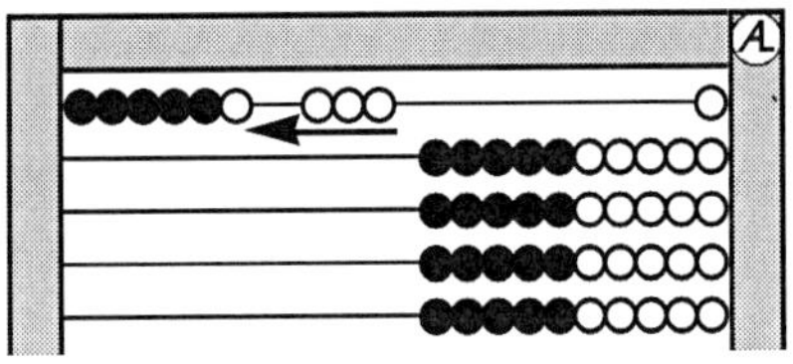

To subtract 9 – 6, think of starting at 6 and going to 9.

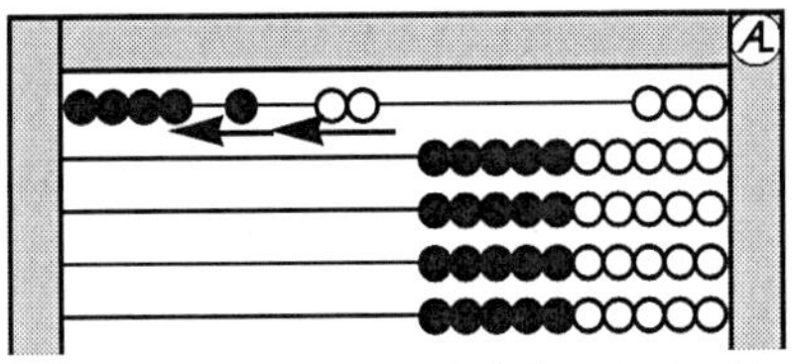

To subtract 7 – 2, think of starting at 2 and first going up to 5 (1) and then continuing to 7 (2 more). [3]

Ask the child to notice if he got the hard ones right or not. Ask him to write the facts he thought were hard. Ask if it helps to see beads in his head in two colors like the abacus he used in Level C. Some examples are given below.

<u>9 – 2.</u> Two strategies work here. Subtracting 2 from an odd number gives the previous odd number. Also since 10 – 2 is 8, then 9 – 2 must be 7.

<u>8 – 6.</u> Subtracting 2 consecutive numbers, such as 9 – 8 gives 1 and subtracting 2 consecutive even or odd numbers gives 2 such as 8 – 6 or 9 – 7.

<u>8 – 5.</u> It could be done by thinking 8 – 6 = 2, so 8 – 5 must be 3. Also show a row of tiles grouped in 5s as shown below. Ask for 8 – 5 [3] and 8 – 3. [5] He could also see it by showing 8 with his hands.

Worksheet. Ask the child to use strategies to complete the second half of the worksheet. The facts and answers are given below.

10	8	3	10	6	4	6	8	5
– 1	– 4	– 2	– 5	– 1	– 3	– 3	– 2	– 1
9	**4**	**1**	**5**	**5**	**1**	**3**	**6**	**4**
10	5	10	9	3	9	7	9	8
– 7	– 2	– 3	– 2	– 1	– 1	– 5	– 7	– 1
3	**3**	**7**	**7**	**2**	**8**	**2**	**2**	**7**
10	9	6	10	7	8	7	4	7
– 2	– 8	– 2	– 4	– 6	– 7	– 1	– 2	– 3
8	**1**	**4**	**6**	**1**	**1**	**6**	**2**	**4**
9	6	7	8	2	9	10	7	8
– 6	– 4	– 4	– 6	– 1	– 3	– 6	– 2	– 5
3	**2**	**3**	**2**	**1**	**6**	**4**	**5**	**3**
5	10	9	8	10	9	5	4	6
– 4	– 9	– 4	– 3	– 8	– 5	– 3	– 1	– 5
1	**1**	**5**	**5**	**2**	**4**	**2**	**3**	**1**

Lesson 35

More Subtraction Strategies

OBJECTIVES
1. To review the subtraction facts
2. To review subtraction strategies for minuends > 10

MATERIALS
An AL abacus, if available
Worksheet 27, "Subtraction Strategies"

WARM-UP
Ask the child to say the multiples of 7. Then ask her to write them and to say the tables. [7 × 1 = 7 and so forth]

7	14	21
28	35	42
49	56	63
70		

Say the following while she writes only the answers:

41 – 2 = [39]   58 – 2 = [56]   47 – 4 = [43]

ACTIVITIES
Subtracting from more than 10. Take out the worksheet and ask the child to do the facts in the first rectangle. Then ask her to circle any problems that were hard for her. Next ask her to correct her answers while you read the answers given below.

12 – 8 = **4**	16 – 7 = **9**	11 – 5 = **6**	12 – 7 = **5**	13 – 9 = **4**
15 – 8 = **7**	14 – 5 = **9**	13 – 7 = **6**	12 – 9 = **3**	15 – 7 = **8**
14 – 9 = **5**	11 – 9 = **2**	16 – 8 = **8**	11 – 7 = **4**	11 – 2 = **9**
14 – 8 = **6**	15 – 6 = **9**	13 – 6 = **7**	12 – 5 = **7**	11 – 8 = **3**
14 – 7 = **7**	12 – 3 = **9**	14 – 6 = **8**	17 – 8 = **9**	16 – 9 = **7**
11 – 6 = **5**	11 – 3 = **8**	12 – 6 = **6**	13 – 8 = **5**	13 – 4 = **9**
15 – 9 = **6**	13 – 5 = **8**	12 – 4 = **8**	17 – 9 = **8**	11 – 4 = **7**
18 – 9 = **9**				

You might tell her that those are all the subtraction facts over 10. How many are there? [36, 5 × 7 + 1]

Strategies for 10 or more. Ask the child to think of as many subtraction strategies as she can. Remind her that counting is not a strategy. Some strategies are listed below.

Going Up. This is the same procedure as used for subtracting from numbers less than 10. For example, in 13 – 9, it takes 1 to get to 10 and 3 to get to 13. So 1 + 3 = 4. See the figures below. The second example shows 14 – 7.

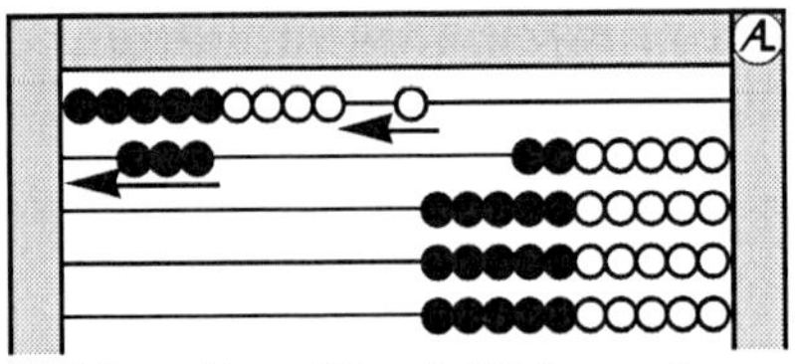

Subtracting 13 – 9 (1) by going up to 10 and then to 13.

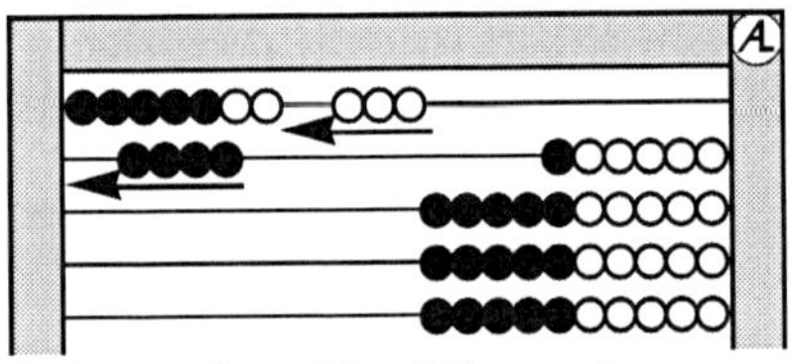

Subtracting 14 – 7 by going up to 10 (3) and then to 14.

Down Over Ten. In this strategy, as much as possible is subtracted from the ones with the remainder being subtracted from the ten. For example in 18 – 9, remove 8 from the 8 and 1 more from the 10. See the figure on the next page for the procedure on the abacus.

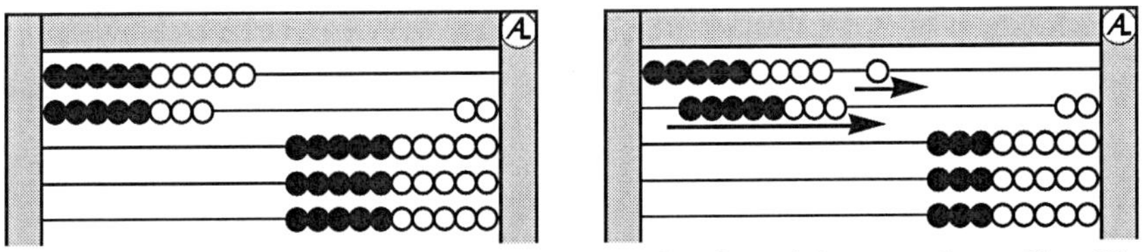

Subtracting 18 – 9 by removing 8 from the 8 and 1 more from the 10.

Enter 15 and ask the child how she could subtract 7, using the same strategy, down-over-ten. [5 is subtracted from the 5 and last 2 from the 10, giving 7.] See the figures below.

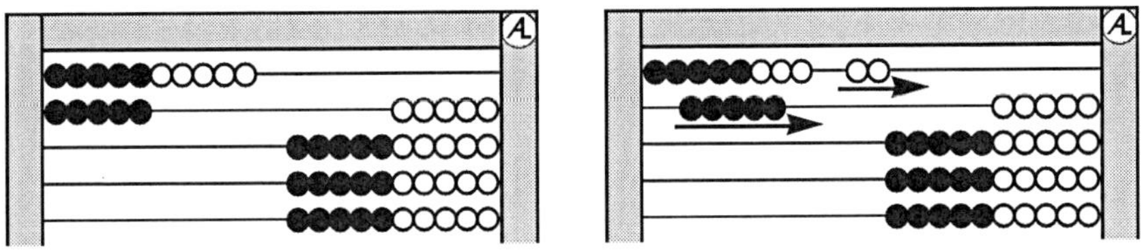

Subtracting 15 – 7 by removing 5 from the 5 and 2 more from the 10.

Doubles. Doubles and near doubles are not as obvious in subtraction as they are in addition.

Subtracting From Ten. Subtracting 15 – 9 can be also thought of as subtracting the 9 from the 10 and adding the result to 5. However, the Going Up strategy is just as effective.

Derived. Of course, there are also derived strategies. For example, if we know that 12 – 6 = 6, then 13 – 6 = 7.

Worksheet. Next ask the child to complete the worksheet. Again ask her to circle any hard ones. Correct her answers as was done earlier. The answers are given below. How many facts are there? [36, 9 × 4]

12	13	11	13	14	11	12	14	13
– 6	– 5	– 3	– 7	– 8	– 2	– 9	– 5	– 9
6	**8**	**8**	**6**	**6**	**9**	**3**	**9**	**4**
12	12	13	11	15	12	13	11	15
– 5	– 8	– 4	– 6	– 9	– 4	– 8	– 4	– 6
7	**4**	**9**	**5**	**6**	**8**	**5**	**7**	**9**
14	12	18	15	12	14	11	14	11
– 6	– 3	– 9	– 8	– 7	– 9	– 8	– 7	– 9
8	**9**	**9**	**7**	**5**	**5**	**3**	**7**	**2**
13	17	16	11	16	15	11	17	16
– 6	– 8	– 8	– 5	– 9	– 7	– 7	– 9	– 7
7	**9**	**8**	**6**	**7**	**8**	**4**	**8**	**9**

The last question asks for strategies for finding 11 – 8. One strategy is to subtract 1 from the 8 and 7 from the 10, giving 3. Another strategy is to go up from 8 to 10, needing 2, and then 1 more to 11, getting 3. A third strategy is deriving the answer from 10 – 8 = 2 by adding 1 more, or 3.

Lesson 36

Review

OBJECTIVE 1. To review and practice

MATERIALS Worksheet 28-A and 28-B, "Review" (2 versions)
Cards for playing games

ACTIVITIES ***Review worksheet.*** Give the children a review sheet.

Note: The first quarter test can be given any time.

The oral problems to be read are as follows:

60 – 56 = 39 – 5 = 53 – 4 =

Review Sheet 28-A.

1. Write only the answers to the oral questions. **4** **34** **49**

4. Write only the answers. 56 – 51 = **5** 61 – 3 = **58** 32 – 10 = **22**

7. Write the multiples of 8.

8 **16** **24** **32** **40**

48 **56** **64** **72** **80**

17. How much is four 8s? **32**

18. How much is ten 8s? **80**

19. How much is eight 8s? **64**

20. How much is six 8s? **48**

21. How much is one 8? **8**

22. What is perimeter?

The distance around the edge.

23. Circle the angles that are 90°. What is another name for these angles? **right angle**

25. Draw $3\frac{1}{4}$ pizzas.

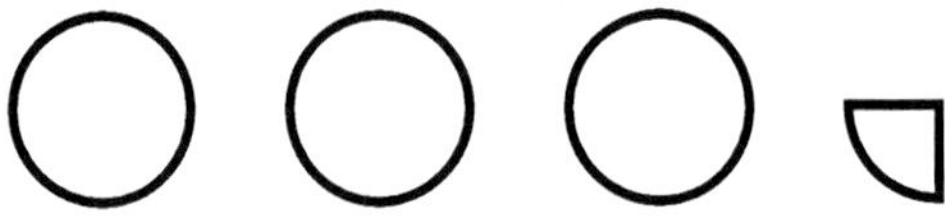

26. David spends 55 minutes from Monday to Friday doing math. How many minutes is that in a school week?

			2	
			5	5
			5	5
			5	5
			5	5
		+	5	5
		2	7	5

275 min.

27. What strategy could you use to find 15 – 9 if you didn't know it?

Subtract 9 from the 10 to get 1 and add it to 5 to get 6.

Subtract 5 from 5 and 4 from 10 to get 6.

The oral problems to be read to the child are as follows:

80 – 73 =   47 – 2 =   37 – 8 =

Review Sheet 28-B.

1. Write only the answers to the oral questions. **7** **45** **29**

4. Write only the answers. 76 – 42 = **34** 32 – 4 = **28** 57 – 10 = **47**

7. Write the multiples of 6.

6 **12** **18** **24** **30**

36 **42** **48** **54** **60**

17. How much is three 6s? **18**

18. How much is six 6s? **36**

19. How much is ten 6s? **60**

20. How much is four 6s? **24**

21. How much is one 6? **6**

22. What is perimeter?

The distance around the edge.

23. Circle the angles that are 90°. What is another name for these angles? **right angle**

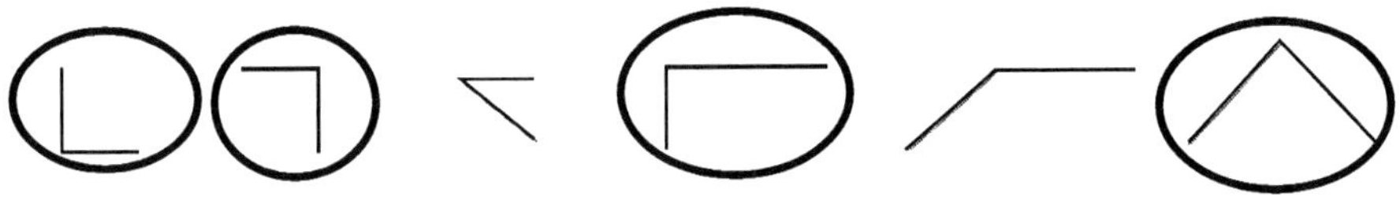

25. Draw $1\frac{1}{4}$ pizzas.

26. Danielle spends 45 minutes from Monday to Friday doing math. How many minutes is that in a school week?

			2	
			4	5
			4	5
			4	5
			4	5
		+	4	5
		2	2	5

225 min.

27. What strategy could you use to find 16 – 9 if you didn't know it?

Subtract 9 from the 10 to get 1 and add it to 6 to get 7.

Subtract 6 from 6 and 3 from 10 to get 7.

Games. Spend the remaining time playing games.

SKIP-COUNTING. Multiples Memory (P2), Mystery Multiplication Card (P4), and Missing Multiplication Cards (P5) *(Math Card Games).*

MULTIPLICATION. Multiplication Memory *(Math Card Games,* P10).

MENTAL SUBTRACTING. Corners variations *(Math Card Games,* S9).

Lesson 37

Adding Hours and Minutes

OBJECTIVE 1. To experience trading with 60 minutes

MATERIALS Worksheet 29, "Adding Hours and Minutes"

WARM-UP Ask the child to say the multiples of 7. Then ask her to write them and to say the tables. [7 × 1 = 7 and so forth]

Say the following and ask the child to write only the answers:

11 – 9 = [2]	14 – 9 = [5]	16 – 9 = [7]
31 – 9 = [22]	44 – 9 = [35]	56 – 9 = [47]

Ask how many minutes are in an hour. [60 min]

ACTIVITIES ***Time problems.*** The following 2 problems can be done mentally.

A half-hour program ended at 8:00. What time did it start? [7:30] Ask the child for an explanation of how she did it.

The second problem is more difficult.

The Green family left at 3:30 for a trip. The trip was an hour and a half long. What time did they arrive? [5:00] (You might need to explain the word *arrive.*)

Ask how she found the answer. [probably first adding the hour, then the half hour]

Problem 1. Ask the child to work on problem 1 on the worksheet.

Alberto wanted to know how long it took him to build a model. The first day he worked 2 hours and 20 minutes. The second day he worked 1 hour and 45 minutes and finished it. How long did it take him? [**4 hr 5 min**]

Suggest she use columns. After she has solved it, ask her to share her solution.

```
  2 hr 20 min
+ 1 hr 45 min
  3 hr 65 min   = 4 hr 5 min
```

Discuss the transition from 3 hr 65 min to 4 hr 5 min. Is 65 minutes greater than 1 hour? [yes] How many minutes more than 1 hour? [5 min]

> **Note:** No particular algorithm is being taught at this point, just common sense.

Next ask the child to read and solve problem 2.

Problem 2. Anna is building a different model. She spent 1 hour 45 minutes for three days. How long did it take her? [**5 hr 15 min**]

```
  1 hr 45 min
  1 hr 45 min
+ 1 hr 45 min
 3 hr 135 min   = 5 hr 15 min
```

Encourage her to discuss the way she arrived at her answer, especially how she went from 135 minutes to 15 minutes.

> **Note:** If the child does not suggest a particular solution, ask, "How about . . . ." to start the discussion.

Problem 3. Amelia practices her violin for 25 minutes each day except on Sunday. How much time does she practice in a week? [**2 hr 30 min**]

Suggest she use columns. After she has solved it, ask her to share her solution.

```
  25 min
  25 min
  25 min
  25 min
  25 min
+ 25 min
 150 min   = 2 hr 30 min
```

To change from minutes to hours, she needs to either subtract 60 twice or think to subtract 120, the number of minutes in 2 hours. She might also solve it by thinking that two 25 minute periods is 10 minutes short of an hour. So the total is 3 hours minus 30 minutes, or 2 hours 30 minutes. It is also possibly to add 2 amounts at a time, change it to hours and minutes and then add the next amount.

Next ask the child to read and solve problem 4.

Problem 4. Adam practices the piano 35 minutes 5 days a week. How much time does he practice in a week? [**2 hr 55 min**]

```
  35 min
  35 min
  35 min
  35 min
+ 35 min
 175 min   = (1 hr 115 min) = 2 hr 55 min
```

Encourage her to discuss the way she arrived at her answers, especially how she went from 175 minutes to 55 minutes. It also could be done by thinking of 30 minutes five times, which is 2 hours and 30 minutes. Then adding the extra 25 minutes to get 2 hours and 55 minutes.

Worksheet. The remaining two problems can be assigned as homework. The solutions are given below.

```
  4 hr 15 min          3 hr 29 min
+ 5 hr 48 min          3 hr 29 min
 10 hr  3 min        + 3 hr 29 min
                      10 hr 27 min
```

Lesson 38

Subtracting Hours and Minutes

OBJECTIVE 1. To practice trading with 60 minutes

MATERIALS Worksheet 30, "Subtracting Hours and Minutes"

WARM-UP Ask the child to say the even numbers backward from 20.

What even number comes before 8. [6] What is 8 – 2? [6]

What even number comes before 10? [8] What is 10 – 2? [8]

What even number comes before 12? [10] What is 12 – 2? [10]

What even number comes before 6? [4] What is 6 – 2? [4]

How many minutes are in an hour? [60]

Ask the child to say the multiples of 8. Does saying the 8s remind you of saying the even numbers backward? [The ones follow this pattern.] Then ask them to write them and to say the tables. [8 × 1 = 8 and so forth]

How many hours are in a day? [24] If he is not sure, ask, How many times does the hour hand go around in a day? [2] How many hours each time? [12]

ACTIVITIES ***Time problems.*** Give the child these two problems to be done mentally.

> A two-hour program ended at 8:30. What time did it start? [6:30] Ask him for an explanation of how he did it.

The second problem is more difficult.

> It takes an hour and a half for the Wang family to drive to the grandparents' home. They arrived at 4:00. What time did they leave? [2:30]

Ask how he found the answer. The most common solution probably is to subtract first the hour and then the 30 minutes.

Problem1. Ask the child to work problem 1 on the worksheet.

> Problem 1. Jane is attending a play that lasts 2 hours and 10 minutes. She has been watching the play for 1 hour and 35 minutes. How soon will it be over? [**35 min**] The solutions are shown below.

2 hr 10 min	+ 10 min
– 1 hr 35 min	+ 25 min
35 min	

Discuss various solutions. Here the simplest solution is to add going up; that is, to think it takes 25 minutes to reach 2 hours and 10 more minutes, giving 35 minutes. The numbers can be written to the side as shown above. The child may prefer to write them in reverse order.

Cover the words "2 hr 10 min" and ask what he gets if he adds the 1 hr 35 min and 35 min. [2 hr 10 min] Ask why that is so.

Problem 2. Jon sleeps 9 hours and 45 minutes a day. How many hours is he awake in a day? **[14 hr 15 min]**

The child first needs to remember how many hours are in a day. [24]

```
  24 hr              + 14 hr
– 9 hr 45 min        + 15 min

  14 hr 15 min
```

Ask the child to explain his solution. The solution above is to think that it takes 15 minutes to get to 10 hours and 14 hours to get to 24 hours.

Ask him to read and solve problem 3.

Problem 3. Jack and Jill built a birdhouse together. It took them 4 hours and 5 minutes. Jill spent 2 hours and 15 minutes. How much time did Jack spend? **[1 hr 50 min]**

```
                 5 min
  4 hr  5 min    1 hr
– 2 hr 15 min    45 min

  1 hr 50 min
```

Note: Underlining the numbers being borrowed makes borrowing easier in other subtraction problems.

Another way to solve this problem is to think that if the total time were 4 hr 15 min, the difference would be 2 hrs. But it is 10 less, so the answer must be 10 min less, or 1 hr 50 min.

Next ask the child to read and solve problem 4.

Problem 4. Jay skied for 3 hours and 45 minutes. Jean skied 4 hours and 20 minutes. Who skied more and how much more? **[Jean, 35 min more]**

```
  4 hr 20 min    20 min
– 3 hr 45 min    15 min
       35 min
```

Discuss various solutions. Cover the words 4 hr 20 min and ask what he gets is he adds the 3 hr 45 min and 35 min. [4 hr 20 min] Ask why that is so.

Worksheet. The remaining six problems can be assigned as homework. The solutions are given below.

```
a.  3 hr 10 min        b.  3 hr 35 min
  – 2 hr 20 min              – 45 min
       50 min              2 hr 50 min

c.  5 hr 10 min        d.  24 hr
  – 2 hr 45 min          – 20 hr 52 min
    2 hr 25 min             3 hr  8 min

e.  5 hr 18 min        f.  10 hr 25 min
  – 1 hr 38 min          –  7 hr 17 min
    3 hr 40 min             3 hr  8 min
```

Lesson 39

Trading Between Inches and Feet

OBJECTIVE 1. To experience trading between feet and inches

MATERIALS 4-in-1 ruler
Worksheet 31, "Trading Between Inches and Feet"

WARM-UP Ask the child to say the multiples of 4. Then ask her to write them and to say the tables. [4 × 1 = 4 and so forth]

Write the following for the child to write only the answers:

11 – 8 = [3]	14 – 8 = [6]	16 – 8 = [8]
41 – 8 = [33]	34 – 8 = [26]	56 – 8 = [48]

Ask how many inches are in a foot. [12]

ACTIVITIES ***Inches problem.*** Show the child the 12" ruler. How long is it in inches? [12 inches] How long is it in feet? [1 foot] Then give her this problem.

An equilateral triangle measures 11 inches on a side. What is the perimeter in feet and inches? [2 ft 9 in.]

Ask the child to draw the figure and to show the measurements. See below.

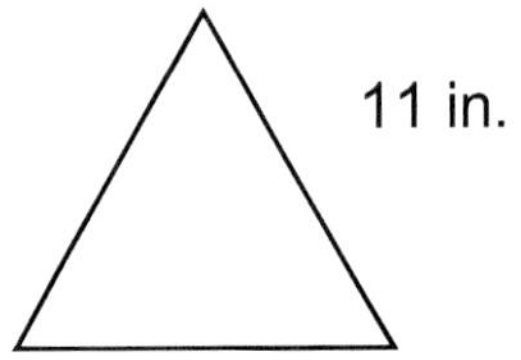

Finding the perimeter of the equilateral triangle.

Ask her to solve it and explain her solutions. One method is to add 11 three times, resulting in 33 in. Next subtract 12 as many times as necessary to find the number of feet. So 33 – 12 = 21; 21 – 12 = 9, so answer must be 2 ft 9 in.

Another solution is to realize 11 in. is 1 inch less than a foot. So it must be 3 ft – 3 in., or 2 ft 9 in.

Next give her this subtraction problem.

Mr. Berg has 5 feet of trim for picture frames. He cut one piece 1 foot 4 inches long. How much is left? [3 ft 8 in.]

5 ft	3 ft
– 1 ft 4 in.	8 in
3 ft 8 in.	

Ask her to show the solution, needing 8 in. to go up to 2 ft and 3 ft to go to 5 ft. Another solution is to think of subtracting the 4 inches from 1 foot.

Worksheet. Ask the child to work the first 4 problems on the worksheet. Ask her to work them alone and then discuss them with you. Tell her that is to explain how she did them. In the process of the explanation, she may change her answers if she is sure they were wrong.

The problems and solutions are given below.

Problem 1. The height of the ceiling in many homes is 8 ft. How many inches is that? [96 in.] One solution is to multiply 12 × 8. Another is to multiply 8 × 12 by first multiplying 8 × 10 to get 80 and then multiplying 8 × 2 to get 16; 80 + 16 = 96.

Note: There is no need for the child to write the measurements on every side of a square.

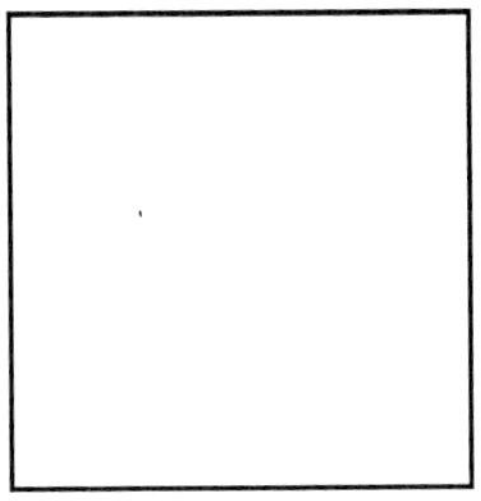

2 ft 9 in.

Robert's rabbit pen.

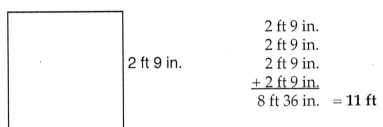

```
   2 ft 9 in.
   2 ft 9 in.
   2 ft 9 in.
 + 2 ft 9 in.
   8 ft 36 in.  = 11 ft
```

Problem 2. Robert is building a square pen for his pet rabbit. One side measures 2 ft 9 in. What is the perimeter? If the rabbit walks next to the fence all the way around the pen, how far would it walk? Draw a picture of the pen.

Both answers are 11 ft. This problem could also be solved by adding 3 ft four times, getting 12 feet and then subtracting 3 in. four times, which is 12 in. or 1 ft, giving the same answer of 11 ft.

Problem 3. Charlotte is 4 feet 2 inches tall. Her little brother, Charles, is 2 feet 10 inches tall. How much taller is Charlotte?

```
                      2 in.
   4 ft  2 in.   1 ft
 – 2 ft 10 in.        2 in.
   1 ft  4 in.
```

Problem 4. Work together with the child to measure each other. How tall are each of you in feet and inches? How tall are each in inches? Explain your work.

The remaining four problems can be assigned as homework. Tell the child to use the space below the problem if she wants to rewrite any of the problems.

The solutions are given below.

```
a) 4 ft 7 in.    b) 3 ft 4 in.    c) 8 ft 4 in.    d) 5 ft  8 in
 + 4 ft 7 in.       3 ft 4 in.     – 3 ft 5 in.     – 1 ft 10 in.
   9 ft 2 in.     + 3 ft 4 in.       4 ft 11 in.      3 ft 10 in..
                    10 ft
```

Lesson 40

Reviewing Place Value Names

OBJECTIVE 1. To review ones, tens, hundreds, and thousands places

MATERIALS Calculator
Worksheet 32, "Place Value Names"

WARM-UP Ask the child to say the multiples of 8. Then ask him to write them and to say the tables. [8 × 1 = 8 and so forth] Also ask for the facts in random order; for example, What is 8 × 3? [24] What is 8 × 7? [56] and so forth. Allow him to look at the patterns.

Write the following for the child to write only the answers:

23 – 5 = [18]	15 – 7 = [8]	12 – 6 = [6]
43 – 5 = [38]	35 – 7 = [28]	52 – 6 = [46]

ACTIVITIES ***Reviewing place value.*** Write

36

368

3682

and ask how many digits are in each of the three numbers. [2, 3, 4]

Note: Deciding the place value of a number by observing the number of digits following it prevents the common error of reading numbers backward by saying, "ones, tens, . . .

What do we call the 3 in 36, thousands, ones, or what? [tens] Ask how he could tell. [One digit after the 3 tells us that it is a ten.] Point to the number 368 and ask: What is the 3 is in this number? [hundred] How do you know? [Two digits after a number means it is hun-dred.]

Refer to 3682 and repeat the questions. [Three digits after a number means it is th-ou-sand.] Then ask what the 6 is [hundreds], 8 is [tens], and 2 is. [ones]

Partitioning numbers. Draw a part-whole set with three parts as shown below. Write 123 in the whole and ask how we could partition the number into hundreds, tens, and ones. Mention the abbreviation "hun" means hundreds. See below on the left.

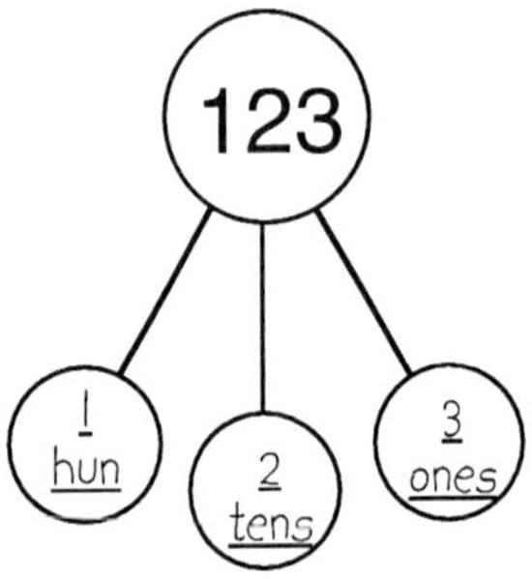

Partitioning 123 into hundreds, tens, and ones.

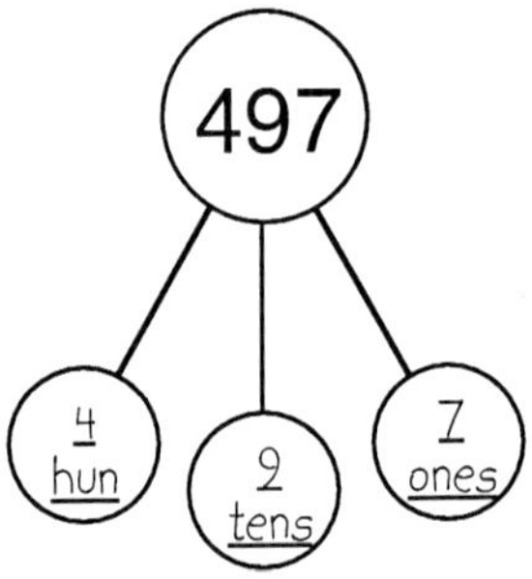

Partitioning 497 into hundreds, tens, and ones.

Replace the whole number with 497and ask the child which numbers go in the parts circles. See above on the right.

Erase the number in the whole circle and change the parts numbers to 6 ones, 2 hun, and 4 tens, as shown on the next page on the left. Tell the child that the numbers might be mixed up and ask him what the new whole number is. [246]

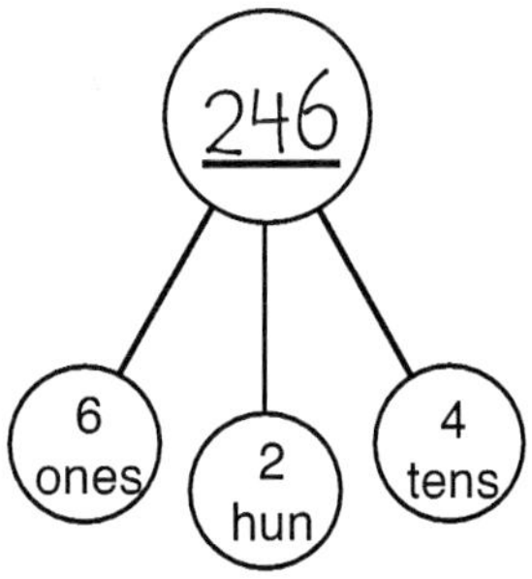

Finding the whole when the parts are known.

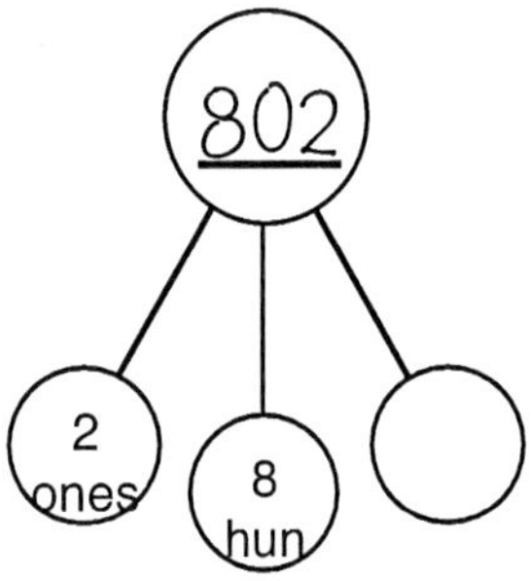

Finding the whole when the parts are known.

This time leave one part-circle blank; write 2 ones and 8 hundreds in the parts circles as shown above on the right. What is the number? [802] See the right figure above.

Add a fourth part to the whole-part set as shown below and write 7 hun, 9 ones, and 8 thou in the parts as shown below. What is the number? [8709] Repeat for other numbers as desired.

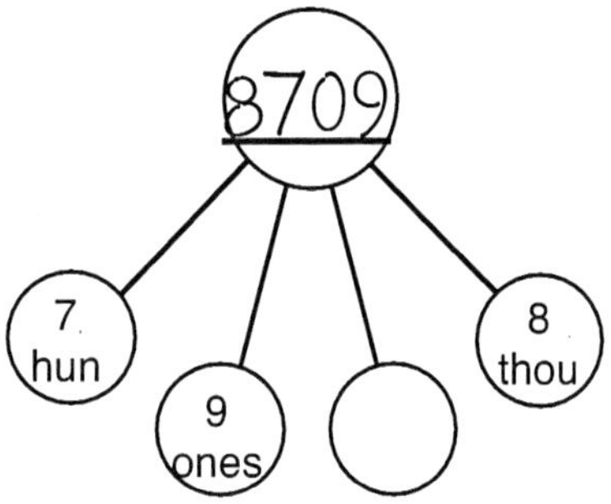

Finding the whole for a four-digit number.

Place value on the calculator. Ask the child to enter the following on the calculator and then change the result without clearing.

Enter 588 and change to 888. [Add 300.]

Enter 5642 and change to 5648. [Add 6.]

Enter 946 and change to 906. [Subtract 40.]

Worksheet. This worksheet asks the child to construct numbers from parts, which are in mixed order. The numbers are on the right side on the worksheet. He also is to solve calculator problems similar to those above. The solutions are as follows.

a. **532**
b. **3422**
c. **3654**
d. **2005**
e. **345**
f. **3240**
g. **3324**
h. **3502**
i. **2023**
j. **subtract 4**
k. **subtract 600**
l. **add 90**
m. **add 300**
n. **subtract 3000**
o. **add 99**
p. **add 440**

Lesson 41

Place Value Problems

OBJECTIVES
1. To review the values of the various denominations
2. To practice quick quantity recognition
3. To translate between ones, tens, hundreds, and thousands

MATERIALS Tiles
A dowel and box representing the roll of candy and box, optional
Worksheet 33, "Place Value Problems"

WARM-UP Ask the child to say the multiples of 8. Then ask her to write them and to say the tables. [8 × 1 = 8 and so forth] Also ask for the facts in random order. For example, <u>What is 8 × 3?</u> [24] <u>What is 8 × 7?</u> [56] and so forth. Allow her to look at the patterns.

Ask how many hours are in a day? [24 hr]

<u>Write the following for the child to write only the answers:</u>

43 – 6 = [37]	54 – 6 = [48]	62 – 8 = [54]
23 – 6 = [17]	94 – 6 = [88]	72 – 8 = [64]

ACTIVITIES ***Review.*** <u>How much is 10 ones?</u> [10] <u>How much is 10 tens?</u> [100] <u>How much is 10 hundreds?</u> [1000]

Ask the child to take between 20 and 50 tiles (about four handfuls). Then ask her to lay them out. Ask if she can tell without counting how many there are. Two examples, 32 and 47, are shown below.

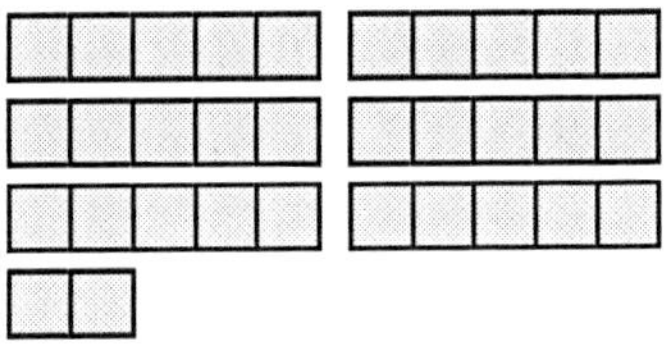

Displaying 32 for instant recognition by grouping.

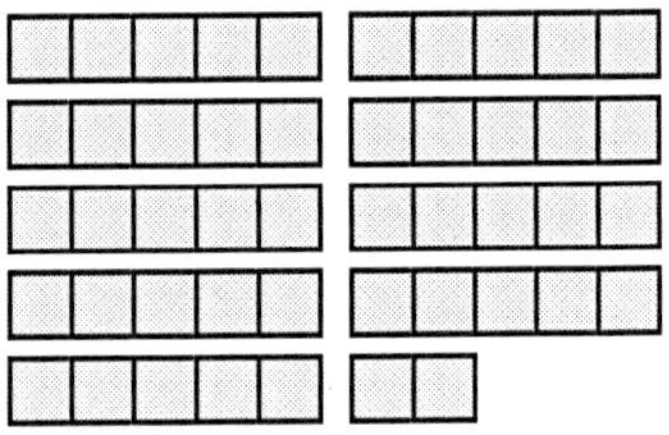

Displaying 47 for instant recognition by grouping.

Ask how she can tell. Discuss the number of tens and ones. Also discuss that most people cannot see more than 5 in a group.

Ask the child to read the quantities in the various groups made by laying out different numbers of tiles.

Visual practice. A box placed between you and the child is the easiest way to conduct this practice. With the box in place, arrange quantities tiles in groups of 5s and 10s as shown above. Then lift the box for 1 second and ask the child to write down how many she saw. Start with quantities 4 to 10 and proceed to quantities up to 50, or even higher.

Problem 1. Tell the child that <u>a certain candy, called Lime Favors, is wrapped with ten in a roll. How many candies are in 24 rolls?</u> [240] Show her the dowel and box, if available as shown in the figure on the next page.

Ask her to think about it for several minutes. Then ask her to solve it. Ask her to share her method and answer with you.

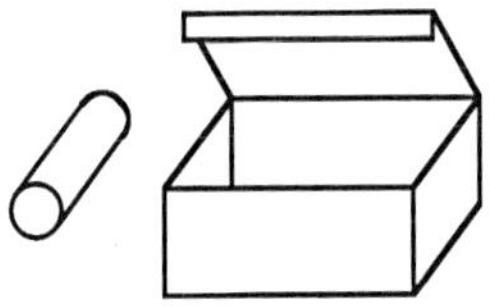

A roll of Lime Favors and the box, which holds 10 rolls.

There are several solutions. The simplest (and most work) is to add 24 ten times. Another solution is to think 24 is 10 + 10 + 4; 10 tens is 100 and 4 tens is 40, giving a total of 240 candies. The most direct solution is to think that 24 tens is 240. Discuss the various solutions. If necessary, introduce other solutions by saying, Could you do it by . . . ?

Problem 2. The Lime Favors are packaged with 10 rolls in a box. How many candies are in a whole box? [100]

How many candies are in 3 boxes and 4 rolls? [340]

Problem 3. In 1 box some of the rolls are broken open and some pieces are missing. There are 8 rolls and 13 single candies left in 1 box. How many pieces of candies are there? [93] How many pieces of candy are missing? [7]

In another box there are 7 rolls and 18 single candies. How many pieces of candies are in that box? [88] How many pieces of candy are missing? [12] To answer the last question, going up is the easiest solution.

Problem 4. Kylie and Kyle work at the Lime Favors factory. Kylie has 65 red candies and 95 green candies. How many rolls can she make? [16]

Kyle has 450 pieces of candy. How many rolls can he make? [45] How many boxes can he fill? [4]

Worksheet. The worksheet has similar problems with the same Lime Favors theme. Ask the child to solve the problems independently. Tell her to solve the problems in the way that makes sense to her. How she solves them is a type of assessment. The solutions follow.

1. **360 pieces**
2. **30 rolls, 3 boxes**
3. **78 pieces**
4. **17 pieces**

a. 100 = **100** ones

b. 200 = **200** ones

c. 110 = **11** tens

d. 180 = **18** tens

e. 100 = **1** hundreds

f. 200 = **2** hundreds

g. 5 tens + 17 ones = **67**

h. 4 tens + 19 ones = **59**

i. 83 = 7 tens + **13** ones

j. 71 = 6 tens + **11** ones

k. 38 = **3** tens + 8 ones

l. 38 = **2** tens + 18 ones

m. 43 tens = 4 hundreds + **3** tens

n. 97 tens = 9 hundreds + **7** tens

Lesson 42

Review

OBJECTIVE 1. To review and practice

MATERIALS Worksheet 34-A and 34-B, "Review" (2 versions)
Cards for playing games

ACTIVITIES ***Review worksheet.*** Give the child about 15 to 20 minutes to do the review sheet. The oral problems to be read are as follows:

60 – 5 =   32 – 5 =   53 – 10 =

Review Sheet 34-A.

1. Write only the answers to the oral questions. 55 27 43

4. Write only the answers. 64 – 41 = 23 72 – 23 = 49 95 – 27 = 68

7. Write the multiples of 6.

6 12 18 24 30

36 42 48 54 60

10. How much is three 6s? 18

11. How much is five 6s? 30

12. How much is ten 6s? 60

13. How much is seven 6s? 42

14. How much is eight 6s? 48

15. This line is 1 inch long.
Draw a line $2\frac{1}{2}$ inches long.

Note: This line is to scale.

16. Explain how you can tell how many squares there are without counting past 10.

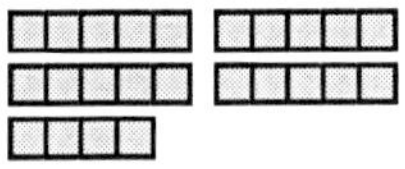

There are 2 tens and 4 ones. That is 24.

17. Write this mixed-up number using digits 0 to 9.
8 hundred 6 ones 9 thousand

9806

18. The Garcia family is traveling to see the grandparents. They drove 2 hours and 45 minutes before a stop and 1 hour 45 minutes after the stop. How long was the drive?

2 hr 45 min
+1 hr 45 min
3 hr 90 min
=4 hr 30 min

19. Chris entered 4972 on a calculator. How could Chris change the number to read 4072 without clearing the calculator?

subtract 900

20.–25.

10	11	13	10	16	9	17	13
–4	–4	–6	–7	–8	–3	–8	–5
6	7	7	3	8	6	9	8

The oral problems to be read to the child are as follows:

80 – 3 =   46 – 8 =   37 – 10 =

Review Sheet 34-B.

Note: This line is to scale.

1. Write only the answers to the oral questions. **77** **38** **27**

4. Write only the answers. 87 – 33 = **54** 66 – 9 = **57** 84 – 36 = **48**

7. Write the multiples of 8.

8 **16** **24** **32** **40**

48 **56** **64** **72** **80**

10. How much is eight 8s? **64**

11. How much is three 8s? **24**

12. How much is six 8s? **48**

13. How much is seven 8s? **56**

14. How much is five 8s? **40**

15. This line is 1 inch long.

Draw a line $1\frac{1}{2}$ inches long.

16. Explain how you can tell how many squares there are without counting past 10.

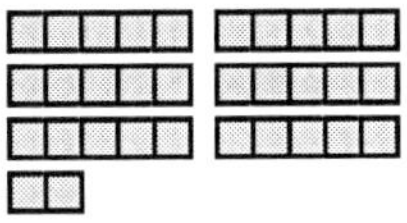

There are 3 tens and 2 ones. That is 32.

17. Write this mixed-up number using digits 0 to 9.

3 thousands 4 ones 7 tens

3074

18. The Silver family is traveling to see an aunt and uncle. It takes 5 hours and to drive. They drove 2 hours and 45 minutes before a stop. How much longer do they need to drive?

5 hr
–2 hr 45 min
2 hr 15 min

19. Jamie entered 8057 on a calculator. How could Jamie change the number to read 8007 without clearing the calculator?

subtract 50

20. 25.

10	11	14	11	17	9	16	12
– 3	– 3	– 6	– 7	– 8	– 4	– 9	– 4
7	**8**	**8**	**4**	**9**	**5**	**7**	**8**

Games. Spend the remaining time playing games.

SKIP-COUNTING. Multiples Memory (P2), Mystery Multiplication Card (P4), and Missing Multiplication Cards (P5), *(Math Card Games).*

MULTIPLICATION. Multiplication Memory (P10), *(Math Card Games.*

SUBTRACTION. Zero Corners (S9), Short Chain Subtraction (S18), *(Math Card Games).*

Lesson 43

Subtracting by Compensating

OBJECTIVES
1. To review adding by compensating
2. To learn subtracting by compensating

MATERIALS
Tiles
Abacus
Worksheet 35, "Adding & Subtracting by Compensating"

WARM-UP Ask the child to say the multiples of 9. Then ask them to write them and to say the tables. [9 × 1 = 9, 9 × 2 = 18, and so forth]

Write the following for her to write only the answers:

27 – 9 = [18]	18 – 9 = [9]	63 – 9 = [54]
54 – 9 = [45]	36 – 9 = [27]	81 – 9 = [72]

What is another name for 12 inches? [foot]

ACTIVITIES ***Reviewing compensating in addition.*** Draw a part-whole circle set and write 15 in the whole circle and 8 and 7 in the parts circles as shown below. Also enter it on the abacus as shown.

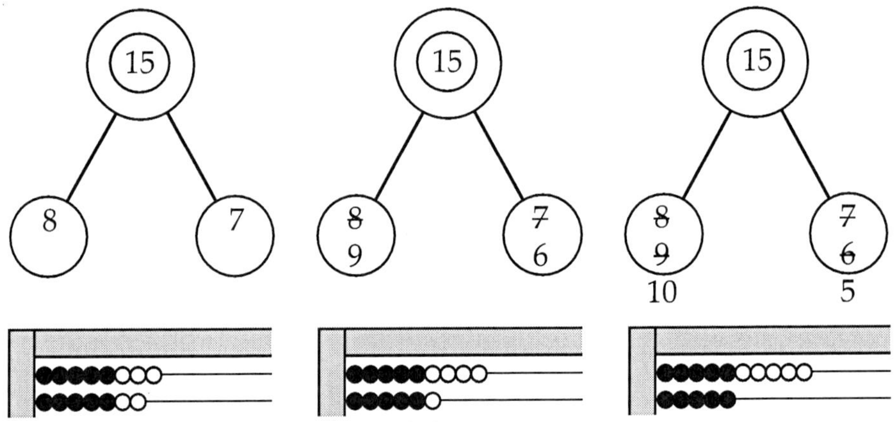

Changing the parts with the whole kept constant.

Note: Use the opportunity to use the mathematical term, *constant,* in an informal situation.

Draw a circle around the 15 and explain that we don't want it to change; we want to keep it constant. Add 1 to the 8 to get 9, as shown above in the middle figures. How must the 7 be changed if the whole is still 15? [Subtract 1 from the 7 to get 6.]

Do it once more; add 1 to the 9 to 10. What must the other number be? [Subtract 1 from the 6 to get 5.] See the figures above.

What pattern do you see? [If a number is added to one number, the same amount must be subtracted from the other number.]

Write

29 + 31 =
[30 + 30 = 60]

How could we use that pattern to add those 2 numbers? [Add 1 to the 29 and subtract 1 from the 31 to get 30 and 30.]

Ask them to use the pattern to add the following.

61 + 59 = _	99 + 101 = _	26 + 24 = _
[60 + 60 = 120]	[100 + 100 = 200]	[25 + 25 = 50]

Compensating in subtraction. Lay out 13 tiles, grouped in 5s as shown below on the left. Below it lay out 7 tiles. What is the difference between the 2 groups? [13 – 7 = 6] If necessary, remind them that *difference* in mathematics means how much greater one quantity is compared to another quantity.

Note: Different models will help the child keep addition and subtraction separate.

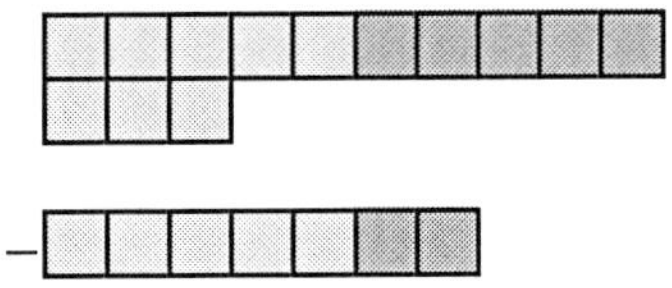

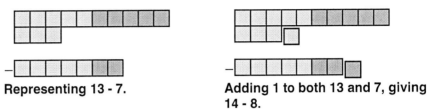

Representing 13 - 7.

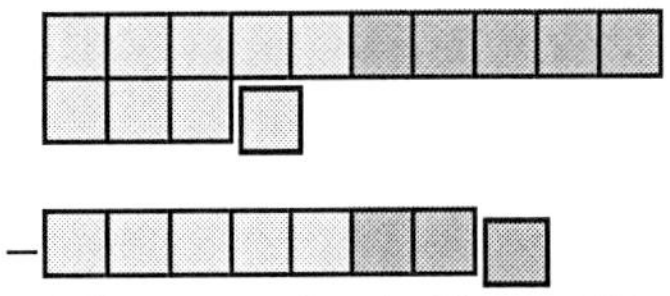

Adding 1 to both 13 and 7, giving 14 - 8.

What happens to the difference if we add 1 to each group? Add 1 tile to the 13 and 1 to the 7. See the right figure above.

What happens to the difference if we add 2 to each group? Add 2 tile to the 14 and 2 to the 8. See the figure below.

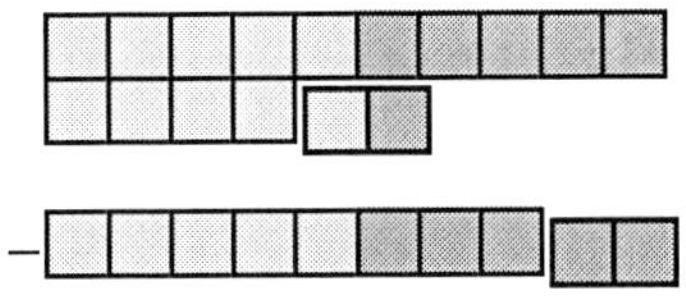

Adding 2 to each quantity, giving 16 – 10.

Write the three equations.

13 – 7 = 6
14 – 8 = 6
16 – 10 = 6

What pattern do you see? [Adding the same number still gives the correct answer.] Which is easiest to subtract? [16 – 10]

Write the following.

18 – 9 = _
[19 – 10 = 9]

What number is easier than 9 to subtract? [10] How can we change the 9 to a 10? [Add 1 to 9 and to the 18.]

Repeat for the following 2 examples.

32 – 28 = _	97 – 89 = _
[34 – 30 = 4]	[98 – 90 = 8]

Worksheet. Give the child the worksheet. They are to write out the new equations. The solutions are shown below.

42 + 38 = _	16 + 14 = _	52 + 49 = _	298 + 198 = _
40 + 40 = 80	**15 + 15 = 30**	**51 + 50 = 101**	**296 + 200 = 496**
64 – 29 = _	32 – 19 = _	55 – 28 = _	60 – 18 = _
65 – 30 = 35	**33 – 20 = 13**	**57 – 30 = 27**	**62 – 20 = 42**
130 – 99 = _	579 – 99 = _	267 – 198 = _	378 – 190 = _
131 – 100 = 31	**580 – 100 = 480**	**269 – 200 = 69**	**388 – 200 = 188**

Lesson 44

Multidigit Subtraction

OBJECTIVE 1. To review or learn the Simplified Subtraction algorithm

MATERIALS Worksheet 36, "Multidigit Subtraction"

WARM-UP Ask the child to write the first five multiples of 6. Then ask him to add 30 to each number to find the second row.

Note: Omit this lesson if the child has done the Transition Lessons.

6	12	18	24	30
36	42	48	54	60

Why does that work? [Six 6s is five 6s, which is 30, and 1 more 6, making 36. Seven 6s is five 6s and two more 6s, 30 + 12, or 36, and so forth.] Then ask for 6s facts in random order.

Next ask the child to look at the multiples and say the tables. [$6 \times 1 = 6$, $6 \times 2 = 12$, and so forth]

ACTIVITIES ***Subtracting 4-digit numbers with no borrowing.*** There are at least 5 different ways to subtract multidigit numbers. Even though a person has learned one method, it is good to explore other methods to enhance understanding.

Note: If a child has an accurate and efficient method of subtraction, do not insist that the child change to a new method.

The following method, called Simplified Subtraction, has several advantages over the decomposition algorithm usually taught in the U.S. (a) The procedure proceeds from left to right, the order children consider "natural" and the same as division (b) All the borrowing is done before any of the actual subtraction, which research has shown is easier for children. (c) Many children will soon be able to avoid writing the extra numbers above the minuend.

Write

$$\begin{array}{r} 3759 \\ -1428 \\ \hline \end{array}$$

and ask the child to look at each denomination in turn and decide if any borrowing is necessary. Point to the 3 and ask: Are there are enough thousands to subtract 1 thousand? [yes] Then point to the 7. Are there are enough hundreds to subtract 4 hundred? [yes] Next point to the 5. Are there enough tens to subtract 2 tens? [yes] Lastly point to the 9. Are there enough ones to subtract 8 ones? [yes]

Next perform the subtraction , starting at the LEFT.

$$\begin{array}{r} 3759 \\ -1428 \\ \hline 2331 \end{array}$$

saying, 3 thousand minus 1 thousand is . . . . [2000] Write 2 in the thousands place. Continue with the remaining denominations.

Subtracting 4-digit numbers with borrowing. Tell the child: That was too easy. So let's make it a little harder. Change the 7 to 2 and the 9 to 7.

$$\begin{array}{r} 6257 \\ -3428 \\ \hline \end{array}$$

Ask the same questions. Point to the 6. Are there enough thousands to subtract 3 thousand? [yes] Then point to the 2. Are there enough hundreds to subtract 4 hundred? [no] How can we can get more? [from the thousands] Underline the 6, explaining it reminds us we needed to trade. How many thousands will we have after the trade? [5] Write the 5 above the 6. See below.

Note: The underlining also helps when the little numbers are no longer written.

```
  5
 6 2 5 7
-3 4 2 8
```

How many hundreds do we have? [12] Will we need to give any to the tens? [no] So write 12 above the 2 as shown.

```
  5 12
 6 2 5 7
-3 4 2 8
```

Next point to the 5 tens. Will we need to give any to the ones? [yes] Underline the tens and write 4 above it. How many ones do we have? [17] Write 17 above the 7.

Note: Underlining the digits being borrowed from is an aid to subtracting without writing the extra numbers.

```
  5 12 4 17
 6 2 5 7
-3 4 2 8
```

Now the actual subtraction can begin. Proceed from LEFT to right.

```
  5 12 4 17
 6 2 5 7
-3 4 2 8
 2 8 2 9
```

Use the following example, the first problem on the worksheet.

```
 8 4 4 5
-5 3 7 2
```

Point to the 8. Will a thousand be needed for the hundreds? [no] Point to the 4 in hundreds place. Will a hundred be needed for the tens? [yes] Underline the 4 and write 3. How many tens do we have? [14] Will a ten be needed for the ones? [no] Write 14 and subtract as shown.

```
    3 14
 8 4 4 5
-5 3 7 2
 3 0 7 3
```

Worksheet. Ask the child to work the problems on the worksheet. The solutions are below.

Note: Do not correct this worksheet. The child check their own answers in the next lesson.

```
  8445      2149      7141      6632
– 5372    – 1524    – 5308    – 4059
  3073       625      1833      2573

  2159      3162      9468      8277
 – 451    – 1094    – 4585    – 5549
  1708      2068      4883      2728
```

Lesson 45

Checking Subtraction by Adding

OBJECTIVE 1. To check subtraction by adding up

MATERIALS Worksheet 37, "Checking Subtraction by Adding"
Worksheet 36 from the previous lesson

WARM-UP Ask the child to say and then write the multiples of 9.

9	18	27	36	45
90	81	72	63	54

Next ask her to look at the multiples and say the tables. [$9 \times 1 = 9$, $9 \times 2 = 18$, and so forth]

What is 9×4? [36] What is 9×5? [45] What is 9×8? [72] What is 9×3? [27] What is 9×9? [81] What is 9×7? [63] What is 9×2? [18]

ACTIVITIES ***Comparing subtraction and addition.*** Present this problem. Lee has 20 cards and gives 7 cards to a friend. How many cards does Lee have now? [13] Draw a part-whole circle set and ask the child to write the numbers in the circles.

Note: The part-whole concept will help the child remember the connection between adding and subtracting.

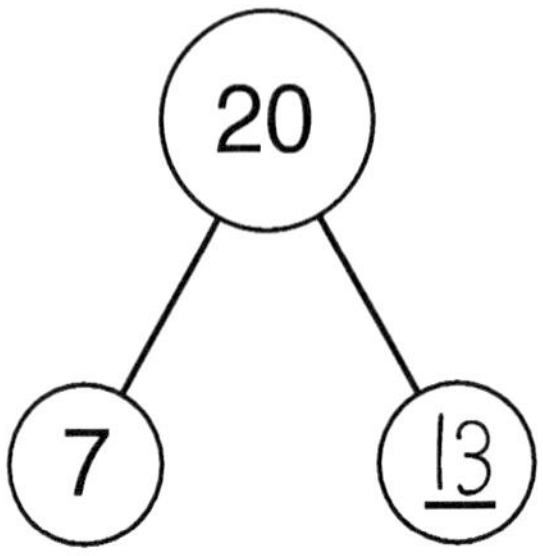

The part-whole circles showing the Lee problem.

Ask her to write the problem vertically and horizontally.

$$20 - 7 = 13$$

$$\begin{array}{r} 20 \\ \underline{-7} \\ 13 \end{array}$$

How many cards do Lee and the friend have? [20] Ask how she found the answer. [add the 7 and 13]

Checking multidigit subtraction. Could we use this idea to see if we subtracted correctly? Write

$$\begin{array}{r} 428 \\ \underline{-157} \\ 331 \end{array}$$

I think the answer might be 331. Let's see if I'm right. Cover the 428 and add the other two numbers.

$$\begin{array}{r} 428 \\ \underline{-157} \\ 331 \\ 488 \end{array}$$

Next uncover the top number. <u>Is the answer right?</u> [no] Ask her to correct the subtraction and to do the addition. Then ask if the subtraction was correct. [yes, presumably]

```
  428
- 157
  271
  428
```

Practice. Write

```
  4620
- 3906
```

Tell the child 1 of the following answers is correct. Ask her to find it by adding up. [714]

1714 1326 714

Worksheet. The worksheet asks the child to find the correct subtraction by adding up. Then she circles the correct subtraction. It might be helpful if she covers the minuend before adding.

8272	8272	8272
− 3446	− 3446	− 3446
5826	5836	4826
9272	**9282**	**8272**
7505	7505	7505
− 1371	− 1371	− 1371
6174	6134	6274
7545	**7505**	**7645**
2642	2642	2642
− 1295	− 1295	− 1295
1447	1357	1347
2742	**2652**	**2642**
7129	7129	7129
− 2736	− 2736	− 2736
4393	4493	4383
7129	**7229**	**7119**
9513	9513	9513
− 7885	− 7885	− 7885
1728	1628	1638
9613	**9513**	**9523**

Checking the previous worksheet. Ask the child to take her copy of that worksheet and check the subtractions by adding. When errors are found, she corrects the subtraction and checks again by adding.

Lesson 46

Subtracting With Doubles and Zeroes

OBJECTIVES 1. To practice multidigit subtraction with *doubles* in a denomination, such as 463 – 264
2. To practice multidigit subtraction with zeroes in the minuend

MATERIALS Worksheet 38, "Subtracting With Doubles and Zeroes"

WARM-UP Ask the child to say and then write the multiples of 9.

9	18	27	36	45
90	81	72	63	54

Next ask her to look at the multiples and say the tables. [$9 \times 1 = 9$, $9 \times 2 = 18$, and so forth]

What is 9 × 7? [63] What is 9 × 9? [81] What is 9 × 2? [18] What is 9 × 3? [27] What is 9 × 5? [45] What is 9 × 6? [54] What is 9 × 8? [72]

ACTIVITIES ***Subtracting with the same digit in a denomination.*** Write

$$\begin{array}{r} 463 \\ -\underline{264} \end{array}$$

and ask the child what he thinks the answer is. [199] Encourage him to use common sense, not an algorithm. If no answer is forthcoming, ask what if the number being subtracted were 263. [200, so the answer must be 1 less, or 199]

Ask if he can subtract using his procedure. Is a hundred needed for the tens? [yes] Explain the way to tell is because 64 is more than 63. Emphasize that when the numbers are the same in a denomination, he needs to look at the remaining numbers. So underline the 4 and write a little 3 as shown.

$$\begin{array}{r} {\scriptstyle 3} \\ \underline{4}63 \\ -\underline{264} \end{array}$$

Continue with the tens. How many tens do we have? [16] Is a ten needed for the ones? [yes] Write a little 15. How many ones do we have? [13] Write a little 13 as shown.

$$\begin{array}{r} {\scriptstyle 3\ 15\,13} \\ \underline{4}\,\underline{6}\,3 \\ -\underline{264} \end{array}$$

Now perform the subtraction.

$$\begin{array}{r} {\scriptstyle 3\ 15\,13} \\ \underline{4}\,\underline{6}\,3 \\ -\underline{264} \\ \mathbf{199} \end{array}$$

Ask if the answer if what he expected. [yes] Ask him to check the answer by adding 264 and 199 and to compare it with 463.

> **Note:** It is not necessary to write the little numbers. When used, write them as each denomination is being prepared.

Example 2. Ask the child to do the following problem and when finished to check it with addition. Then ask him to explain it to you.

$$\begin{array}{r} 7529 \\ -\underline{2553} \end{array}$$

The steps are shown below.

$$\begin{array}{r} {\scriptstyle 6\ 14\ 12} \\ \underline{7}\,\underline{5}\,2\,9 \\ \underline{-2553} \\ \mathbf{4976} \end{array}$$

Example 3. This time double numbers occur twice.

$$\begin{array}{r} \underline{5}\,\underline{4}\,2\,7 \\ \underline{-1429} \end{array}$$

The solution is below.

$$\begin{array}{r} {\scriptstyle 4\ 13\ 11\ 17} \\ \underline{5}\,\underline{4}\,\underline{2}\,7 \\ \underline{-1429} \\ \mathbf{3998} \end{array}$$

Example 4. Zeroes in the minuend pose no special problem. Give him this example.

$$\begin{array}{r} 9020 \\ \underline{-3725} \end{array}$$

The solution is below.

$$\begin{array}{r} {\scriptstyle 8\ 9\ 11\ 10} \\ \underline{9}\,\underline{0}\,\underline{2}\,0 \\ \underline{-3725} \\ \mathbf{5295} \end{array}$$

Worksheet. Give the child the worksheet. Tell him to check his answers by adding. The solutions follow.

8344	6188	5628	6686
– 2381	– 5686	– 2643	– 6589
5963	**502**	**2985**	**97**
5819	7291	9605	8800
– 4949	– 1799	– 3609	– 4981
870	**5492**	**5996**	**3819**
5613	9721	8272	7581
– 5424	– 3354	– 5579	– 4515
189	**6367**	**2693**	**3066**
5128	9805	7447	9720
– 1837	– 5890	– 1589	– 2364
3291	**3915**	**5858**	**7356**
6062	3900	8488	8000
– 4263	– 865	– 4789	– 89
1799	**3035**	**3699**	**7911**

Lesson 47

Using Check Numbers

OBJECTIVES 1. To give the child a way to check computation
2. To introduce check numbers

MATERIALS Worksheet 39, "Using Check Numbers"

WARM-UP Ask the following orally. What is 74 + 20? [94] What is 74 + 25? [99] What is 37 + 30? [67] What is 37 + 38? [75] What is 53 + 29? [82]

Note: The multiples of 3 and 9 play an important role in check numbers.

Ask the child to write the multiples of 3; remind her to write only three numbers in each row. Next ask her to look at the multiples and say the tables from 3 × 1 = 3 to 3 × 10 = 30.

Then ask the child to find and write the following: What is 3 × 3? [9] What is 3 × 6? [18] What is 3 × 9? [27] What is 3 × 5? [15] What is 3 × 2? [6] What is 3 × 8? [24] What is 3 × 10? [30]

Ask the child to subtract 8709 – 2746 in her journal. [5963]

ACTIVITIES ***Introducing check numbers.*** Tell the child that you are going to show her a new way to check her answers, using check numbers. Tell her every number has a special number, called a check number. When you add numbers, the check numbers of the numbers will be the same as the check number of the answer.

Take out Worksheet 39 and explain that it has a table of the check numbers up to 108. Demonstrate how to use them with the following example, 38 + 49. Ask the child to write it and the sum. [87]

Note: Writing the check numbers in parentheses keeps them separate from the numbers being operated on.

$$\begin{array}{r} 38 \\ +\underline{49} \\ 87 \end{array}$$

Now ask her to find 38 on the table and read the check number in the parentheses. [2] Write it after the 38 in parentheses as shown below. Repeat for 49 [4] and 87. [6]

$$\begin{array}{rl} 38 & \mathbf{(2)} \\ +\underline{49} & \underline{\mathbf{(4)}} \\ 87 & \mathbf{(6)} \end{array}$$

Ask her to add the check numbers. Does it equal 6? [Yes!] That means the sum checks and is probably correct. Write a check mark, meaning it checks.

Note: When check numbers do not agree, the answer must be wrong, but even when check numbers do agree, the answer could still be wrong.

$$\begin{array}{rl} 38 & (2) \\ +\underline{49} & \underline{(4)} \\ 87 & (6) \checkmark \end{array}$$

Finding an adding error with check numbers. Write the problem 29 + 29 with 2 different answers, 59 and 58. Tell the child to find the wrong answer using check numbers.

First she finds the check numbers for 29 [2] and 59. [5]

$$\begin{array}{rl} 29 & (2) \\ +\underline{29} & \underline{(2)} \\ 59 & (5) \end{array} \qquad \begin{array}{rl} 29 & (2) \\ +\underline{29} & \underline{(2)} \\ 58 & (4) \checkmark \end{array}$$

Then ask which check numbers add up correctly. [the problem on the right] Place a check mark near the answer as shown. Ask if 58 is the correct answer. [yes]

Second example. Repeat for 59 + 27, using sums 76 and 87.

59 (5)	59 (5)
+ 27 (0)	+ 27 (0)
~~76 (4)~~	~~87 (6)~~
	86 (5)

This time neither answer checks. Ask the child for the correct answer [86] and to find its check number. [5]

Finding a subtracting error with check numbers. Write the following subtraction problem with 2 different answers as shown.

62	62
– 49	– 49
27	13

Ask how she checks subtraction problems. [by adding up] Tell her that is the easiest way to use check numbers. Ask her to find the check numbers for all numbers as shown below.

62 (8)	62 (8)
– 49 (4)	– 49 (4)
27 (0)	13 (4) ✓

In the first example, does 0 + 4 = 8? [no] So that answer must be wrong. Ask her to cross it out. In the second example, does 4 + 4 = 8? [yes] So that answer might be right. Does 13 and 49 equal 62? [yes]

Note: The table on the worksheet will be needed in a future lesson.

Worksheet. Ask the child to read the instructions. Then ask if she has any questions. The solutions are given below.

39 **(3)**	28 **(1)**	75 **(3)**	24 **(6)**	65 **(2)**	83 **(2)**	38 **(2)**
+ 21 **(3)**	+ 57 **(3)**	+ 29 **(2)**	+ 47 **(2)**	+ 29 **(2)**	+ 31 **(4)**	+ 39 **(3)**
60 **(6)**	~~75 (3)~~	~~105 (6)~~	~~72 (0)~~	94 **(4)**	~~124 (7)~~	~~79 (7)~~
	85 (4)	**104 (5)**	**71 (8)**		**114 (6)**	**77 (5)**

52 **(7)**	71 **(8)**	34 **(7)**	106 **(7)**	38 **(2)**	82 **(1)**	103 **(4)**
– 13 **(4)**	– 48 **(3)**	– 29 **(2)**	– 87 **(6)**	– 19 **(1)**	– 45 **(0)**	– 94 **(4)**
39 **(3)**	~~37 (1)~~	5 **(5)**	~~21 (3)~~	~~27 (0)~~	~~47 (2)~~	9 **(0)**
	23 (5)		**19 (1)**	**19 (1)**	**37 (1)**	

Lesson 48

Review

OBJECTIVE 1. To review and practice

MATERIALS Worksheet 40-A and 40-B, "Review" (2 versions)
Cards for playing games

ACTIVITIES ***Review worksheet.*** Give the child about 15 to 20 minutes to do the review sheet. The oral problems to be read are as follows:

40 – 8 =   32 + 58 =   53 – 10 =

Review Sheet 40-A.

1. Write only the answers to the oral questions. **32** **90** **43**

4. Write only the answers. 64 – 56 = **8** 72 – 23 = **49** 95 – 43 = **52**

7. Write the multiples of 4.

4 **8** **12** **16** **20**

24 **28** **32** **36** **40**

17. 4 × 6 = **24** 4 × 2 = **8**

19. 4 × 3 = **12** 4 × 5 = **20**

21. 4 × 4 = **16** 4 × 10 = **40**

23. 4 × 1 = **4** 4 × 8 = **32**

25. 4 × 9 = **36** 4 × 7 = **28**

27. This is 1 triangle.
Draw $1\frac{1}{2}$ triangles.

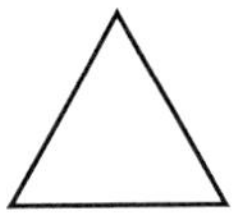

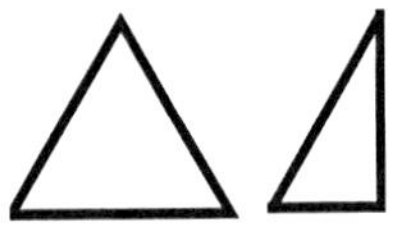

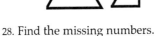

28. Find the missing numbers.

8927	**8430**
– 6578	– 5476
2349	2954

29. Use digits and write this number in the proper order: 2 hundred, 8 ones, and 7 thousand.

7208

30. Jamie is buying four small gifts for $1.26 each. What is the total cost?

			1	2	
		$	1.	2	6
		$	1.	2	6
		$	1.	2	6
	+	$	1.	2	6
		$	5.	0	4

31. A library has 5 thousand 6 hundred 6 books. Of these 1368 books are checked out. How many books does the library have in?

		5	6	0	6
	–	1	3	6	8
		4	2	3	8

The oral problems to be read to the child are as follows:

100 – 3 =   46 + 25 =   46 – 11 =

Review Sheet 40-B.

1. Write only the answers to the oral questions. 97   71   35

4. Write only the answers. 82 – 27 = 55   98 – 50 = 48   98 – 55 = 43

7. Write the multiples of 9.

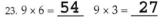

9 18 27 36 45

90 81 72 63 54

17. 9 × 5 = 45   9 × 8 = 72

19. 9 × 7 = 63   9 × 9 = 81

21. 9 × 1 = 9   9 × 4 = 36

23. 9 × 6 = 54   9 × 3 = 27

25. 9 × 2 = 18   9 × 10 = 90

27. This is 1 triangle.
Draw $2\frac{1}{2}$ triangles.

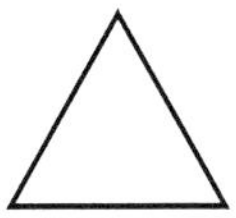

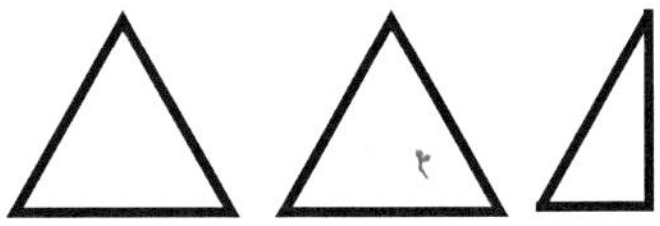

28. Find the missing numbers.

7851	9477
– 4654	– 5793
3197	3684

29. Use digits and write this number in the proper order: 7 tens, 9 hundred, 8 thousand.

8970

30. Morgan is buying five small gifts for $2.38 each. What will they cost?

			1	4	
		$	2.	3	8
		$	2.	3	8
		$	2.	3	8
		$	2.	3	8
	+	$	2.	3	8
	$	1	1.	9	0

31. Six thousand eighty-one children live in the town of Rockville. Of these 3086 are girls. How many are boys?

		6	0	8	1
	-	3	0	8	6
		2	9	9	5

4000
– 115
3885
– 210
3675
– 215
3360

Subtraction Corners in the Thousands game. A good game for practicing subtraction skills is the Corners game in which the players start with a score of 4000. In addition to the usual points, players receive an extra 100 points if the colors are green or black and 200 points if the colors are red or blue. All these points are subtracted from their score.

The subtractions are done on paper, preferably in his math journal. The player with the lowest score is the winner. A sample of the subtraction is shown on the left.

Lesson 49

Finding Check Numbers

OBJECTIVES
1. To help the child discover ways of finding check numbers
2. To check multidigit addition and subtraction with check numbers

MATERIALS Worksheet 39, "Using Check Numbers"
Worksheet 41, "Finding Check Numbers"

WARM-UP Give the following orally and ask the child to add mentally: 35 + 15 [50], 78 + 15 [93], 48 + 15 [63], 46 + 15 [61], and 19 + 15. [34]

Note: For mental work, be sure the child adds the tens and then the ones. For example, 35 + 15 = 35 + 10 [45] + 5 [50].

Ask the child to write the multiples of 3 and the multiples of 9.

Next ask the child to look at the multiples and say the tables of 3. Then ask her for the following: What is 3 × 6? [18] What is 3 × 3? [9] What is 3 × 9? [27] What is 3 × 5? [15] What is 3 × 2? [6] What is 3 × 8? [24] What is 3 × 10? [30]

ACTIVITIES ***Discovering how to find check numbers.*** Ask the child to look at her check number tables (Worksheet 39); the first four rows are shown below.

1 (1)	2 (2)	3 (3)	4 (4)	5 (5)	6 (6)	7 (7)	8 (8)	9 (0)
10 (1)	11 (2)	12 (3)	13 (4)	14 (5)	15 (6)	16 (7)	17 (8)	18 (0)
19 (1)	20 (2)	21 (3)	22 (4)	23 (5)	24 (6)	25 (7)	26 (8)	27 (0)
28 (1)	29 (2)	30 (3)	31 (4)	32 (5)	33 (6)	34 (7)	35 (8)	36 (0)

The first four rows of the check numbers table.

Then ask the following questions to help her discover how to find check numbers.

1. Do you see the multiples of 9 somewhere on the table? [the right column]
2. What is the *range*, that is, what values, or numbers, can check numbers have? [0 to 8]
3. Look at the first, or top, row. Which number doesn't seem to follow a pattern and what is different? [The check number for 9 is not 9, but 0.]
4. Look at the second row. How could you find those check numbers? [by adding the digits together]
5. What about 18? [Adding 1 and 8 equals 9, but 9 is the same as 0 with check numbers.]
6. Look at the third row. Which number seems harder? [19] How could you find the check number for 19? [The easiest way is to ignore the 9, so it is 1.]

An observant child might notice the check number is the same as how far the number is past a multiple of 9. For example, 19 is 1 more than 18, so the check number is 1.

7. How could you find the check number for 28? There are 2 ways other than the one just above. a) Add 2 + 8, getting 10; then add 1 + 0, getting 1. b) Or, think if the 2 is partitioned into

1 + 1 and a 1 combined with the 8, then we have 1 and 9, which is 1. Be sure the child understands both ways.

Practice. Tell the child she is ready for bigger numbers. Write

2745

and ask her to find the check number. [0] Ask how she did it. The easiest way is to note that 2 + 7 = 9 and 4 + 5 = 9, so the check number is 0.

Second example. Write

5635

and ask her to find the check number. [1] Ask how she did it. The easiest way is to see that 6 + 3 = 9 and 5 + 5 = 10, so the check number is 1. It could also be done by adding 5 + 6 = 11, which is 2, then 2 + 3 + 5 = 10, or 1.

Third example. Write

6974

and ask her to find the check number. [8] Ask how she did it. The 9 can be ignored; 6 + 7 = 13, which is 4; and 4 + 4 = 8. Or, 6 + 4 = 10; 10 (or 1) + 7 = 8.

Checking multidigit addition and subtraction. Ask the child to add the following and to check them with check numbers

Note: Just as subtraction problems can be checked by adding up, so can check numbers.

2796	2796 (6)	7146	7146 (0)
+ 5192	+ 5192 (8)	– 3207	– 3207 (3)
	7988 (5) ✓		3939 (6) ✓

Ask her if the check numbers check. [Yes, for the addition 6 + 8 = 14, which is 5.] Add up the check numbers for subtraction. [6 + 3 = 9, which is 0, which matches the check number for 7146.]

Worksheet. Give her the worksheet for practice in finding check numbers and in finding errors. The solutions are below.

4214 **(2)**	6936 **(6)**	7892 **(8)**	4968 **(0)**
2588 **(5)**	5674 **(4)**	3895 **(7)**	6375 **(3)**

7832 **(2)**	8863 **(7)**	5737 **(4)**	5326 **(7)**
+ 1446 **(6)**	+ 687 **(3)**	+ 2546 **(8)**	+ 1788 **(6)**
9278 (8)	**9550 (1)**	**8283 (3)**	**7114 (4)**

4136 **(5)**	2557 **(1)**	972 **(0)**	4145 **(5)**
120 **(3)**	3271 **(4)**	1785 **(3)**	819 **(0)**
+ 5186 **(2)**	+ 2582 **(8)**	+ 8557 **(7)**	+ 2733 **(6)**
9442 (1)	**8410 (4)**	**11314 (1)**	**7697 (2)**

3641 **(5)**	6565 **(4)**	5232 **(3)**	7900 **(7)**
– 1262 **(2)**	– 3542 **(5)**	– 5053 **(4)**	– 5073 **(6)**
2379 (3)	**3023 (8)**	**179 (8)**	**2827 (1)**

5586 **(6)**	3628 **(1)**	8405 **(8)**	5067 **(0)**
– 2689 **(7)**	– 2811 **(3)**	– 1687 **(4)**	– 3368 **(2)**
2897 (8)	**817 (7)**	**6718 (4)**	**1699 (7)**

Lesson 50

Check Numbers and Multiples of Three

OBJECTIVES

1. To emphasize the multiples of three
2. To practice using check numbers

MATERIALS

A set of Corners cards
Instructions for Corners Game (Lesson 8)

WARM-UP

Give the following orally and ask the child to subtract mentally: 35 – 15 [20], 41 – 15 [26], 48 – 15 [33], 76 – 15 [61], and 150 – 15. [135]

Note: For mental work, be sure the children subtract the tens and then the ones. For example, 35 – 15 = 35 – 10 [25] – 5. [20]

Ask the child to subtract 4379 – 2590 in his journal. [1789] Tell him to use check numbers to check his answer. [(5) – (7) = (7) Here the 7 and 7 are added to get 14; then the 1 and 4 are added to get 5.]

Ask the child to write the multiples of 3, but to leave space after each multiple. Next ask him to look at the multiples and say the tables of 3. Then ask him for the following: <u>What is 3 × 6?</u> [18] <u>What is 3 × 7?</u> [21] <u>What is 3 × 5?</u> [15] <u>What is 3 × 3?</u> [9] <u>What is 3 × 4?</u> [12] <u>What is 3 × 9?</u> [27] <u>What is 3 × 8?</u> [24]

ACTIVITIES

Discovering another pattern with the 3s. Ask the child to write the check numbers for the multiples of threes he just wrote. The results are shown below.

3 **(3)**	6 **(6)**	9 **(0)**
12 **(3)**	15 **(6)**	18 **(0)**
21 **(3)**	24 **(6)**	27 **(0)**
30 **(3)**		

Ask what patterns he sees. [They are either 3, 6, or 0.]

Note: The child may have played the addition version of this game, Corners Three, in Level C.

Corners Three Subtraction game. Tell the child that today he will play Corners game a different way. Instead of matches being multiples of 5, they will be multiples of 3.

Ask him to list all the possible combinations he can make with the Corners cards, including 10s. Do it in an organized way; start by asking for the sums that total 3. See below.

Note: There are other ways to organize this.

3 = 1 + 2	6 = 1 + 5	9 = 1 + 8
	6 = 2 + 4	9 = 2 + 7
12 = 2 + 10	6 = 3 + 3	9 = 3 + 6
12 = 3 + 9		9 = 4 + 5
12 = 4 + 8	15 = 5 + 10	
12 = 5 + 7	15 = 6 + 9	18 = 9 + 9
12 = 6 + 6	15 = 7 + 8	18 = 10 + 8

Corners Three Subtraction is played like the regular subtraction Corners Game, but with 2 exceptions.

1) The cards joining must be a multiple of 3.

2) Matching without scoring is *not* allowed, such as 1 + 1 and 4 + 4. Play must be to the last card played or to a Corner.

Note: As usual, play must be to the last card played or to a corner.

Ask him to start with a score of 150 and tell him he must use check numbers.

Check number patterns. A sample game is shown below along with the check numbers.

```
 150 (6)
– 12 (3)
 138 (3)
– 15 (6)
 123 (6)
– 12 (3)
 111 (3)
  – 6 (6)
 105 (6)
– 15 (6)
  90 (0)
– 18 (0)
  72 (0)
  – 9 (0)
  63 (0)
  – 3 (3)
  60 (6)
```

Reserve the last 5 minutes or so of the lesson time and ask the child to discuss what check number patterns he sees. [They are all either 0s, 3s, or 6s.]

ENRICHMENT What would the check numbers be if he started the game with a score of 100? [1, 4, and 7]

What would the check numbers be if he started the game with a score of 200? [2, 5, and 8]

EXTRA HELP ***Practicing finding check numbers.*** If the child needs help finding check numbers, play the game, Check Numbers, found in *Math Card Games* (A63).

Lesson 51

The Almost Subtraction Strategy

OBJECTIVES 1. To practice the *almost* subtraction strategy
2. To practice using check numbers

MATERIALS Worksheet 42, "The 'Almost' Subtraction Strategy"

WARM-UP Give the following orally and ask the child to add mentally: 35 + 26 [61], 78 + 26 [104], 48 + 26 [74], 46 + 26 [72], and 99 + 26. [125]

Ask the child to subtract 5008 – 2452 in her journal. [2556] Tell her to use check numbers to check her answer. [(4) – (4) = (0)]

Ask the child to say the multiples of three; remind her to write only three numbers in each row.

3	6	9
12	15	18
21	24	27
30		

Next ask the child to look at the multiples and say the tables from 3 × 1 = 3 to 3 × 10 = 30.

ACTIVITIES ***Problem 1.*** Give the child this problem to solve without paper or pencil. Mindy is buying some bread that costs 1 dollar and 1 cent ($1.01). She pays for it with a five-dollar bill. How much money does she get back? [$3.99]

Note: Reasoning, not an algorithm, should be the final authority on the correctness of a answer.

Ask her to think about and then explain her solution.

The preferred solution is to think that $1.01 is almost one dollar and if it were, the answer would be $4. But since it is 1 cent more, the answer must be $3.99. Tell her we will call this strategy the *almost* strategy.

Problem 2. Mike is a clerk at a fruit store. A customer buys some fruit that costs 2 dollars and 99 cents ($2.99). The customer pays for it with a 10-dollar bill. How much change does Mike need to give the customer? [$7.01]

Again ask her to think about and then to explain her solution.

The preferred solution is to think that $2.99 is almost 3 dollars and if it were, the answer would be $7. But since it is one cent less, the answer must be 1 cent more, $7.01. What we can call this strategy? [the almost strategy]

Problem 3. Amanda and Zachary and their family left at 1:32 to drive to their cousins' place. They arrived there at 3:30. How long did it take them? [1hr 58 min]

Let her solve it without making any additional comments. After she has explained her solution, discuss various other solutions. Stress the almost solution: if they had left at 1:30, it would have been 2 hours. But it is 2 minutes less, so the answer is 1 hour 58 minutes.

Almost strategy for subtraction facts. Write the following

15	15
– 5	– 6

How could you use the almost strategy to help in remembering 15 – 6? [We know 15 – 5 = 10, so 15 – 6 must be 1 less, or 9.]

Write another set of facts.

14	14
– 4	– 5

How could you remember 14 – 5? [one less than 14 – 4, or 9]

Write the following

18	15	12	13	14
– 9	– 6	– 4	– 4	– 5

Which fact does not have 9 for an answer? [12 – 4]

How could you remember 12 – 4? [It is 2 away from 12 – 2, so it must be 8.]

Worksheet. The worksheet has four similar problems. It also has a row of facts and 2 rows of computations to be done mentally. Ask the child to use check numbers for the last row of problems. The answers are as follows.

1. **$1.99** 2. **$2.01**

3. **8:58** 4. **198**

14	11	13	16	14	12	17	13	15
– 5	– 2	– 5	– 7	– 8	– 3	– 9	– 4	– 6
9	**9**	**8**	**9**	**6**	**9**	**8**	**9**	**9**

23	38	65	72	53	97	34	46	80
– 4	– 9	– 7	– 3	– 5	– 8	– 6	– 7	– 1
19	**29**	**58**	**69**	**48**	**89**	**28**	**39**	**79**

425 (2)	187 (7)	247 (4)	105 (6)	636 (6)	300 (3)	312 (6)
– 26 (8)	– 88 (7)	– 49 (4)	– 7 (7)	– 37 (1)	– 1 (1)	– 13 (4)
399 (3)	**99 (0)**	**198 (0)**	**98 (8)**	**599 (5)**	**299 (2)**	**299 (2)**

Lesson 52

Terry's Subtraction Strategy

Note: Omit this lesson if the child has done the Transition Lessons.

OBJECTIVES
1. To practice subtraction by *Terry's* method
2. To practice using check numbers
3. To learn mathematics by reading

MATERIALS Worksheet 43, "Terry's Way to Subtract"

WARM-UP Give the following (for which the almost strategy works) orally and ask the child to compute mentally: 45 – 26 [19], 78 – 29 [49], 45 – 26 [19], 100 – 26 [74], and 100 – 49. [51]

Ask the child to say the multiples of three; remind him to write only three numbers in each row.

3	6	9
12	15	18
21	24	27
30		

Next ask the child to look at the multiples and say the tables from $3 \times 1 = 3$ to $3 \times 10 = 30$.

Then ask the child to write the even multiples of three in a row. What do you have? [multiples of 6 to 30]

6	12	18	24	30

Then ask the child to add 30 to each multiple and write it below each multiple. What do you have? [multiples of 6 to 60]

6	12	18	24	30
36	42	48	54	60

ACTIVITIES ***Terry's method.*** Take out the worksheet. Ask the child to read it carefully and then to work to figure it out. Give him about 10 to 20 minutes.

When he thinks he understands it, write the following problem and ask him to solve it.

$$\begin{array}{r} 835 \\ \underline{-\ 639} \end{array}$$

The solution according to Terry is:

$$\begin{array}{r} 835 \\ \underline{-\ 639} \\ 200 \\ 0 \\ \underline{-\ 4} \\ 196 \end{array}$$

Ask for an explanation.

Write the next example and ask the child to tell you what to write.

5627
– 2818
3000
– 200
10
– 1
2809

Terry's method for facts. Write 13 – 7. Do you think it will work for this?

13
– 7
10
– 4
6

One way to understand why it works is to think that if the problem were 17 – 7, we know the answer would be 10. But since 13 is 4 less than 17, the answer must be 4 less than 10, or 6.

Repeat for other facts, such as 14 – 8 and 12 – 4.

Worksheet. Ask the child to complete the worksheet. The solutions are as follows.

76	77	51	62	64	610	859
– 13	– 69	– 9	– 35	– 48	– 112	– 375
60	**10**	**50**	**30**	**20**	**500**	**500**
3	**– 2**	**– 8**	**– 3**	**– 4**	**0**	**– 20**
63	**8**	**42**	**27**	**16**	**– 2**	**4**
					498	**484**

792	821	6076	3843	7083	6005
–139	–626	–4753	–3346	–2249	–1009
600	**200**	**2000**	**0**	**5000**	**5000**
60	**0**	**– 700**	**500**	**–200**	**0**
– 7	**– 5**	**20**	**0**	**40**	**0**
653	**195**	**3**	**– 3**	**– 6**	**– 4**
		1323	**497**	**4834**	**4996**

Lesson 53

Working With Twos

Note: Omit this lesson if the child knows the 2s facts.

OBJECTIVES
1. To learn the table of 2s
2. To review the meaning of multiplication
3. To review the commutative law

MATERIALS Tiles
Worksheet 44, "Working With Twos"

WARM-UP Ask the child to subtract 3152 – 209 in her journal. [2943] Tell her to use check numbers to check her answer. [(2) – (2) = (0)]

Then ask the child to write the 2s multiples to 20:

2	4	6	8	10
12	14	16	18	20

Ask the child to look at the multiples and answer together. What is 2×6? [12] What is 2×7? [14] What is 2×4? [8] What is 2×9? [18] What is 2×8? [16] What is 2×3? [6] and What is 2×2?

ACTIVITIES ***Preliminary practice.*** Put away the multiples of 2. Take out Worksheet 44 and tell the child she is to do only the **first** row, inside the rectangle, but not to begin until you say Start. Tell her to raise her hand when she finishes and to write on her paper the number of seconds that you tell her.

Note: This type of timed testing is less stressful than requiring the children to stop at a certain time.

Ask her to correct her paper as you read the answers.

8 18 14 6 16 2 4 20 10 12

Ask her to write the number correct at the top on the line.

Reviewing one meaning of multiplication. Lay out 12 tiles (or use an abacus) in the arrangement shown below on the left.

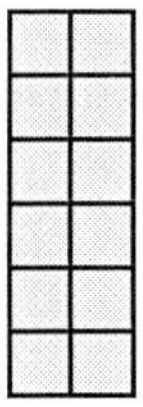

Showing 2 × 6 tiles.

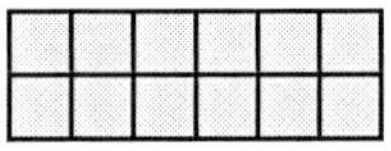

Turning the tiles to show and 6 × 2.

Ask the child to write the addition equation. [2 + 2 + 2 + 2 + 2 + 2 = 12] Next ask her to write the multiplication equation. [$2 \times 6 = 12$] Then write the following equation

$$2 + 2 + 2 + 2 + 2 + 2 = 2 \times 6$$

and ask if it is okay to write it that way. [yes] Why? [Both sides are equal.]

Now turn the 12 tiles 90 degrees as shown above on the right. Ask the child to write the addition [6 + 6 = 12] and multiplication equations. [$6 \times 2 = 12$]

Ask her to write the combined equation.

$$6 + 6 = 6 \times 2$$

Reviewing the commutative law. Write

$$2 \times 7 = 7 \times 2$$

and ask if that is true. [yes] Ask the child to prove it to you. [the arrangement below]

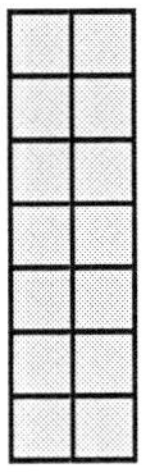

Showing 2 × 7 tiles.

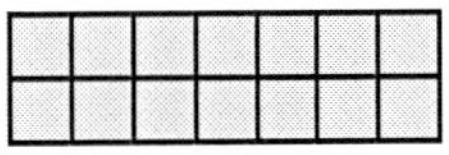

Turning the tiles to show and 7 × 2.

Repeat for 3 × 5.

Ask the her, Which is more, 5 dimes or 10 nickels? [same]

The 2s table. Write

$$2 \times 7$$

and ask the child how she could remember it. [possibly 7 + 7]

Next ask her to look over her wrong answers on the worksheet and think how she could remember them.

Note: A good time is 40 seconds or less.

Concluding practice. Repeat the procedure for the time test done earlier using the last row on the worksheet. The answers are:

20 8 18 12 14 10 2 6 4 16

Ask if she had more correct. Ask if she did it faster.

Worksheet. The rest of the worksheet asks the child to write addition and multiplication equations for groups of tiles. There is also more 2s practice. The answers follow.

$2 + 2 + 2 + 2 + 2 = \mathbf{10}$
$2 \times 5 = \mathbf{10}$

$8 + 8 = \mathbf{16}$
$8 \times 2 = \mathbf{16}$

$10 + 10 + 10 = \mathbf{30}$
$10 \times 3 = \mathbf{30}$

$3 + 3 + 3 + 3 + 3 + 3 = \mathbf{18}$
$3 \times 6 = \mathbf{18}$

$1 + 1 + 1 + 1 + 1 + 1 = \mathbf{6}$
$1 \times 6 = \mathbf{6}$

$2 \times 3 = \mathbf{6}$
$2 \times 6 = \mathbf{12}$
$2 \times 1 = \mathbf{2}$
$2 \times 8 = \mathbf{16}$
$2 \times 10 = \mathbf{20}$
$2 \times 7 = \mathbf{14}$
$2 \times 5 = \mathbf{10}$
$2 \times 9 = \mathbf{18}$
$2 \times 4 = \mathbf{8}$
$2 \times 2 = \mathbf{4}$

EXTRA HELP ***Practicing multiplication facts.*** If the child needs help learning the multiplication facts, play the game Multiplication Memory found in *Math Card Games* (P10).

Lesson 54

Review

OBJECTIVE 1. To review and practice

MATERIALS Worksheet 45-A and 45-B, "Review" (2 versions)
Cards for playing games

ACTIVITIES ***Review worksheet.*** Give the child about 15 to 20 minutes to do the review sheet. The oral problems to be read are as follows:

61 – 18 = 75 + 26 = 53 – 15 =

Review Sheet 45-A.

1. Write only the answers to the oral questions. 43 101 38

4. Write only the answers: 67 – 28 = 39 43 – 29 = 14 62 – 18 = 44

7. Write the multiples of 3.

3 6 9
12 15 18
21 24 27
30

17.
3 × 1 = 3 3 × 2 = 6 3 × 3 = 9
3 × 4 = 12 3 × 5 = 15 3 × 6 = 18
3 × 7 = 21 3 × 8 = 24 3 × 9 = 27
3 × 10 = 30

27. Write the check numbers.

14 (5) 99 (0)
457 (7) 852 (6)

31. Jamie is buying three notebooks for $2.58 each. What is the total cost?

$ 2.58	(6)
2.58	(6)
+ 2.58	(6)
$ 7.74	(0)

32. A pair of boots costs $19.85. A coat costs $35.37 and a cap costs $7.47. What is the total cost?

$ 19.85	(5)
35.37	(0)
+ 7.47	(0)
$ 62.69	(5)

33. 569 (2) − 284 (5) = 285 (6)

34. 756 (0) − 157 (4) = 599 (5)

35. 7823 (2) − 4905 (0) = 2918 (2)

36. 1365 (6) − 846 (0) = 519 (6)

37. 2 × 5 = 10 2 × 8 = 16 2 × 4 = 8 2 × 9 = 18 2 × 6 = 12

The oral problems to be read to the child are as follows:

97 – 38 =   51 + 49 =   46 – 19 =

Review Sheet 45-B.

1. Write only the answers to the oral questions. 59 100 27

4. Write only the answers: 54 – 35 = 19   100 – 29 = 71   88 – 37 = 51

7. Write the multiples of 4.

4 8 12 16 20

24 28 32 36 40

17. 4 × 8 = 32   4 × 4 = 16

19. 4 × 3 = 12   4 × 5 = 20

41. 4 × 4 = 16   4 × 10 = 40

43. 4 × 1 = 4   4 × 6 = 24

45. 4 × 7 = 28   4 × 9 = 36

27. Write the check numbers.

25 (7)   63 (0)   178 (7)   467 (8)

31. Jamie is buying three notebooks for 4 dollars and 9 cents each. What is the total cost?

	$	4.	0	9	(4)
		4.	0	9	(4)
	+	4.	0	9	(4)
$	1	2.	2	7	(3)

32. A pair of gloves costs $4.27. A scarf costs $8.99 and a hat costs $13.59. What is the total cost?

	$	4.	2	7	(4)
		8.	9	9	(8)
+	1	3.	5	9	(0)
$	2	6.	8	5	(3)

33. 346 (4) − 271 (1) = 75 (3)

34. 532 (1) − 133 (7) = 399 (3)

35. 6539 (5) − 2632 (4) = 3907 (1)

36. 2864 (2) − 975 (3) = 1889 (8)

37. 3 × 1 = 3   3 × 10 = 30   3 × 2 = 6   3 × 5 = 15   3 × 3 = 9

Games. Spend the remaining time playing games.

CHECK NUMBERS. Check Numbers *(Math Card Games* A63).

SUBTRACTION. Corners Three Subtraction (Lesson 50), Short Chain Subtraction *(Math Card Games,* S18-S23*).*

Treasure Hunt *(Math Card Games,* P6).

Lesson 55

Working With Fives

Note: Omit this lesson if the child has done the Transition Lessons.

OBJECTIVES

1. To learn the table of 5s
2. To apply the multiples of 5 to clocks
3. To read a scale when some numbers are not shown

MATERIALS

A large clock with hour numbers, but not minute numbers
The scale shown on the right page
Worksheet 46, "Working With Fives"

WARM-UP

Give the following orally and ask the child to subtract mentally: 35 – 28 [7], 41 – 28 [13], 48 – 28 [20], 76 – 28 [48], and 150 – 28. [122]

Ask the child to subtract 9063 – 4084 in her journal. [4979] Tell her to use check numbers to check her answer. The check numbers for this problem are (0) – (7) = (2).

While the child is counting, write the 5s multiples:

5	10
15	20
25	30
35	40
45	50

Ask the child to look at the multiples and answer 5 × 6 [30], 5 × 7 [35], 5 × 4 [20], 5 × 9 [45], 5 × 8 [40], 5 × 3 [15], and 5 × 2. [10]

ACTIVITIES

Preliminary practice. Put away the multiples. Take out the worksheet and tell the child she is to do the first row, inside the rectangle, when you say <u>Start.</u> Tell her to raise her hand when she finishes and to write the number of seconds that you tell her.

When she is finished, ask her to correct them as you read the answers.

15 50 30 25 45 40 5 10 35 20

Ask her to write the number correct at the top.

Practicing the 5s. Write out the facts in the following order.

5 × 1 = 5	5 × 2 = 10
5 × 3 = 15	5 × 4 = 20
5 × 5 = 25	5 × 6 = 30
5 × 7 = 35	5 × 8 = 40
5 × 9 = 45	5 × 10 = 50

<u>What pattern do you see with 5 times an even number?</u> [multiples are 10s, actually half of the multiplier.] <u>Why is that so?</u> [Two 5s make a ten, so you need half as many.] Practice with the child by asking for the even facts in mixed order, with the answers showing and then with the answers covered. Looking at the list helps with memory because it gives an organized visual picture.

<u>What pattern do you see with 5 times an odd number?</u> [end in 5] Practice the odd numbers, first with the answers, then without. Also practice with both the evens and odds. Repeat the ones she has the most trouble with.

Note: A child may be helped with her fives by correlating them with the clock.

Fives on the clock. Show the clock. How many hour numbers does a clock have? [12] How many minutes are between each number? [5] Remind the child a minute is the *space* between the marks.

Ask her to count the minutes around the clock while you point to each number in turn, starting with 1. [oh 5, 10, 15, . . . 55, o'clock]

Now tell her you are going to mix up the numbers. First point to quarter hours and ask for the minutes: 3 [15], 6 [30], and 9. [45]

Next point to the even numbers in turn, 2 [10], 4 [20], 6 [30], 8 [40], 10. [50] What pattern did you notice? [multiples of 10] Repeat for the odd numbers.

Concluding practice. Repeat the procedure for the time test done earlier using the left column inside the rectangle on the worksheet. The answers are:

15 30 5 40 50 35 25 45 20 10

Ask is she had more correct. Ask if she did it faster.

Reading a scale. Draw the scale and the arrow as shown below.

Note: A paper arrow could be used.

This scale will be needed for the next lesson.

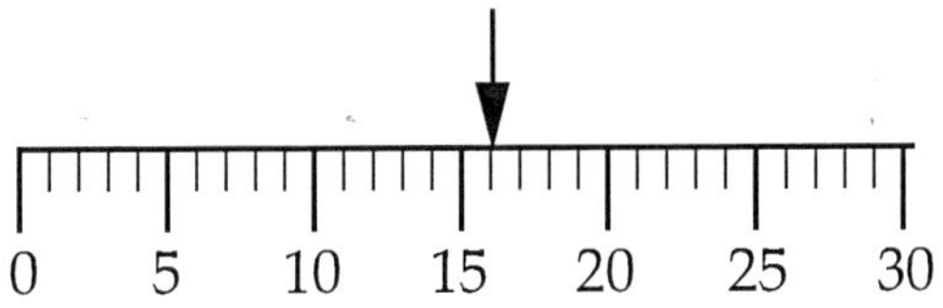

A scale where each line represents 1. The arrow points to 16.

How much is each line worth? [1] What number does the arrow point to? [16] How do you know? [first line after 15]

Move the arrow to other locations such as, 3, 11, 29 and 17.

Worksheet. The worksheet asks the child to read a scale and to draw the arrows. Answers are below.

1.

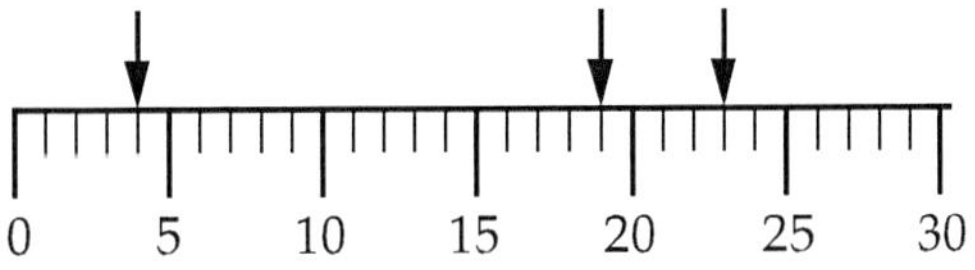

2.

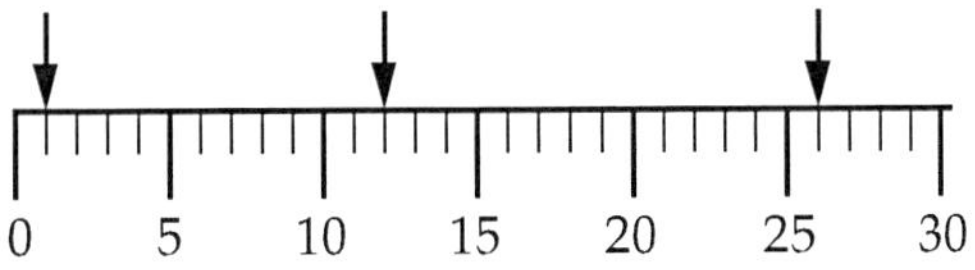

3. **6, 21**
4. **13, 22**
5. **9, 19**
6. **17, 28**

Lesson 56

Telling Time to the Minute

Note: Omit this lesson except the Facts practice if the child can tell time to the minute.

OBJECTIVES
1. To review telling time to the five-minute interval
2. To practice telling time to the minute

MATERIALS
A geared clock with minute marks (If the clock has the minute numbers, cover them with opaque tape.)
The scale from the previous lesson or a copy of Worksheet 46
Worksheet 47, "Quick Practice-1"
Worksheet 48 "Telling Time to the Minute"

WARM-UP
Give the following orally and ask the child to subtract mentally: 73 – 35 [38], 92 – 56 [36], and 100 – 34. [66]

Ask the child to subtract 7093 – 6275 in his journal. [818] Tell him to use check numbers to check his answer. [(1) – (2) = (8)] To check the answer *add* the check numbers, (2) + (8) = (1).

Note: Quick practice sheets will be part of the daily lessons from now on.

Facts practice. Take out the quick practice sheet. Explain he is to do what is in the **top** rectangle when you say start. Correct it when finished. The answers are:

14	**4**	**18**	**12**	**8**	**2**	**20**	**6**	**10**	**16**
30	**25**	**15**	**5**	**50**	**20**	**45**	**40**	**35**	**10**

ACTIVITIES
Reviewing telling time to five-minutes. Show the clock; ask where the hour starts. [at the top, at the 12] Set the hands to 8:00. Why do we count by fives? [There are 5 minutes between each number.] Ask the child to move the hands as you say the time from 8:00 to 9:00 in five-minute increments. [eight oh-five, eight ten, . . . , eight fifty-five, nine o'clock]

Where is the minute hand when the hour is half over? [at the 6] Where will the hour hand be? [halfway between the 8 and 9] Ask him to move the hands so he can see for himself. Where is the hour hand at 9:30? [halfway between 9 and 10]

Spend some time practicing telling time. Ask him to move the hands to various positions and ask for oral responses. For extra practice on difficult numbers, such as :25, set the hands at 2:25, 3:25 and so. Also name times for him to set on the clock.

Reviewing finding numbers. Refer to the previous day's scale. Ask how he could find 28 on the scale. [3 after 25 or 2 before 30] Repeat for 17, 2 and other numbers. Place the arrow at various positions for him to name.

Time to the minute. Tell the child that now they are going to tell the time between the numbers. Ask him to set the hands to 2:00. Then ask him to move to 2:01 (two oh-one). Then ask him to move it to the next minute. What time does the clock say? [two oh-two] Continue minute by minute asking him to respond.

Where is the minute hand for 6:21? [1 mark after 6:20] Ask him to set the clock. Repeat for 6:24. [1 mark before 6:25 or 4 after 6:20] Repeat for other times.

Set the clock for him to read the times.

Drawing hands. On a blank paper clock, demonstrate how to draw the hands. Start near the numbers and draw toward the center. First draw the minute hand, starting at the edge of the circle and drawing toward the center. Then draw the hour hand, starting near the numbers and taking in consideration where the minute hand is. Lastly, draw the arrows by starting a short ways from the point and drawing toward the point. See the example below for drawing the hands at 3:20.

Drawing the hands for 3:20. First draw the minute hand, starting at the number and drawing toward the center. Repeat for the hour hand. Lastly, draw the arrows.

Worksheet. The worksheet requires the child to draw hands for the 12 clocks, followed by asking him to read the time on 8 clocks. Those times are given below.

4:53	**10:19**	**6:27**	**2:43**
8:37	**1:06**	**9:24**	**3:13**

Lesson 57

Telling Time Practice

Note: Omit this lesson except the Facts practice if the child can tell time to the minute.

OBJECTIVES
1. To apply the multiples of 5 to clocks
2. To practice telling time to the minute

MATERIALS
Worksheet 47, "Quick Practice-1"
A geared clock
40 to 60 number cards 0 to 9 for Name That Time Game
Sets of clock cards for a child not knowing the minute numbers

WARM-UP
Ask the child to subtract 7771 – 7429 in her journal. [342] The check numbers are: [(4) – (4) = (0)]

Write 5 × 4 and 4 × 5 ask which is more. [both 20] Repeat for 5 × 9 and 9 × 5. [both 45] Ask the child to explain. [same] This can easily be demonstrated on the abacus as shown below.

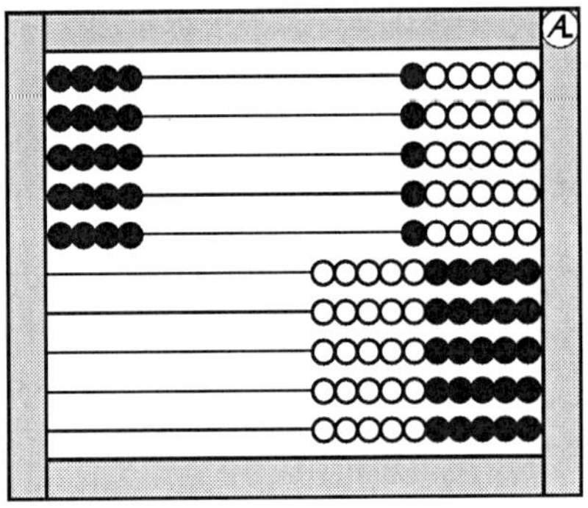

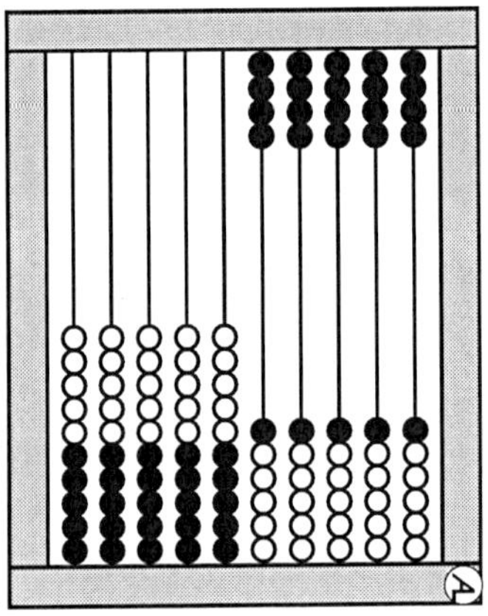

Representing 4 × 5.

Then turning the abacus 90° to see it 5 × 4.

What is 5 × 8? [40] What is 3 ×5? [15] What is 2 × 7? [14] What is 7 × 2? [14] What is 5 × 9? [45] What is 9 × 5? [45]

Facts practice. Ask the child to take out her Quick Practice sheet. Explain she is to do those in the **second** rectangle when you say start. Correct it when she is finished. The answers are:

8	20	6	12	10	16	2	14	18	4
25	10	30	20	15	50	5	35	45	40

ACTIVITIES
Minute practice. Use a large clock and either point to or set the minute hand to 3. What is the minute number? [15] Point to the other numbers in random order and ask for the minute numbers.

Clocks and multiplying by 5. Help the child see the correlation with the numbers on the clock and multiplying by 5. Point to 3 on the clock and ask what is 5 × 3. [15] Point to the other numbers from 1 to 11 and ask what is 5 times that number.

Name That Time Game. Demonstrate the following game to the child. Tell her it is to help her practice telling time.

Players choose the top three cards and form a time with them. Forming the time requires knowing that there cannot be more 59 minutes in an hour.

After the game ask What sets of three numbers could not be a possible time? [all 0s or all 3 numbers greater than 5]

NAME THAT TIME

Objective To practice telling time to the nearest minute

Manipulatives The geared clock or a paper clock (Appendix of *Math Card Games*, use pencils for hands).

Cards About 60 basic number cards 0 to 9 (Fewer cards can be used twice.)

Number of players 2 or 3

Object of game To read the time entered on the clock

Layout The stock with the cards face down

Play The first person draws 3 cards from the stock and forms a time with them without letting the other person see it. The same player then sets the clock that same time. The other player(s) reads the time, writes it down, and compares the answer with the 3 cards.

Then the next player takes a turn setting the clock with 3 new cards. Play continues until the cards are exhausted; the cards may be reused.

Very rarely, a player might have three cards that cannot be made into a time, such as 9, 8, 9. In this case the player can return 1 or 2 cards and take new cards to form a another time.

Note For players still memorizing the minute numbers, the clock cards could be set around the clock in the correct order.

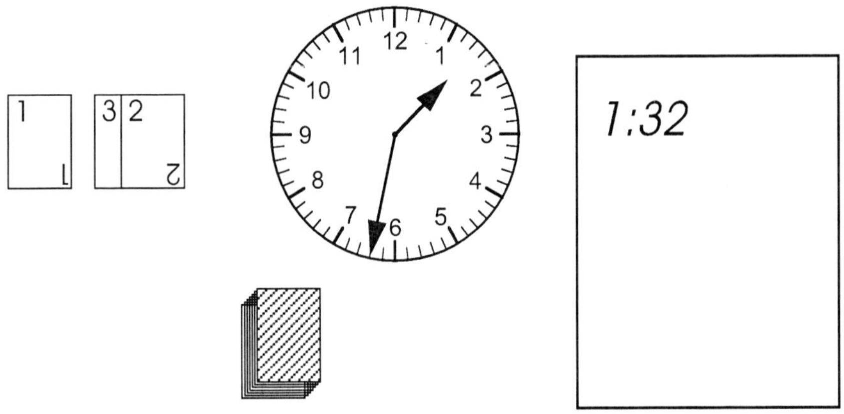

EXTRA HELP ***Learning the minutes.*** For the child needing help learning the minutes, ask them to play the games Minute Memory and Minute Solitaire, found in *Math Card Games* (C11-12).

Lesson 58

Multiplying With Money

OBJECTIVES
1. To apply the multiples of 5 to money
2. To write equations for groups of coins

MATERIALS
Worksheet 47, "Quick Practice-1"
A geared clock
Coins: 10 each of pennies, nickels, dimes, and quarters
Tiles
Worksheet 49, "Multiplying With Money"

WARM-UP
Ask the child to subtract 8096 – 5618 in his journal. [2478] The check numbers are: [(5) – (2) = (3)]

Say the following orally for the child to compute mentally. What is 77 + 39? [116] What is 77 – 39? [38]

Set the hands of the clock to various times for the child to name. Then ask the child to set the clock at various times you specify.

Facts practice. Ask the child to take out his Quick Practice sheet. Explain he is to do those in the **third** rectangle when you say start. Correct it when he is finished. The answers are:

10	**12**	**8**	**6**	**4**	**16**	**20**	**14**	**18**	**2**
35	**10**	**45**	**15**	**30**	**40**	**50**	**25**	**20**	**5**

ACTIVITIES
Multiplication equations with nickels. Review the value of the coins by showing the child the various coins and asking what their values are.

Next lay out 6 nickels in groups of 2s as shown below on the left. What is their value? [30¢] Ask the child to write the addition and multiplication equations. [5¢ + 5¢ + 5¢ + 5¢ + 5¢ + 5¢ = 30¢ and 5¢ × 6 = 30¢]

Note: A child may be helped by reading the multiplication equationas 5¢ *taken 6 times* equals 30¢.

5¢ × 6 = 30¢
Finding the value of 6 nickels.

5¢ × 7 = 35¢
Finding the value of 7 nickels.

5¢ × 4 = 20¢
Finding the value of 4 nickels.

Repeat for other patterns, such as 7 and 4, as shown above. Ask him to write or say the multiplication equations. Do you see any patterns to help you? [2 nickels in each row equal 10¢]

Multiplication equations with other coins. Lay out dimes in groups of 2s and ask for the values and the equations. Several examples are shown below.

10¢ × 8 = 80¢
Finding the value of 8 dimes.

25¢ × 3 = 75¢
Finding the value of 3 quarters.

1¢ × 5 = 5¢
Finding the value of 5 pennies.

Repeat for pennies and quarters.

Tens facts. Write

3 × 10 = _ 10 × 3 = _

Which is more? Show me with tiles. [both 30] See below. Then ask for other ten facts. What is 10 × 2? [20] What is 5 ×10? [50] What is 10 × 7? [70] What is 7 × 10? [70] What is 10 × 10? [100]

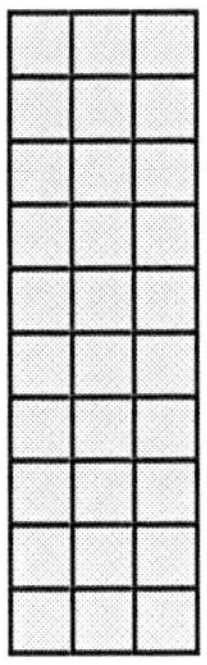

Showing 3 × 10 = 30.

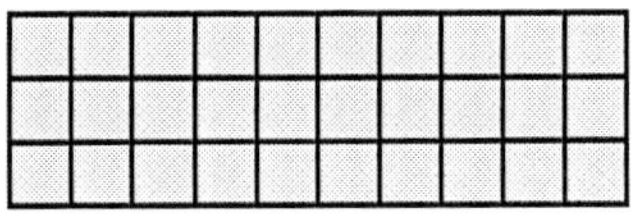

Showing 10 × 3 = 30.

Writing equations. Ask the child to look around the room and to write equations for things he sees. Tell him to explain his numbers.

Worksheet. The worksheet asks for equations for money. It also provides practice in the 10s facts. The answers are below.

10¢ × 9 = 90¢ **5¢ × 5 = 25¢** **1¢ × 7 = 7¢**

5¢ × 8 = **40¢** 10¢ × 6 = **60¢** 25¢ × 4 = **100¢**

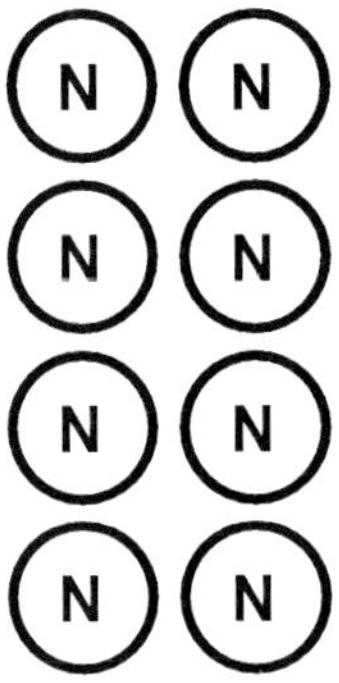

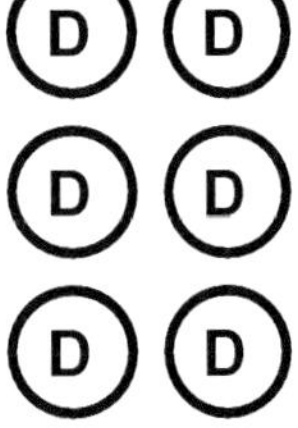

Lesson 59

Multiplying With 1s and 0s

OBJECTIVES 1. To understand multiplying with 1s and 0s
2. To learn the facts with multiplying 1s and 0s

MATERIALS Worksheet 47, "Quick Practice-1"
A geared clock
Tiles
Worksheet 50, "Multiplying With 1s and 0s"

WARM-UP Ask the child to subtract 6512 – 1089 in her journal. [5423] The check numbers are: [(5) – (0) = (5)]

Say the following orally for the child to compute mentally. What is 37 + 19? [56] What is 37 – 19? [18]

Set the hands of the clock to various times for the child to name. If desired, ask the child to set the clock at various times you specify.

Facts practice. Ask the child to take out her Quick Practice sheet. Ask her to do those that are in the **fourth** rectangle when you say start. Correct them when she is finished. The answers are:

45	**40**	**10**	**14**	**40**	**10**	**30**	**2**	**15**	**10**
9	**5**	**40**	**3**	**100**	**80**	**20**	**1**	**6**	**70**

ACTIVITIES ***Multiplying with 1s.*** Ask the child, Can you show 1 × 6 with tiles? See the left figure below.

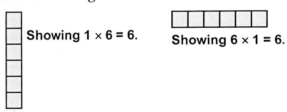

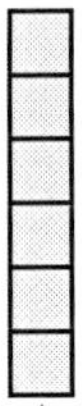

Showing 1 × 6 = 6.

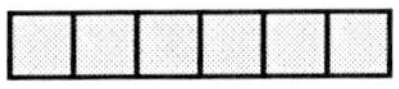

Showing 6 × 1 = 6.

Now can you show 6 × 1? See the right figure above. Write or ask her to write the equations.

Proceed with other examples, if helpful. The abacus is very effective for teaching this concept. Write

$$1 \times 7 = \_$$

How many do we need to enter at a time? [1] How many times do we enter it? [7] See the left figure below.

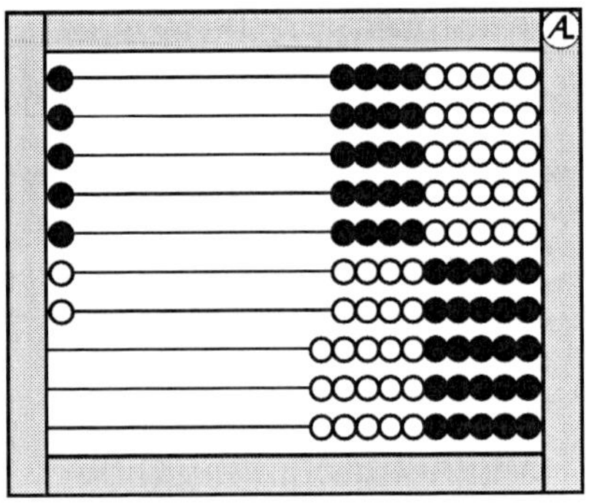

Showing 1 × 7 on the abacus: enter 1 seven times.

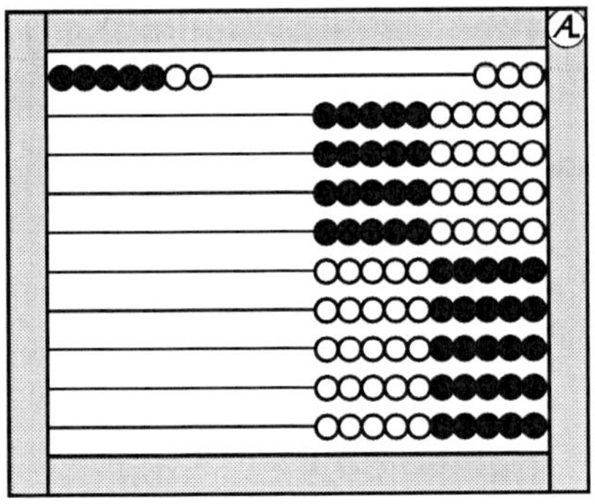

Showing 7 × 1 on the abacus: enter 7 one time.

Repeat for 7 × 1. How many do we need to enter at a time? [7] How many times do we enter it? [1] See the right figure on the previous page.

Then ask in a light tone, Some children think 7 × 1 = 8. Why is that? [They're confusing it with adding.]

Next give the child examples, both orally and written, such as, What is 1 × 4? [4] What is 9 × 1? [9] What is 1 × 1? [1] What is 1 × 30? [30] What is 51 × 1? [51] What is 104 × 1? [104]

Multiplying with 0s. The abacus is a good way to concretely demonstrate multiplying with zeroes. Write

$$0 \times 6 = \_$$

How many do we need to enter at a time? [0] How many times do we enter it? [6] So, move your empty hand, as if to move an imaginary bead, across the top wire and the next five wires. I've entered 0 six times: how much is it? [0]

Repeat for 6 × **0**. How many do we need to enter at a time? [6] How many times do we enter it? [0] This time select 6 beads, but freeze your hand because you cannot enter it any times.

Problems. With the remaining time, use some of the following problems. Ask the child to write the equations.

1. Yolanda practiced 1 hour every day for 2 weeks. How many hours was that? [1 × 14 = 14 hours] She could write an equation to find the number of days in 2 weeks: 2 × 7 = 14.

2. Yuri shoveled sidewalks at the rate of $2 an hour. Yuri worked 9 hours. How much money did Yuri make? [$2 × 9 = $18]

3. In your journal, draw a rectangle that is 4 cm wide by 5 cm high. (Each square measures 1 cm.) How many square centimeters are inside? [4 × 5 = 20 sq cm]

4. Repeat problem 3 for a rectangle 1 cm by 6 cm. [1 × 6 = 6 sq cm]

Worksheet. The worksheet provides practice in the 1s and 0s facts. The answers are below.

10¢ × 1 = 10¢ **1¢ × 1 = 1¢** **25¢ × 1 = 25¢**

3	9	5	5	2	7	8	6	1	7
6	9	1	4	10	0	4	10	8	3
0	0	0	0	0	0	0	0	0	0
0	0	0	0	0	0	0	0	0	0

Lesson 60

Review

OBJECTIVE 1. To review and practice

MATERIALS Worksheet 51-A and 51-B, "Review" (2 versions)
Cards for playing games

ACTIVITIES ***Review worksheet.*** Give the child about 10 to 15 minutes to do the review sheet. The oral problems to be read are as follows:

70 – 13 = 86 + 86 = 73 – 37 =

Review Sheet 51-A.

1. Write only the answers to the oral questions. 57 172 36

4. Write only the answers. 64 – 47 = 17 72 – 23 = 49 95 + 47 = 142

7. Write the multiples of 6.

6 12 18 24 30

36 42 48 54 60

17. 3691 − 2837 = 854

18. 3082 − 984 = 2098

19. Draw the hands for these two clocks.

5:13 8:41

21. Write the time on these two clocks.

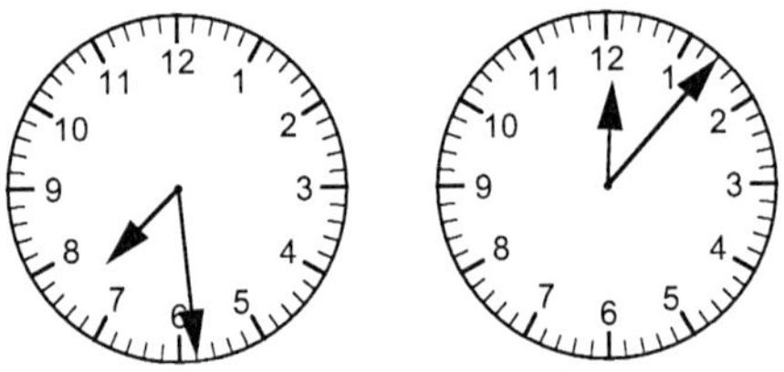

7:29 12:07

23. Explain how you can tell how many squares there are without counting each one.

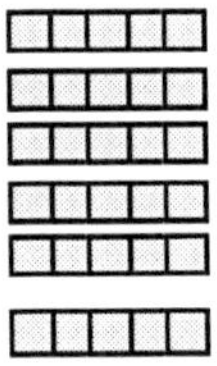

There are 6 fives. That makes 30.

24. How many trees are still growing in North Park? They planted 54 walnut trees, 638 oak trees and 89 maple trees. But storms killed 10 trees.

		²5	4	(0)	7	8	1
	6	3	8	(8)	-	1	0
	+	8	9	(8)	7	7	1
	7	8	1	(7)			

There are 771 trees.

26. 7 × 5 = 35 10 × 2 = 20 2 × 8 = 16 4 × 10 = 40 5 × 8 = 40 9 × 5 = 45 0 × 212 = 0 1 × 486 = 486

The oral problems to be read to the children are as follows:

100 – 29 = 78 + 78 = 55 – 38 =

Review Sheet 51-B.

1. Write only the answers to the oral questions. **71** **156** **17**

4. Write only the answers. 82 – 29 = **53** 94 – 46 = **48** 96 + 57 = **153**

7. Write the multiples of 8.

8 16 24 32 40

48 56 64 72 80

17. $\begin{array}{r} 6431 \\ -4916 \\ \hline \mathbf{1515} \end{array}$

18. $\begin{array}{r} 9550 \\ -955 \\ \hline \mathbf{8595} \end{array}$

19. Draw the hands for these two clocks.

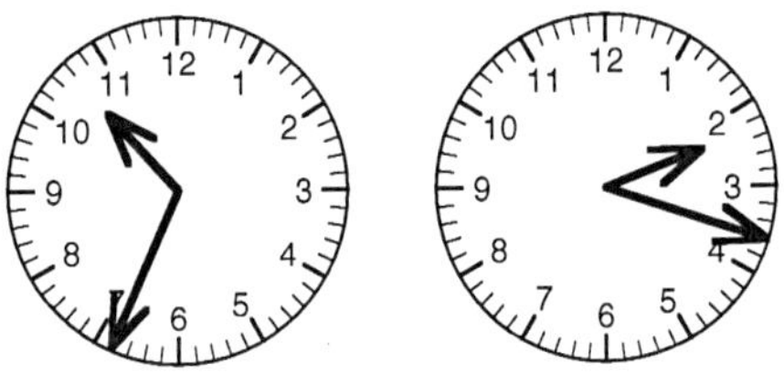

10:34 2:18

21. Write the time on these two clocks.

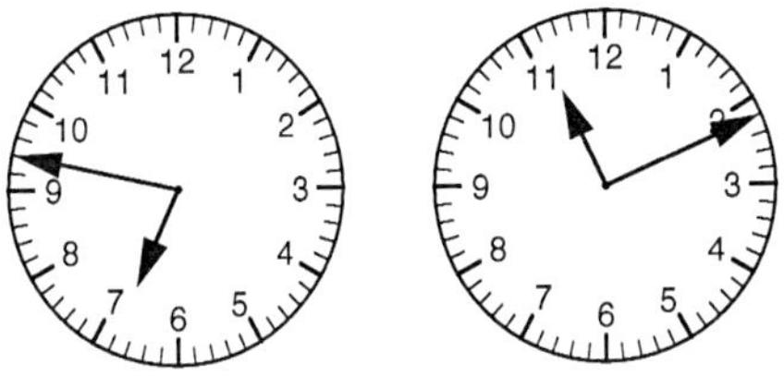

6:47 **11:11**

23. Explain how you can tell how many squares there are without having to count each one.

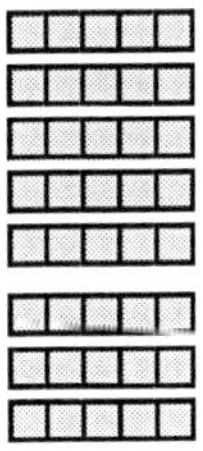

There are 8 fives. That makes 40.

24. Lauren had 10 dollars and bought some food for $3.28 and a ticket for a ride for $2.75. How much money does Lauren have left?

$3.28	**$10.00**
+ 2.75	**– $6.03**
$6.03	**$3.97**

Lauren had $3.97 left.

26. $\begin{array}{r} 9 \\ \times 5 \\ \hline \mathbf{45} \end{array}$ $\begin{array}{r} 4 \\ \times 2 \\ \hline \mathbf{8} \end{array}$ $\begin{array}{r} 2 \\ \times 7 \\ \hline \mathbf{14} \end{array}$ $\begin{array}{r} 4 \\ \times 5 \\ \hline \mathbf{20} \end{array}$ $\begin{array}{r} 5 \\ \times 6 \\ \hline \mathbf{30} \end{array}$ $\begin{array}{r} 7 \\ \times 5 \\ \hline \mathbf{35} \end{array}$ $\begin{array}{r} 578 \\ \times 0 \\ \hline \mathbf{0} \end{array}$ $\begin{array}{r} 267 \\ \times 1 \\ \hline \mathbf{267} \end{array}$

Games. Spend the remaining time playing games.

SUBTRACTION. Corners Three Subtraction (Lesson 50), Short Chain Subtraction *(Math Card Games,* S18*).*

TELLING TIME. Between the Numbers (Lesson 56) and Name That Time *(Math Card Games,* C28*).*

MULTIPLICATION . Mixed-Up Multiplication Cards *(Math Card Games,* P7*).*

Lesson 61

Multiplication Problems

OBJECTIVES
1. To understand multiplying with 1s and 0s
2. To learn the facts with multiplying 1s and 0s

MATERIALS
Worksheet 52, "Quick Practice-2"
A geared clock
Tiles
Worksheet 53, "Multiplication Problems"

WARM-UP
Ask the child to subtract 6512 – 1089 in her journal. [5423] Tell her to use check numbers to check her answers. [(5) – (0) = (5)]

Say the following orally for the child to compute mentally. What is 39 + 54? [93] What is 139 + 54? [193]

Set the hands of the clock to various times and ask the child to say them. If desired, ask the child to set her clock at various times you specify.

Facts practice. Take out the new Quick Practice sheets. Ask the child to do those in the **top** rectangle when you say start. Ask her to correct them when she finishes. The answers are:

1	18	25	15	8
16	12	14	8	45
2	5	0	15	30
40	18	6	7	8
12	35	50	16	9
10	50	35	3	20
6	10	40	4	20
8	14	10	5	30
0	20	7	45	3
20	6	0	4	9

ACTIVITIES
Problems. Each of the eight problems on the worksheet is a multiplication problem, but additional information or an extra step is often necessary.

1. Megan practiced handwriting for 10 minutes every day last week. How much time was that? [10 × 7 = 70 min] The answer should also be changed into 1 hr 10 min.

2. Matt and his brothers have 9 pairs of shoes altogether. How many shoes is that? [9 × 2 = 18] For this problem the child needs to think that a pair is 2.

3. Mario set the table for 8 people. He set 1 knife, 2 forks, and 2 spoons at each place. How many utensils did he set? [5 × 8 = 40 utensils] It is simpler to first add the number of utensils, 1 + 2 + 2 = 5 If she decides to do it piecemeal, 8 × 1 = 8, 8 × 2 = 16, and then add the sums, discuss both methods.

4. Marva wants to buy a notebook for 50¢. She has 9 nickels. Is that enough money? [9 × 5¢ = 45¢, No, she does not have enough money.]

5. Mike earns two dollars a day for delivering papers. How much money does he make in a week? [$2 × 7 = $14] Here she must think about the days in a week.

6. Madison saved $1 every day in March. How much did she save during the month? [$1 × 31 = $31] She needs to remember the number of days in March.

7. Maria bought four books at four dollars each. She also bought zero books at three dollars each. How much did she pay? [4 × $4 = $16 and 0 × $3 = 0, total is $16]

8. Mark bought 2 dozen eggs. How many eggs is that? [12 × 2 = 24 eggs]

Lesson 62

The Multiplication Table

OBJECTIVES 1. To begin constructing a multiplication table
2. To become aware of the number of multiplications facts

MATERIALS Worksheet 52, "Quick Practice-2"
A large multiplication table (Appendix pg. 2), optional
Worksheet 54, "Multiplication Table"
Abacus

WARM-UP Ask the child to subtract 7761 – 4582 in his journal. [3179] The check numbers are (3) – (1) = (2).

Facts practice. Ask the child to take out his Quick Practice sheet. Ask him to do those in the **bottom** rectangle when you say start. Ask him to correct them. The answers are:

45	6	8	4	14
45	16	7	0	18
8	25	16	20	50
40	15	5	9	20
40	35	50	8	12

0	3	10	2	6
1	8	35	0	5
12	9	15	10	4
20	30	10	6	3
7	20	30	18	14

ACTIVITIES ***2s on the multiplication table.*** Take out the worksheet. What do you think a multiplication chart is? [a place to write the multiplication facts]

Which tables have you been working with recently? [2s, 5s, 10s, and 1s (The 0s are not on the chart.)] Tell the child he will enter *only* those on his multiplication table. Explain that to find the place to write 2 × 1 is to find 2 in the top row and then slide a piece of paper, a ruler, or a finger down to the 1 column. See the left figure below. Ask him to write the 2s as shown below on the right.

Note: In Level C the child constructed addition and subtraction tables over the period of several days.

×	1	2	3	4	5	6	7	8	9	10
1		2								

Writing 2 × 1 in the table.

×	1	2	3	4	5	6	7	8	9	10
1		2								
2		4								
3		6								
4		8								
5		10								
6		12								
7		14								
8		16								
9		18								
10		20								

The 2s written.

Note: Tell the child to write neatly because he will be making a project with the charts when it is filled in. Show him a sample.

5s, 10s, and 1s on the multiplication table. Ask the child to write the 5s on the table. Continue with the 10s and 1s. See the table on the next page.

> **Note:** The recent additions are shown in boldface.

×	1	2	3	4	5	6	7	8	9	10
1	**1**	2			**5**					**10**
2	**2**	4			**10**					**20**
3	**3**	6			**15**					**30**
4	**4**	8			**20**					**40**
5	**5**	10			**25**					**50**
6	**6**	12			**30**					**60**
7	**7**	14			**35**					**70**
8	**8**	16			**40**					**80**
9	**9**	18			**45**					**90**
10	**10**	20			**50**					**100**

Entering the 5s, 10s, and 1s.

×	1	2	3	4	5	6	7	8	9	10
1	1	2			5					10
2	2	4	**6**	**8**	10	**12**	**14**	**16**	**18**	20
3	3	6			15					30
4	4	8			20					40
5	5	10			25					50
6	6	12			30					60
7	7	14			35					70
8	8	16			40					80
9	9	18			45					90
10	10	20			50					100

Entering the 2s as multipliers.

Multiplying by 2s. Write

$$3 \times 2 =$$

Where do you write it? [in the third row] Ask him to write 4 × 2, 6 × 2 and the others to 10 × 2. See above on the right.

Multiplying by 5s, 10s, and 1s. Proceed the same way for numbers times 5, 10, and 1. The table with these facts are shown below. What is special about the numbers written in? [facts he knows]

> **Note:** This multiplication table will be needed for several more lessons.

×	1	2	3	4	5	6	7	8	9	10
1	1	2	**3**	**4**	5	**6**	**7**	**8**	**9**	10
2	2	4	6	8	10	12	14	16	18	20
3	3	6			15					30
4	4	8			20					40
5	5	10	**15**	**20**	25	**30**	**35**	**40**	**45**	50
6	6	12			30					60
7	7	14			35					70
8	8	16			40					80
9	9	18			45					90
10	10	20	**30**	**40**	50	**60**	**70**	**80**	**90**	100

Entering the 5s, 10s, and 1s as multipliers.

Questions about the table. Ask the following questions:

1. How many multiplication facts are there? [100] Ask how the child figured it out. [10 rows by 10 columns = 100]

2. How many blank spots are left on the table? [36] Again ask how he found his answer. If he counts, ask if he could do it a different way. [seeing 9 groups of 4s, or 6 groups of 6] If he doesn't think of it, suggest it by saying you see the groups.

3. How many facts have you entered on the table? [64] Ask how he figured it out. They could be counted, but it is easier to subtract from 100.

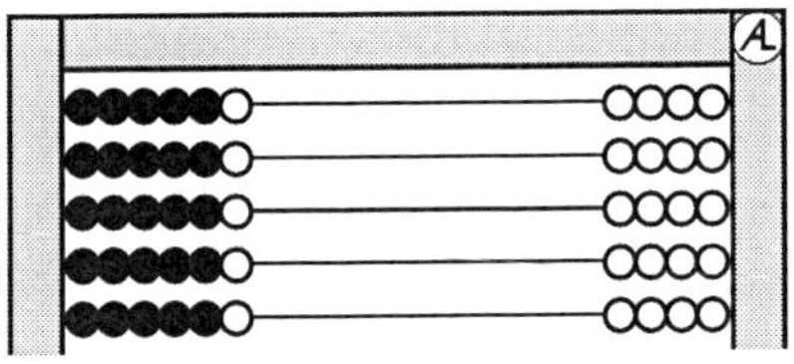

> **Note:** Seeing such arrays mentally helps the children improve their visual skills as well as learn the multiplication facts.

Abacus practice. Enter on the abacus the array of 5 × 5. Explain that you will show the abacus briefly. Flash it for 2-3 seconds. What is the equation? [5 × 5 = 25] Discuss the groups of 5s on the abacus. Repeat for other 5s facts and the 1s and 10s facts.

Lesson 63

Working With Threes

OBJECTIVES
1. To learn the 3s table
2. To learn the facts with 3 as the multiplier

MATERIALS
Worksheet 55, "Quick Practice-3"
Abacus
Worksheet 54, "Multiplication Table" from previous lessons
Worksheet 56, "Working With Threes"

WARM-UP
Ask the child to subtract 8403 – 657 in her journal. [7746] The check numbers are (6) – (0) = (6).

Say the following orally for the child to compute mentally. What is 43 + 72? [115] What is 120 – 45? [75]

ACTIVITIES
Facts practice. Take out the new Quick Practice sheet. Ask the child to do those in the **top** rectangle when you say start. Ask her to correct them. The answers are:

20	14	40	16	2	10	25	40	35	12
50	45	6	20	15	18	10	50	0	16
6	0	5	20	1	4	30	35	8	70
12	15	20	8	2	30	18	45	10	14

Note: A good time is 3 seconds per fact, 120 seconds for this group.

Ask her to practice the ones that were incorrect. You can also discuss strategies to remember those facts. She can practice orally.

3s practice. Ask the child to write the multiples of 3; ask how many numbers she writes in a row for the 3s. [3] Write the following on the drawing board. (Leave sufficient space between the columns to write later, for example, "3 × 2 =" in front of it.)

3	6	9
12	15	18
21	24	27
30		

Ask her to say the 3s table in order. [3 × 1 = 3, 3 × 2 = 6, . . . , 3 × 10 = 30] Then ask the following questions:

1. Where is 3 × 3? [last number in first row, 9]
2. Where is 3 × 6? [last number in second row, 18]
3. Where is 3 × 9? [last number in third row, 27]
4. Where is 3 × 5? [in the middle, 15]
5. Where is 3 × 4? [first number in second row, 12]
6. Where is 3 × 7? [first number in third row, 21]
7. Where is 3 × 8? [second number in third row, 24]

What are the check numbers? [0, 3, 6] Is 16 a multiple of 3? [No, the check number of 16 is 7.] Is 63 a multiple of 3? [Yes, the check number of 63 is 0.]

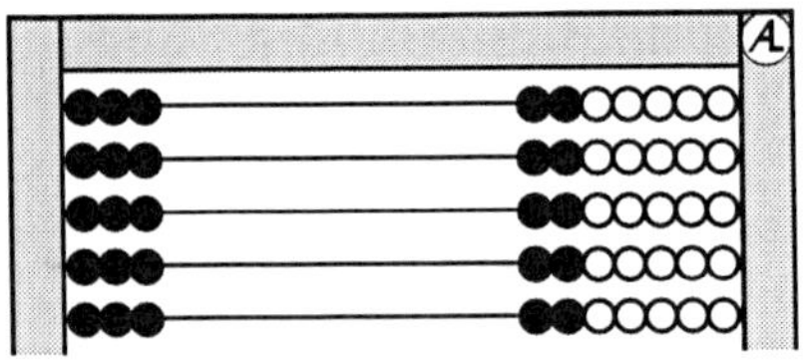

Abacus practice. Enter 3 × 5 on an abacus and flash it for 2-3 seconds. What is the equation? [3 × 5 = 15] Encourage her to see groups of 5s, rearranging beads mentally, if necessary. Repeat for other facts.

Practice with written symbols. Write the following in front of the corresponding multiples.

$3 \times 1 = [3]$ $3 \times 2 = [6]$ $3 \times 3 = [9]$
$3 \times 4 = [12]$ $3 \times 5 = [15]$ $3 \times 6 = [18]$
$3 \times 7 = [21]$ $3 \times 8 = [24]$ $3 \times 9 = [27]$
$3 \times 10 = [30]$

<u>If you know that 3 × 4 is 12, how could you find 3 × 8?</u> [twice 3 × 4, or 24] <u>Kayla knows that 3 × 6 is 18, how could she find 3 × 7?</u> [add 3 to 18] <u>Ken knows 3 × 10 is 30, how can he find 3 × 9?</u> [subtract 3 from 30 to get 27]

Ask for the facts in mixed order. Erase the answers and repeat.

3s on the multiplication table. Ask the child to write these facts on her multiplication table. See the table below on the left.

Note: The recent additions are shown in boldface.

×	**1**	**2**	**3**	**4**	**5**	**6**	**7**	**8**	**9**	**10**
1	1	2	3	4	5	6	7	8	9	10
2	2	4	6	8	10	12	14	16	18	20
3	3	6	**9**		15					30
4	4	8	**12**		20					40
5	5	10	15	20	25	30	35	40	45	50
6	6	12	**18**		30					60
7	7	14	**21**		35					70
8	8	16	**24**		40					80
9	9	18	**27**		45					90
10	10	20	30	40	50	60	70	80	90	100

Entering the 3s.

×	**1**	**2**	**3**	**4**	**5**	**6**	**7**	**8**	**9**	**10**
1	1	2	3	4	5	6	7	8	9	10
2	2	4	6	8	10	12	14	16	18	20
3	3	6	9	**12**	15	**18**	**21**	**24**	**27**	30
4	4	8	12		20					40
5	5	10	15	20	25	30	35	40	45	50
6	6	12	18		30					60
7	7	14	21		35					70
8	8	16	24		40					80
9	9	18	27		45					90
10	10	20	30	40	50	60	70	80	90	100

Entering the 3s as multipliers.

Write

$$\begin{array}{r} 4 \\ \underline{\times 3} \end{array}$$

Ask for the answer and how she knows. [same as 3 × 4] Ask for other facts with 3 as the multiplier. Then ask her to enter those facts on her multiplication table as shown above on the right.

Worksheet. The answers to the worksheet follow.

27 21 6 9 3 24 30 15 18 12
9 3 24 18 6 15 12 21 30 27

3 × 4 = **12** 8 × 3 = **24** **21 days**
3 × 2 = **6** 6 × 3 = **18** **180 min**
3 × 1 = **3** 5 × 3 = **15** **96 oz**
3 × 10 = **30** 2 × 3 = **6** **72 hr**
3 × 6 = **18** 3 × 3 = **9** **1095 days**
3 × 3 = **9** 10 × 3 = **30**
3 × 8 = **24** 1 × 3 = **3**
3 × 5 = **15** 7 × 3 = **21**
3 × 9 = **27** 4 × 3 = **12**
3 × 7 = **21** 9 × 3 = **27**

Lesson 64 (1 or 2 days)

Representing One Thousand

OBJECTIVES
1. To review the term *cube*
2. To represent the thousands
3. To construct a thousand-cube

MATERIALS
Worksheet 55, "Quick Practice-3"
Centimeter cubes
Base-10 picture cards
Thousand Cube, up to 100 copies (available separately or on Appendix pg. 3.)
Scissors

WARM-UP Ask the child to subtract 8222 – 5353 in his journal. [2869] The check numbers are (5) – (7) = (7).

Say the following orally for the child to compute mentally. What is 56 + 57? [113] What is 130 – 27? [103]

Ask the child to write the multiples of 3s with only 3 in a row. Then ask for the multiples of 3 in random order. What is 3 × 5? [15] Continue with the other 3s facts.

ACTIVITIES ***Facts practice.*** Ask the child to take out his Quick Practice sheet and ask him to do the facts in the **bottom** rectangle when you say start. Ask him to correct them. The answers are:

12	12	24	50	24	35	18	18	16	30
40	12	14	5	10	30	30	2	18	70
15	16	20	21	35	12	10	3	27	3
18	45	40	21	30	20	25	4	6	45
15	8	14	10	20	9	5	27	6	50

Note: It would be good to invite others to help in the cube building effort.

Ask him to practice the ones that were incorrect. You can also discuss strategies to remember those facts.

Reviewing 1, 10, 100, and 1000. Show the child the centimeter cube. How much is it? [1] Ask him to make a row 5 times greater. Then ask him to make a row 10 times greater than the single cube. See the figures below. How much is it? [10]

A row 5 times greater than the single cube.

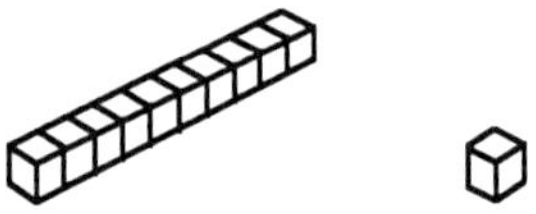

A row 10 times greater than the single cube.

Show a 10-row and ask what it represents. [10] Ask him to make a square 10 times greater than the 10-row. See the figure on the next page. How much is it? [100]

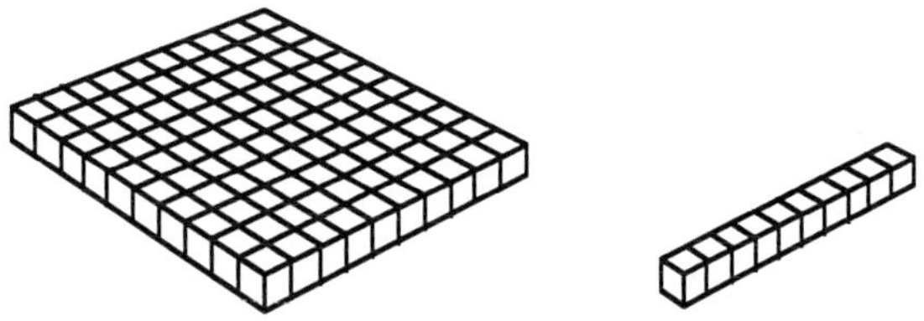

A square 10 times greater than the 10-rod.

Show the 100-square and ask what it represents. [100] Ask him to make a cube 10 times greater than the 100-flat. See below. How much is it? [1000]

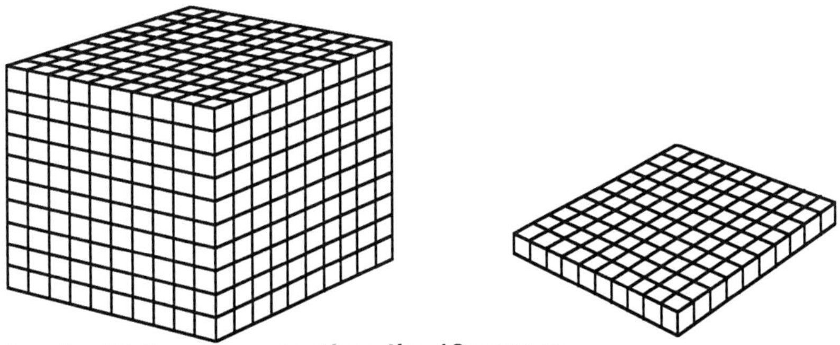

A cube 10 times greater than the 10-square.

What is the name of the shape ? [cube] What is the difference between a cube and a square? [A square fits on a sheet of paper, but a cube doesn't. A cube has six sides, each in the shape of a square.]

Note: It is helpful to construct a cube beforehand to show the child.

The cubes show quantities grouped by 5s to correlate with work done in earlier levels.

Show the child the 1000-cube and ask if he sees another cube–a very small cube. [the 1 centimeter cube] How many small cubes would it take to build the large cube? [1000] How many times greater is the large cube compared to the small cube? [1000]

Building thousand cubes. Explain that you want to talk about many thousands, but since you don't have many thousand cubes, you may make 100 or you may make enough for the child to picture 100 in his head.

Note: The 100 thousand-cubes will be used in the future lessons.

Give templates or copies of the Thousand Cube to the child. No glue or tape is necessary because the flaps interlock. Ask him to build the cubes. He may need assistance at the final assembly stage when an extra hand might be helpful.

Ask him to make more cubes.

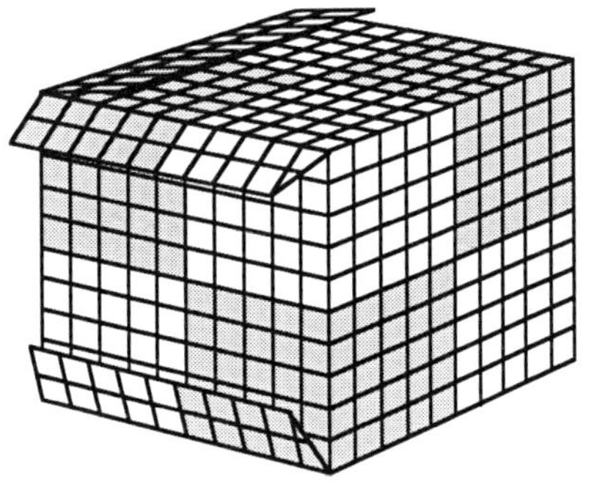

A constructed thousand cube.

Lesson 65

Reviewing Place Value

OBJECTIVES
1. To represent quantities in terms of ones, tens, hundreds, and thousands
2. To review such quantities

MATERIALS
Worksheet 57, "Quick Practice-4"
Base-10 picture cards
A thousand cube made during the previous lesson
Worksheet 58, "Reviewing Place Value"

WARM-UP
Ask the child to subtract 5711 – 543 in her journal. [5168] The check numbers are (5) – (3) = (2).

Say the following orally for the child to compute mentally. What is 68 + 72? [140] What is 130 – 114? [16]

8s practice. Ask the child to write the multiples of 8 in two rows. Write the following on the board. (Leave sufficient space between the columns to write later, for example "8 × 1 =" in front of it.)

8	16	24	32	40
48	56	64	72	80

ACTIVITIES
Facts practice. Take out Quick Practice-4 and ask the child to do the facts in the **top** rectangle when you say start. Ask her to correct them. The answers are:

5	35	20	50	14	10	5	24	35	9
8	21	18	12	15	70	40	10	15	10
3	3	14	25	16	30	12	30	12	4
50	20	24	18	20	6	18	30	21	18
6	40	16	45	2	12	27	30	27	45

Ask her to practice the ones that were incorrect.

Reviewing the thousand cube. Ask the child to take a thousand cube and to think about it when answering the following questions.

1. How many little cubes are in a row, or line, at the bottom of your thousand cube? [10] How many zeroes do we need to write the number 10? [1]

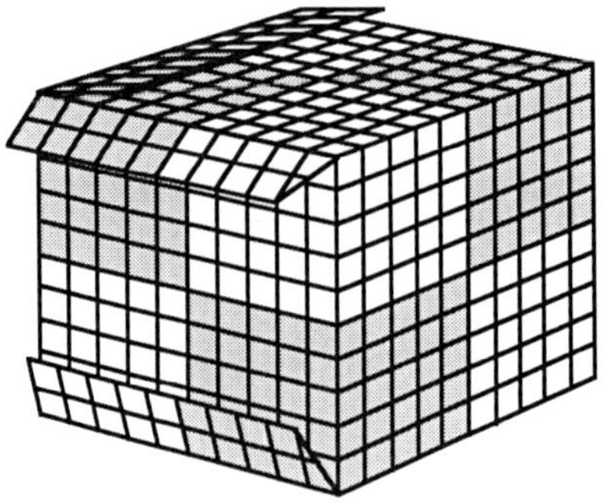

A thousand cube.

2. How many rows of 10 are at the bottom of your thousand cube? [10] What do we call 10 tens. [100] How many little cubes are on the bottom of your cube? [100] What is the equation? [10 × 10 = 100] In how many directions do you see tens? [2] How many zeroes do we need to write the number 100? [2]

3. How many layers of 100 cubes are in your thousand cube? [10] Write the equation using only 10s. [10 × 10 × 10 = 1000] In how many directions do you see tens? [3] How many zeroes do we need to write the number 1000? [3]

Comparing number of 10s with number of digits. Tell the child there is an interesting pattern between the number of lines of tens and the number of zeroes we need to write the number. Ask her to fill in the chart shown below.

	Rows of tens	Number of 0s	Geometric name
1	0	0	cube
10	1	1	line
100	2	2	square
1000	3	3	cube

Making quantities. Ask the child to show 2 hundred [2 hundred squares] and to write it. [200] How can you tell that number is 2 hundred and not 2 thousand? Stress we know because 2 digits come after the 2. Do **not** start at the right and name columns, ones, tens, and so forth.

Note: We read numbers from left to right. So it is important to identify a thousand by starting at the left.

Repeat for quantities such as 6-ten [60], 5 thousand [5000], and 12 hundred. [1200] Discuss both ways of reading it: 12 hundred and 1 thousand 2 hundred. Also write some quantities for her to build.

Continue with two quantities, such as 6 hundred 8 ones [608], three quantities, such as 2 thousand 6 hundred 5, and four quantities.

Naming quantities. Lay out various quantities with the rods, or the pictures, for the child to write the numbers.

Worksheet. The child is to determine quantities from figures and to add various combinations. The answers are as follows.

Note: This worksheet will be needed in Lesson 68.

8	21	60
100	300	
700	1,000	
29	68	160
108	121	120
308	400	321
800	360	1,000
1,008	2,000	2,189

Lesson 66

Review

OBJECTIVE 1. To review and practice

MATERIALS Worksheet 59-A and 59-B, "Review" (2 versions)
Cards for playing games

ACTIVITIES ***Review worksheet.*** Give the child about 15 to 20 minutes to do the review sheet. The oral problems to be read are as follows:

47 + 17 =   100 – 89 =   65 – 28 =

Review Sheet 59-A.

1. Write only the answers to the oral questions. **64** **11** **37**

4. Write only the answers. 64 – 47 = **17** 72 – 48 = **24** 95 + 93 = **188**

7. Write the multiples of 8.

8 **16** **24** **32** **40**

48 **56** **64** **72** **80**

17. Kim has three bags. Each has nine books. How many books does Kim have? Write the equation.

3 × 9 = 27 books

19. Draw the hands for these two clocks.

6:07

1:28

21. Write the time on these two clocks.

9:11

3:54

23. Each square is a box with 2 mittens. How many mittens are there altogether? Write the equation.

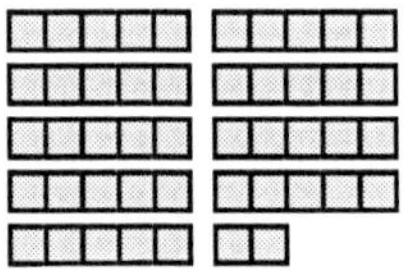

47 × 2 = 94 mittens (47 + 47 = 94) or 40 × 2 = 80 and 7 × 2 = 14, 80 + 14 = 94

24. A 4-H club is planting 300 trees. On Monday they planted 57 trees. On Tuesday they planted 116 trees. How many trees do they have left to plant?

		5	7			3	0	0	
+	1	1	6		-	1	7	3	
	1	7	3			1	2	7	

127 trees left

26.

3	189	428	9	5	3	3	8
× 5	× 1	× 0	× 5	× 8	× 6	× 7	× 3
15	**189**	**0**	**45**	**40**	**18**	**21**	**24**

The oral problems to be read to the child are as follows:

100 – 29 = 78 + 78 = 55 – 38 =

Review Sheet 59-B.

1. Write only the answers to the oral questions. **71** **156** **17**

4. Write only the answers. 82 – 54 = **28** 43 – 29 = **14** 94 + 98 = **192**

7. Write the multiples of 9.

9 18 27 36 45

90 81 72 63 54

17. Kerry has three packages. Each has 100 cards. How many cards does Kerry have altogether? Write the equation.

3 × 100 = 300 books

19. Draw the hands for these two clocks.

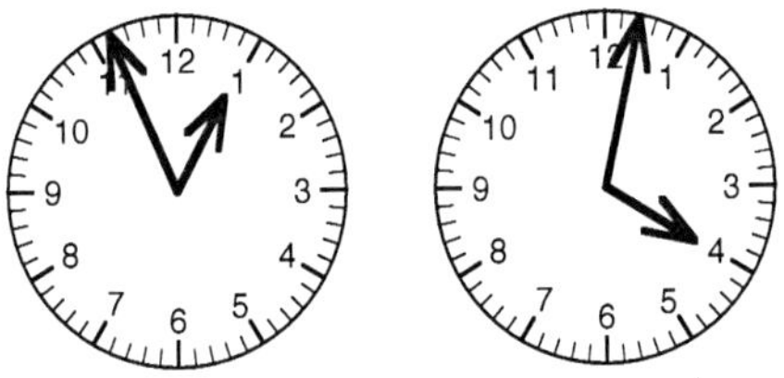

21. Write the time on these two clocks.

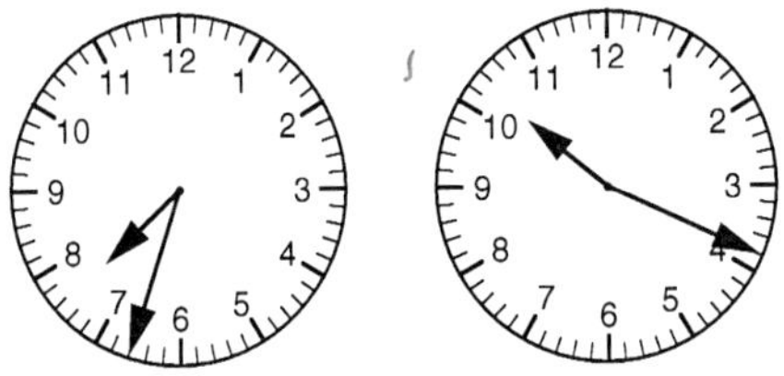

7:33 **10:19**

23. Each square is a box with 2 shoes. How many shoes are there altogether? Write the equation.

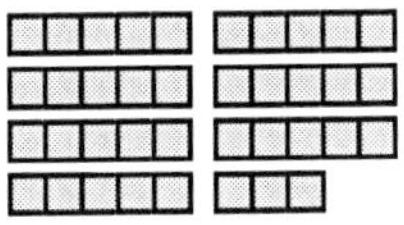

38 × 2 = 76 shoes (38 + 38 = 76) or 30 × 2 = 60 and 8 × 2 = 16, 60 + 16 = 76

24. A group of scouts is trying to raise \$100. They made \$36.49 on Friday and \$52.86 on Saturday. How much do they have left to raise?

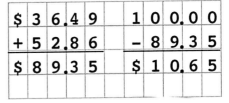

\$ 36.49	**100.00**
+ 52.86	**– 89.35**
\$ 89.35	**\$ 10.65**

26. $3 \times 6 =$ **18** $1 \times 76 =$ **76** $3 \times 5 =$ **15** $0 \times 3 =$ **0** $3 \times 8 =$ **24** $5 \times 6 =$ **30** $5 \times 7 =$ **35** $3 \times 3 =$ **9**

Games. Spend the remaining time playing games.

TELLING TIME: Between the Numbers (Lesson 56) and Name That Time (*Math Card Games*, C28).

MULTIPLICATION. Multiplication Old Main (P14), Slide-a-Thon Solitaire (P9), and Multiples Solitaire (P19) *(Math Card Games)*.

Note: *Multiples Solitaire* was formerly called *Threesome Solitaire.*

Lesson 67

Working With Fours

OBJECTIVES
1. To learn the 4s table
2. To learn the facts with 4 as the multiplier

MATERIALS
Worksheet 57, "Quick Practice-4"
Abacus
Worksheet 54, "Multiplication Table" from previous lessons
Worksheet 60, "Working With Fours"

WARM-UP
Ask the child to subtract 8403 – 657 in her journal. [7746] The check numbers are (6) – (0) = (6).

Say the following orally for the child to compute mentally. What is 83 + 19? [102] What is 150 – 35? [115]

ACTIVITIES
Facts practice. Ask the child to take out her Quick Practice sheet and ask her to do the facts in the **bottom** rectangle when you say start. Ask her to correct them. The answers are:

6	**30**	**10**	**12**	**12**	**18**	**10**	**20**	**15**	**12**
50	**25**	**14**	**30**	**5**	**30**	**35**	**20**	**50**	**14**
27	**70**	**40**	**8**	**3**	**20**	**18**	**12**	**27**	**18**
18	**45**	**9**	**2**	**4**	**24**	**35**	**6**	**16**	**10**
24	**40**	**5**	**21**	**21**	**3**	**16**	**30**	**15**	**45**

Ask her to practice the ones that were incorrect. You can also discuss strategies to remember those facts.

Note: This is another lesson emphasizing one of the multiples, 4 in this case, and the associated facts.

4s practice. Ask the child to write the multiples of 4 in two rows. Write the following on the board. (Leave sufficient space between the columns to write later, for example "4 × 1 =" in front it.)

4	8	12	16	20
24	28	32	36	40

Ask her to say the 4s table in order. [4 × 1 = 4, 4 × 2 = 8, . . . , 4 × 10 = 40] Then ask the following questions.

Where is 4 × 4? [fourth number in first row, 16]
Where is 4 × 6? [first number in second row, 24]
Where is 4 × 9? [fourth (second to last) in second row, 36]
Where is 4 × 5? [last in first row, 20]
Where is 4 × 3? [third number in first row, 12]
Where is 4 × 7? [second number in second row, 28]

Practice orally by asking, for example, what is 4 × 5 [20], 4 × 10 [40], and so forth, by using the multiples.

Comparing 4s to 2s. Ask the child to write the multiples of 2 in a row. Then ask her to double each one. What is the results? [multiples of 4] See below.

2	4	6	8	10	12	14	16	18	20
4	8	12	16	20	24	28	32	36	40

If you know 2 × 7 = 14, how can that help to find 4 × 7? [Double it to get 28.]

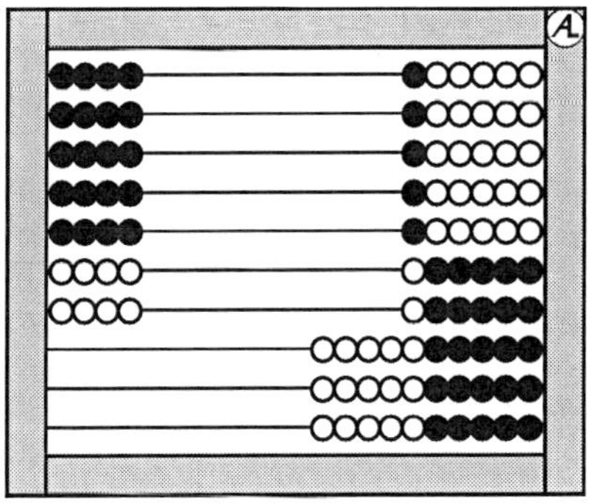

Abacus practice. Enter 4 × 7 on an abacus and flash it for 2-3 seconds. What is the equation? [4 × 7 = 28] Repeat for other facts.

Practice with written symbols. Write the following in front of the multiples of 4, written earlier.

4 × 1 = [4]   4 × 2 = [8]   4 × 3 = [12] 4 × 4 = [16] 4 × 5 = [20]

4 × 6 = [24] 4 × 7 = [28] 4 × 8 = [32] 4 × 9 = [36] 4 × 10 = [40]

If you know 4 × 5 was 20, how can you find 4 × 6? [20 + 4, or 24]
Kayla knows that 4 × 6 is 24, how could she find 4 × 7? [add 4 to 24]

Ask for the facts in mixed order. Erase just the answers and practice again. They can use either their mental abacus or the patterns.

Note: The recent additions are shown in boldface.

×	**1**	**2**	**3**	**4**	**5**	**6**	**7**	**8**	**9**	**10**
1	1	2	3	4	5	6	7	8	9	10
2	2	4	6	8	10	12	14	16	18	20
3	3	6	9	12	15	18	21	24	27	30
4	4	8	12	**16**	20					40
5	5	10	15	20	25	30	35	40	45	50
6	6	12	18	**24**	30					60
7	7	14	21	**28**	35					70
8	8	16	24	**32**	40					80
9	9	18	27	**36**	45					90
10	10	20	30	40	50	60	70	80	90	100

Entering the 4s.

×	**1**	**2**	**3**	**4**	**5**	**6**	**7**	**8**	**9**	**10**
1	1	2	3	4	5	6	7	8	9	10
2	2	4	6	8	10	12	14	16	18	20
3	3	6	9	12	15	18	21	24	27	30
4	4	8	12	16	20	**24**	**28**	**32**	**36**	40
5	5	10	15	20	25	30	35	40	45	50
6	6	12	18	24	30					60
7	7	14	21	28	35					70
8	8	16	24	32	40					80
9	9	18	27	36	45					90
10	10	20	30	40	50	60	70	80	90	100

Entering the 4s as multipliers.

4s on the multiplication table. Ask the child to write the 4s facts on her multiplication table. See the table above on the left.

Write

$$\begin{array}{r} 7 \\ \underline{\times 4} \end{array}$$

Ask for the answer and how she knows. [same as 4 × 7] Ask for other facts with 4 as the multiplier. Then ask her to enter those facts on her multiplication table as shown above on the right. How many spaces are left to be filled? [16]

Worksheet. The answers for the worksheet follow.

24	**40**	**20**	**8**	**32**	**36**	**16**	**4**	**28**	**12**
28	**8**	**20**	**24**	**32**	**4**	**40**	**16**	**36**	**12**

4 × 4 = **16**	2 × 4 = **8**	1. **28 days, less**
4 × 1 = **4**	5 × 4 = **20**	2. **240 min**
4 × 7 = **28**	1 × 4 = **4**	3. **48 in.**
4 × 5 = **20**	4 × 4 = **16**	4. **20 slices**
4 × 10 = **40**	10 × 4 = **40**	5. **96 hr**
4 × 8 = **32**	7 × 4 = **28**	
4 × 6 = **24**	3 × 4 = **12**	
4 × 3 = **12**	6 × 4 = **24**	
4 × 9 = **36**	9 × 4 = **36**	
4 × 2 = **8**	5 × 4 = **20**	
4 × 5 = **20**	8 × 4 = **32**	

Lesson 68

Representing Many Thousands

OBJECTIVE 1. To represent numbers to the millions

MATERIALS Worksheet 61, "Quick Practice-5"
The 100 thousand cubes made during Lesson 65 if available
Worksheet 62, "Representing Many Thousands"
Worksheet 63, "Building Thousands"

WARM-UP Ask the child to subtract 8000 – 5426 in his journal. [2574] The check numbers are (8) – (8) = (0).

Ask the child to write the multiples of 4s with 5 numbers in a row. Then ask for the multiples of 4 in random order.

ACTIVITIES ***Facts practice.*** Take out Quick Practice-5 and ask the child to do the facts in the **top** rectangle when you say start. Ask him to correct them. The answers are:

Note: Quick Practice-5 is more difficult since it has very few facts involving 1, 2, or 10.

20	**32**	**9**	**45**	**24**	**16**	**6**	**6**	**36**	**40**
3	**12**	**18**	**21**	**12**	**45**	**16**	**15**	**35**	**24**
14	**12**	**27**	**36**	**24**	**27**	**20**	**8**	**18**	**25**
4	**18**	**16**	**30**	**56**	**25**	**15**	**18**	**28**	**14**
24	**0**	**8**	**12**	**30**	**28**	**10**	**40**	**32**	**10**

Ask him to practice the ones that were incorrect.

Reviewing the thousand cube. Ask the child to take a thousand cube and to think about it when answering the following:

1. <u>How many little cubes does this large cube represent?</u> [1000]

2. <u>In how directions do you see 10?</u> [3, two across the top and one down]

3. <u>Can you think of a figure that shows 10s in four directions?</u> [no] Explain that there are only 3 dimensions–that is what we mean by 3-D.

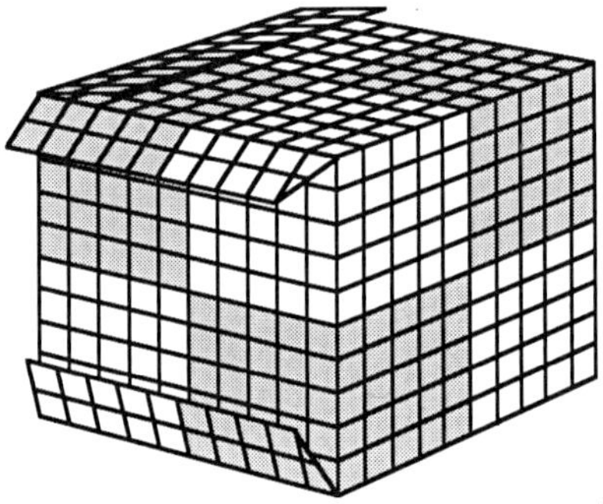

A thousand cube.

Multiple thousands. Ask the child to show 2 thousand with the cubes. [2 cubes in a row] Ask him to write it. [2000] <u>How can you tell that the number is 2 thousand and not 2 hundred?</u> [3 zeroes]

Repeat for 5 thousand [5000] and 9 thousand. [9000]

Add 1 more thousand-cube and ask <u>How many do we have now?</u> [10,000] Write it

10,000

and explain that when we write numbers starting at 10 thousand, we usually add a comma where we say the word *thousand.*

Now ask him to construct 34 thousand. Remind him that no more than 10 thousand-cubes can be in a row. See the figure on the next page.

Ask the child to write it. [34,000] <u>What is 1000 × 34?</u> [34 thousand] Repeat for 47 thousand [47,000] and 52 thousand. [52,000]

Now write

62,000

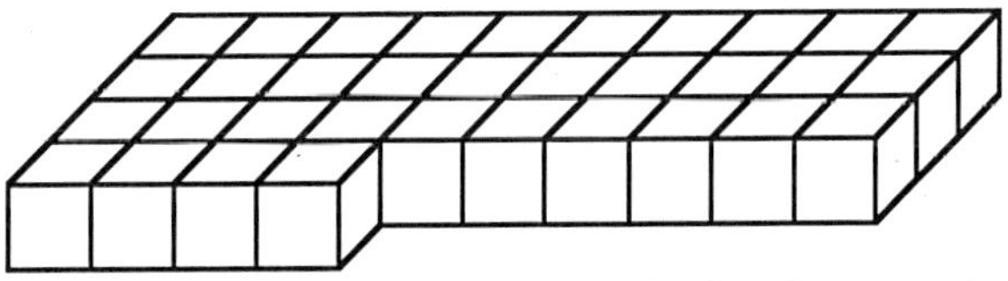

Constructing 34 thousand with the thousand cubes.

and ask him to read it and construct it. [62 thousand-cubes] Repeat for 79,000. [79 thousand-cubes]

Next challenge him to construct and write 100 thousand. [100,000] Does that look like 100,000?

What would 200 thousand look like if there were enough thousand-cubes to construct it? [a second layer of 100 thousand-cubes] Ask him to write it. [200,000] Repeat for 500,000.

What would 1000 thousands look like? [10 layers of 100 thousand-cubes. What is the shape? [a cube] What do you think its name is? [1 million] Write it

1,000,000

and explain it needs 2 commas. How many zeroes does 1 million need? [6] What is 1 thousand thousands? [1 million] What is 1000 × 1000? [1 million]

Worksheet. In the first worksheet the child determines quantities from figures and adds various combinations. Be sure he realizes each cube in the pictures is a thousand-cube. The answers are as follows:

8,000	**21,000**	**60,000**
100,000	**300,000**	
700,000	**1,000,000**	
29,000	**68,000**	**160,000**
108,000	**121,000**	**120,000**
308,000	**400,000**	**321,000**
800,000	**360,000**	**1,000,000**
1,008,000	**2,000,000**	**2,189,000**

When he has completed the worksheet, ask him to compare the results with Worksheet 58. [all answers on this worksheet are one thousand times greater than the earlier worksheet.]

For the second worksheet, he decides what he needs to construct the numbers on the left. The answers follow:

		500,000	100,000	50,000	10,000	5,000	1,000
g.	8,000					**1**	**3**
h.	36,000				**3**	**1**	**1**
i.	301,000		**3**				**1**
j.	910,000	**1**	**4**		**1**		
k.	1,000,000	**2**					
l.	666,000	**1**	**1**	**1**	**1**	**1**	**1**
m.	217,000		2		1	**1**	**2**
n.	480,000		**4**	**1**	**3**		
o.	1,005,000	**2**				**1**	

Lesson 69

Reading and Writing Large Numbers

OBJECTIVE 1. To read and write numbers of more than four digits

MATERIALS Worksheet 61, "Quick Practice-5"
Worksheet 64-1, "Digits and Commas," cut apart
Worksheet 64-2, "Reading and Writing Large Numbers"

WARM-UP Ask the child to subtract 7944 – 7448 in her journal. [496] The check numbers are (6) – (5) = (1)

Ask the child to count by 9s to 90 and to write them. Remind her to write the second row going from right to left as shown below.

9	18	27	36	45
90	81	72	63	54

Ask the child to look at the multiples and to recite the 9s table. [9 × 1 = 9, 9 × 2 = 18, . . . 9 × 10 = 90]

ACTIVITIES ***Preliminary practice.*** Ask the child to take out her Quick Practice sheet and ask her to do the facts in the **bottom** rectangle when you say start. Ask her to correct them. The answers are:

Note: Quick Practice-5 is more difficult since it has very few facts involving 1, 2, or 10.

3	25	24	18	36	9	24	4	16	18
14	40	6	30	18	28	0	27	45	24
10	35	15	12	45	18	12	32	12	16
27	8	36	8	16	4	10	32	40	24
35	15	20	28	21	14	12	30	21	16

Ask her to practice the ones that were incorrect.

Constructing numbers. Take out the digits 1 to 5, 7, 9, four 0s, and two commas. Write the following numbers.

Note: Mathematically speaking, *and* does not belong in these numbers, even though it is often heard in everyday speech.

203,719 450,000

Ask the child to arrange the cards to make the following numbers. See below.

2	0	3	,	7	1	9

4	5	0	,	0	0	0

Using the cards to make the two numbers.

Next give her the digits 2 to 7 and a comma and ask her to make the largest number that she can. [765,432] Then ask her to make the smallest number that she can. [234,567] See below.

7	6	5	,	4	3	2

2	3	4	,	5	6	7

Making the largest and smallest numbers from 2 to 7.

Repeat for 1 3 5 7 9 0 and a comma. The number may not start with a 0. [975,310 and 103,579]

Reading numbers. Give the child 7 8 9 and ask her to form seven hundred eighty-nine. Then give her the 6 and ask her to place it to the left of the 7. See the left figure on the next page. Can you read the new number? [six thousand seven hundred eighty-nine]

Next give the 5 to the child; ask her to place it to the left of the 6. Also give a comma to her and tell her to place the comma where it belongs. See the right figure below. Can you read the new number? [fifty-six thousand seven hundred eighty-nine]

6	7	8	9

5	6	,	7	8	9

Continue in the same way by giving the 4 to the child and then the 3 and another comma [3,456,789] and ask her to read it. [three million four hundred fifty-six thousand seven hundred eighty-nine]

Finally give her the 2 and the 1. [123,456,789] See below.

1	2	3	,	4	5	6	,	7	8	9

Ask her to read it. [one hundred twenty-three million four hundred fifty-six thousand seven hundred eighty-nine]

Which numbers are the millions? [123] Which are the thousands? [456] Which are just ones? [789]

Replace various digits with zeroes and ask the child to read it.

Writing the number words. Write

72,072

and ask the child to write it in words. [seventy-two thousand seventy-two] Remind her to write a dash between the words *seventy* and *two.*

Also write 11, 580 and ask her to write the words. [eleven thousand five hundred eighty] Repeat for 806, 241. [eight hundred six thousand two hundred forty-one] You might point out that the word *forty* does not have a *u* in it like the word *four.*

Worksheet. The worksheet is similar to the above activities. The solutions are as follows.

a. **10,000** b. **72,000**

c. **102,000** d. **245,000**

e. **950,000** f. **1,000,000**

g. **8,000,000** h. **12,000,000**

i. **620,602**

j. **62,062**

k. **6,262**

l. **602,622**

m. **622,062**

n. **622,602**

o. **forty-one thousand six hundred thirty-two**

p. **five hundred nineteen thousand eight hundred three**

q. **three hundred three thousand seventy-three**

Lesson 70 (1 or 2 days)

Working With Large Numbers

OBJECTIVES 1. To solve problems with numbers having more than 4 digits
2. To measure strides in inches

MATERIALS Worksheet 65, "Quick Practice-6"
At least 2 yardsticks or a tape measure
Worksheet 66, "Working With Large Numbers"

WARM-UP Ask the child to count by 9s to 90 and to write them. Remind him to write the second row going from right to left as shown below.

9	18	27	36	45
90	81	72	63	54

Ask the child to look at the multiples and answer 9 × 6 [54], 9 × 7 [63], 9 × 4 [36], 9 × 9 [81], 9 × 8 [72], 9 × 3 [27], and 9 × 1. [9]

ACTIVITIES ***Facts practice.*** Take out Quick Practice-6 and ask the child to do the facts in the **top** rectangle when you say start. Ask him to correct them. The answers are:

24	**9**	**36**	**21**	**3**	**10**	**30**	**28**	**40**	**15**
18	**28**	**27**	**40**	**8**	**35**	**16**	**6**	**24**	**12**
20	**18**	**30**	**18**	**16**	**36**	**8**	**32**	**14**	**35**
0	**4**	**12**	**20**	**12**	**45**	**12**	**27**	**12**	**25**
24	**15**	**24**	**14**	**45**	**18**	**10**	**21**	**32**	**6**

Ask him to practice any that were incorrect.

Population problems. Write the following cities and their populations (or choose other cities with widely varying populations).

Chicago	2,783,726
Hutchinson	11,473
Minneapolis	368,383
Saint Cloud	48,812
Saint Paul	272,235
Silver Lake	782

Tell the child these are 6 cities and their populations. What is population? [the number of people living in a certain place] What order are the cities written in? [alphabetical] Ask him to write them in order of increasing population, that is, in order from lowest population to highest.

Silver Lake	782
Hutchinson	11,473
Saint Cloud	48,812
Saint Paul	272,235
Minneapolis	368,383
Chicago	2,783,726

Then ask the following questions:

1. Which city has the largest population? What is it? [Chicago, two million seven hundred eighty-three thousand seven hundred twenty-six]

2. Which city has the lowest population and what is it? [Silver Lake]
3. Is the population of the first 4 cities (revised list) less than or more than that of Minneapolis? [less] Ask him to add them to find out for sure. [333,302] Ask him read his answer.

4. Do more people live in the first five cities or more in Chicago? [in Chicago] Ask him to add them to find out for sure. [701,685] Discuss his solution. The easiest way is to add the population of Minneapolis to the sum of the first four cities.

"Mile" problems. Discuss with the child how far a mile is. Tell him that originally a mile was 1000 strides. Explain that a stride is the distance a person covers when the heel of one foot touches the ground until that heel touches the ground again. It is 2 steps.

Ask the child how he could measure your stride. Decide on a starting line. If using 2 yardsticks, he will discover he needs to add the totals if the stride exceeds 36 inches. Repeat the procedure with him measuring your stride.

A mile used to be 1000 strides. Ask him how long the miles in inches are for both of you. Discuss what it would be like if everyone's mile was different.

Tell him that the official mile since the year 1500 is 63,360 inches. Ask if each of you will need more or less than 1000 strides to walk an actual mile. [probably more for him]

Perimeter problem. Give this problem to the child. The largest office building in the world is called The Pentagon because it is in the shape of a pentagon. The distance around one side is about 1056 feet. What is the perimeter? [5280 ft] Draw a pentagon on the board.

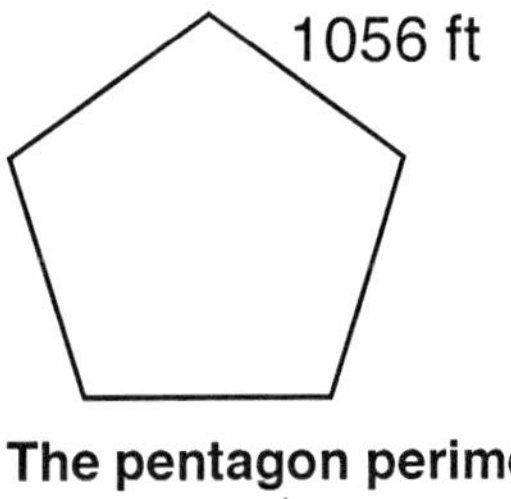

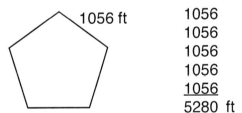

1056
1056
1056
1056
1056
5280 ft

The pentagon perimeter problem.

Worksheet. There are seven problems for the child to read and solve. You might want to discuss them after the child has worked on them. The solutions are as follows:

a. **42 mi** b. **324 ft**

c. **31,255 mi** d. **18,112 mi**

e. **372,564 mi, yes** f. **745,128 mi**

g. **21,120 ft**

Lesson 71

Working With Nines

OBJECTIVES
1. To learn the 9s table
2. To learn the facts with 9 as the multiplier

MATERIALS Worksheet 65, "Quick Practice-6"
Abacus
Worksheet 54, "Multiplication Table"
Worksheet 67, "Working With Nines"

WARM-UP Ask the child to add 364 thousand and 777 thousand in her journal. [1,141,000] The check numbers are (4) + (3) = (7) Ask her to read the answer. [1 million 1 hundred forty-one thousand]

Ask what is needed with 7 to make 9? [2] with 6? [3], with 8? [1] with 5? [4] with 3? [6] with 1? [8] with 4? [5] with 2? [7]

ACTIVITIES ***Facts practice.*** Ask the child to take out her Quick Practice sheet and ask her to do the facts in the **bottom** rectangle when you say start. Ask her to correct them. The answers are:

30	**32**	**8**	**16**	**27**	**6**	**6**	**27**	**45**	**3**
12	**36**	**15**	**12**	**20**	**10**	**8**	**14**	**0**	**20**
18	**12**	**10**	**45**	**35**	**24**	**21**	**25**	**18**	**18**
24	**4**	**15**	**40**	**21**	**20**	**9**	**14**	**24**	**16**
24	**35**	**16**	**12**	**21**	**18**	**36**	**40**	**32**	**30**

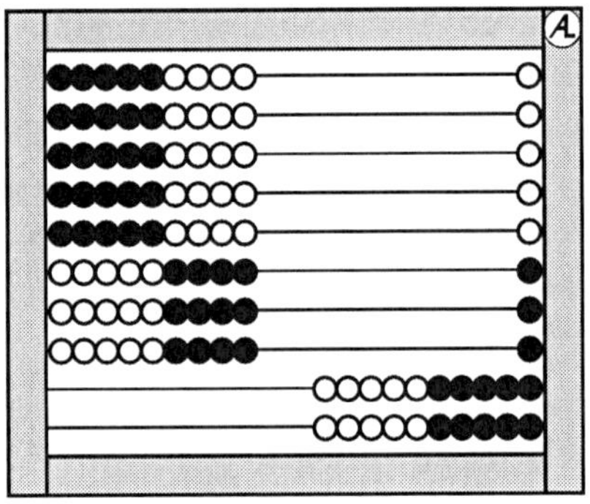

Ask her to practice the ones that were incorrect.

Abacus practice. Enter 9 × 8 on an abacus. How could you find the product? Discuss seeing it as 80 – 8 or as groups of 5s and 10s. Enter other 9 facts, such as 9 × 4 and 6 × 9 and flash the abacus for several seconds.

9s multiples. Ask the child to write the multiples of 9 in two rows with the second row reversed as shown below.

9	18	27	36	45
90	81	72	63	54

What are some special things you notice about these 9s multiples? [The second row is the same as the first row with digits reversed.] What happens when you add the digits for the these multiples? [The sum is always 9.] What are the check numbers for any multiple of 9? [0]

Practice orally by asking, for example, What is 9 × 9? [81] What is 9 × 7? [63] and so forth. Ask the child to see their mental abacus or use the skip counting patterns. Concentrate especially on the new facts.

Comparing 9s to 10s. Ask her to write the multiples of 10 in a row and the multiples of 9 directly below as shown.

10	20	30	40	50	60	70	80	90	100
9	18	27	36	45	54	63	72	81	90

What patterns do you notice; how could you remember the 9s facts? Point to the 27 and say, 9 × 3 is how much less than 30? [3] Repeat for 9 × 8. [8 less than 80, or 72] Repeat for other 9s facts.

An even easier way, using for example, 9 × 4, is to think that the answer must be the decade below the number being multiplied, the 30s in this case. The second digit must be what is needed with 3 to make 9, that is 6, so 9 × 4 = 36. Another example is 9 × 6: it must be in the 50s and the second digit must 4, so the answer is 54.

Practice with written symbols. Write 9 × 1 to 9 × 10 in order. Point to the various facts in random order and ask the child for the products.

9s on the multiplication table. Ask the child to write the 9s facts on her multiplication table. See the tables below.

Note: The recent additions are shown in boldface.

×	1	2	3	4	5	6	7	8	9	10
1	1	2	3	4	5	6	7	8	9	10
2	2	4	6	8	10	12	14	16	18	20
3	3	6	9	12	15	18	21	24	27	30
4	4	8	12	16	20	24	28	32	36	40
5	5	10	15	20	25	30	35	40	45	50
6	6	12	18	24	30				**54**	60
7	7	14	21	28	35				**63**	70
8	8	16	24	32	40				**72**	80
9	9	18	27	36	45				**81**	90
10	10	20	30	40	50	60	70	80	90	100

Entering the 9s.

×	1	2	3	4	5	6	7	8	9	10
1	1	2	3	4	5	6	7	8	9	10
2	2	4	6	8	10	12	14	16	18	20
3	3	6	9	12	15	18	21	24	27	30
4	4	8	12	16	20	24	28	32	36	40
5	5	10	15	20	25	30	35	40	45	50
6	6	12	18	24	30				54	60
7	7	14	21	28	35				63	70
8	8	16	24	32	40				72	80
9	9	18	27	36	45	**54**	**63**	**72**	81	90
10	10	20	30	40	50	60	70	80	90	100

Entering the 9s as multipliers.

Worksheet. The worksheet concentrates on 9s. The answers follow.

9	9	9	9	9	9	9	9	9	9
× 3	× 9	× 6	× 5	× 1	× 10	× 8	× 4	× 2	× 7
27	**81**	**54**	**45**	**9**	**90**	**72**	**36**	**18**	**63**

2	10	7	6	5	1	8	4	3	9
× 9	× 9	× 9	× 9	× 9	× 9	× 9	× 9	× 9	× 9
18	**90**	**63**	**54**	**45**	**9**	**72**	**36**	**27**	**81**

9 × 8 = **72**	5 × 9 = **45**	a. **63 days, more**
9 × 2 = **18**	2 × 9 = **18**	b. **540 min**
9 × 1 = **9**	10 × 9 = **90**	c. **288 oz**
9 × 6 = **54**	9 × 9 = **81**	d. 72
9 × 10 = **90**	3 × 9 = **27**	e. **3,285 days**
9 × 9 = **81**	6 × 9 = **54**	
9 × 5 = **45**	8 × 9 = **72**	
9 × 7 = **63**	4 × 9 = **36**	
9 × 3 = **27**	1 × 9 = **9**	
9 × 4 = **36**	7 × 9 = **63**	

Lesson 72

Review

OBJECTIVE 1. To review and practice

MATERIALS Worksheet 68-A and 68-B, "Review" (2 versions)
Cards for playing games

ACTIVITIES ***Review worksheet.*** Give the child about 15 to 20 minutes to do the review sheet. The oral problems to be read are as follows:

60 × 6 = 25¢ × 6 = 70 – 12 =

Review Sheet 68-A.

1. Write only the answers to the oral questions. **360** **$1.50** **58**

4. Write only the answers. 76 + 47 = **123** 64 – 48 = **16** 92 – 89 = **3**

7. Write the multiples of 9.

9 18 27 36 45

90 81 72 63 54

17.
9 × 8 = **72** 6 × 9 = **54**
9 × 4 = **36** 7 × 9 = **63**
9 × 1 = **9** 3 × 9 = **27**

23. Explain how you could find 9 × 6 if you didn't know it.
10 × 6 = 60, so 9 × 6 = 60 – 6 = 54
OR 6th multiple of 9
OR 9 × 5 = 45 and 45 + 9 = 54

24. The population of Indiana is 6,080,000. Write this number in words.

six million eighty thousand

25. In City Bank, there are 28 bags with money. Each bag has 1 thousand dollars. How much money is in the bags?

$28,000

26. A machine is packing seeds by putting one thousand seeds in a bag. How many bags will the machine fill for 17,000 seeds?

17 bags

27. Three million four hundred eighty-five thousand three hundred ninety-eight people live in Los Angeles. Write this number using digits.

3,485,398

28. In 1990, the population of Saint Paul was 272,235. The population of Minneapolis was 368,383. How many people lived in those 2 cities that year?

		2	7	2,	2	3	5
	+	3	6	8,	3	8	3
		6	4	0,	6	1	8

29.

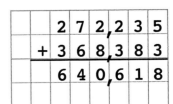

$$\begin{array}{r} 808{,}791 \\ +\,352{,}240 \\ \hline \mathbf{1{,}161{,}031} \end{array} \qquad \begin{array}{r} 560{,}567 \\ -\,63{,}945 \\ \hline \mathbf{496{,}622} \end{array}$$

The oral problems to be read to the child are as follows:

$40 \times 4 =$ $25¢ \times 5 =$ $80 - 13 =$

Review Sheet 68-B.

1. Write only the answers to the oral questions. 160 $1.25 67

4. Write only the answers. $55 + 88 =$ 143 $75 - 39 =$ 36 $81 - 76 =$ 5

7. Write the multiples of 4.

4 8 12 16 20

24 28 32 36 40

17. $4 \times 7 =$ 28 $9 \times 4 =$ 36

$4 \times 5 =$ 20 $6 \times 4 =$ 24

$4 \times 8 =$ 32 $4 \times 4 =$ 16

23. Explain how you could find 4×9 if you didn't know it.

10 × 4 = 40, so 4 × 9 = 40 − 4 = 36
OR 9th multiple of 4
OR 4 × 5 + 4 × 4 = 20 + 16 = 36

24. The population of Colorado is 4,300,000. Write this number in words.

four million three hundred thousand

25. State Bank has $49,000 in bags. Each bag has one thousand dollars. How many bags do they have?

49

26. A seed corn dealer has 83 bags of seed. Each bag has about 1000 seeds. About how many seeds is that all together?

83,000

27. One million six hundred thirty thousand five hundred fifty-three people live in Houston. Write this number using digits.

1,630,553

28. In 1990, the population of Dallas was 1,006,877. The population of Fort Worth was 447,619. How many people lived in those 2 cities that year?

	1,	0	0	6,	8	7	7
	+	4	4	7,	6	1	9
	1,	4	5	4,	4	9	6

29.

$$\begin{array}{r} 622{,}199 \\ +\,414{,}249 \\ \hline 1{,}036{,}448 \end{array} \qquad \begin{array}{r} 875{,}208 \\ -\,35{,}487 \\ \hline 839{,}721 \end{array}$$

Games. Spend the remaining time playing games.

TELLING TIME: Between the Numbers (Lesson 57) and Name That Time (C27) *(Math Card Games).*

MULTIPLICATION. Multiplication Old Main (P14), Slide-a-Thon Solitaire (P9) *(Math Card Games),* and Multiples Solitaire (P19) *(Math Card Games).*

Note: *Multiples Solitaire* was formerly called *Threesome Solitaire.*

Lesson 73

Multiplying and Adding

OBJECTIVE 1. To prepare for the single-digit multiplication algorithm

MATERIALS Worksheet 69, "Quick Practice-7"
Tiles
Worksheet 70, "Multiplying and Adding"

WARM-UP Ask the child to write the multiples of 9s to 90.

9	18	27	36	45
90	81	72	63	54

Ask the child to look at the multiples and answer 9 × 7 [63], 9 × 6 [54], 9 × 4 [36], 9 × 9 [81], and 9 × 8. [72]

What is 2 × 2 + 2? [6] What is 5 × 5 + 5? [30] What is 4 × 4 + 4? [20] What is 3 × 3 + 3? [12] What is 1 × 1 + 1? [2] What is 1 × 2 + 3? [5]

ACTIVITIES ***Facts practice.*** Take out Quick Practice-7 and ask the child to do the facts in the **top** rectangle when you say start. Ask her to correct them. The answers are:

12	28	36	54	18
72	81	6	15	6
45	72	24	27	20
9	24	30	0	54
32	36	18	12	16

63	8	18	16	18
21	9	27	8	21
24	20	45	35	90
12	24	28	40	63
12	9	30	32	15

Ask her to practice the ones that were incorrect.

Reviewing area equations. Either lay out a 4 by 6 array with tiles as shown below on the left, or ask the child to do so. Use a second color after 5 to minimize counting.

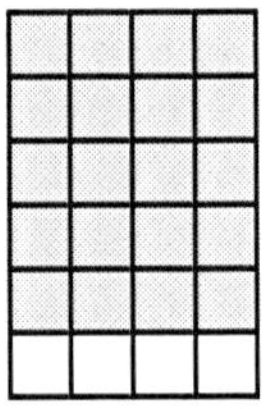

4 × 6 = 24

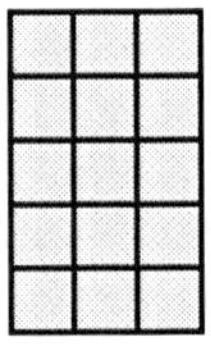

3 × 5 = 15

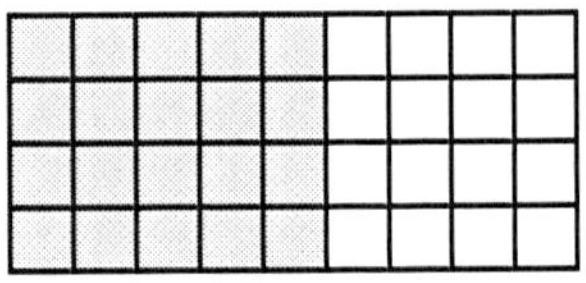

9 × 4 = 36

Ask her to write the equation that shows the number of tiles. [4 × 6 = 24] If she writes 4 + 4 + 4 + 4 + 4 + 4 = 24, ask her to try it with multiplication.

Repeat for 3 by 5 [3 × 5 = 15] and 9 by 4 [9 × 4 = 36] arrays. See above center and right.

> **Note:** Encourage the child to write the multiplication part first. On a simple calculator it must be done first to get the correct answer.

Using multiplication and adding. Lay out a 4 by 5 array in a row; place 2 extra tiles on top as shown below on the left. Ask the child to write the equation. [$5 \times 4 + 2 = 22$]

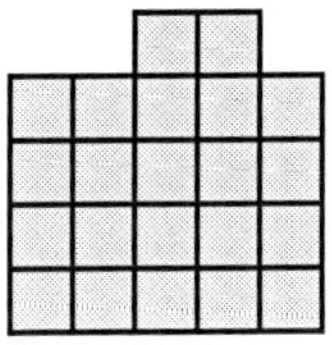

5 × 4 + 2 = 22

Then make the arrays shown, and ask for the equations.

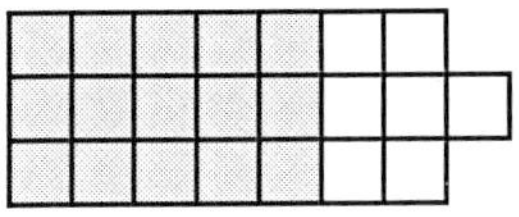

7 × 3 + 1 = 22

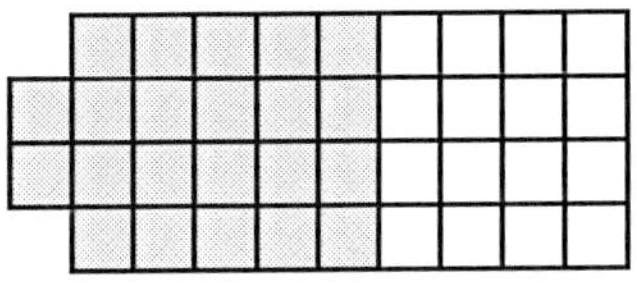

9 × 4 + 2 = 38

Multiplying to simplify adding. Write 246 four times as shown below on the left. Ask the child to copy it and find the sum. Then ask her to show her solution. It probably will look like that shown below.

	2	1 2	1 2	
246	246	246	246	
246	246	246	246	
246	246	246	246	**246**
+ 246	+ 246	+ 246	+ 246	**× 4**
	4	**84**	984	**984**

Encourage using multiplication. For example, in adding the ones, 6×4 is 24. Also encourage her to write the trades at the top of the columns. Write this problem as a multiplication problem.

For a second example, write 542 six times and ask her to find the sum. See below.

	1	2 1	2 1	
542	542	542	542	
542	542	542	542	
542	542	542	542	
542	542	542	542	
542	542	542	542	**542**
+ 542	+ 542	+ 542	+ 542	**× 6**
	2	**52**	**3,252**	**3,252**

Worksheet. The worksheet has both arrays and adding the same number. The solutions are:

a. **4 × 7 + 2 = 30**	b. **2 × 9 + 1 = 19**	c. **5 × 6 + 3 = 33**
d. **2,600**	e. **15,125**	f. **3,516**
g. **104,156**	h. **274,389**	i. **24,680**
3,025	**879**	**2,468**
× 5	**× 4**	**× 10**
15,125	**3,516**	**24,680**

Lesson 74

Multiplying by a One-Digit Number

OBJECTIVE 1. To learn the single-digit multiplication algorithm

MATERIALS Worksheet 69, "Quick Practice-7"
Worksheet 71, "Multiplying by a One-Digit Number"

WARM-UP Ask what is 2 × 2 + 3? [7] What is 10 × 10 + 10? [110] What is 4 × 2 + 2? [10] What is 5 × 3 + 3? [18] What is 7 × 4 + 4? [32]

Ask the child to count by 8s to 80 and to write them.

8	16	24	32	40
48	56	64	72	80

Point to the multiples and ask the child to recite the corresponding fact. For example, point to 24 and the child says 8 × 3 = 24. Repeat for other multiples

ACTIVITIES ***Facts practice.*** Ask the child to take out his Quick Practice sheet and ask him to do the facts in the **bottom** rectangle when you say start. Ask him to correct them. The answers are:

27	**16**	**12**	**12**	**72**
6	**24**	**20**	**12**	**18**
81	**21**	**0**	**15**	**24**
72	**54**	**27**	**9**	**8**
18	**90**	**30**	**40**	**36**

6	**45**	**9**	**45**	**18**
8	**30**	**32**	**36**	**63**
20	**16**	**24**	**12**	**15**
21	**63**	**28**	**54**	**35**
24	**18**	**28**	**32**	**9**

Note: The term equation refers to an equality written horizontally and includes an "=." Numbers written in columns are not equations.

Ask him to practice the ones that were incorrect.

Multiplying by one digit. Write the following:

$$\begin{array}{r} 175 \\ \underline{\times 5} \end{array}$$

[875] and ask the child to find 175 times 5. Suggest he can write it out like it was done in the previous lesson.

Ask him to explain his solution. Ask him if there is another way to solve the problem.

$$\begin{array}{r} {\scriptstyle 3\,2} \\ 175 \\ \underline{\times\ 5} \\ 875 \end{array} \qquad \begin{array}{r} {\scriptstyle 3\,2} \\ 175 \\ 175 \\ 175 \\ 175 \\ \underline{+\ 175} \\ 875 \end{array}$$

$$\begin{array}{r} 531 \\ \underline{\times 6} \end{array}$$

Ask him to solve it first and then to explain it to you.

$$\begin{array}{r} {\scriptstyle 1} \\ 531 \\ \underline{\times 6} \\ 3{,}186 \end{array} \qquad \begin{array}{r} {\scriptstyle 1} \\ 531 \\ 531 \\ 531 \\ 531 \\ 531 \\ \underline{+ 531} \\ 3{,}186 \end{array}$$

Tell him to write out the addition method if he needs to.

Write and ask him to multiply

$$\begin{array}{r} 9{,}088 \\ \underline{\times 3} \end{array}$$

[27,264] Ask him to solve it and then explain it to you.

Check numbers. Ask the child how he thinks check numbers would work for multiplying. [The check numbers of the 2 numbers are multiplied together.] In the above example, the check number for 9088 is (7) and (7) × (3) = (21), which is (3), as is the check number for 27,264.

Worksheet. The worksheet provides practice in multiplying. There is space, especially for the first 2 rows, to write the problems out as multiplication problems. The answers are given below.

96	**84**	**220**
125	**2,439**	**878**
940	**800**	**966**
2,574	**10,876**	**4,145**
22,557	**10,899**	**30,420**
18,010	**15,120**	**77,571**

Lesson 75

Introducing Area

OBJECTIVES
1. To introduce the term *area*
2. To introduce *square inches*

MATERIALS
Worksheet 72, "Quick Practice-8"
Worksheet 73, "Introducing Area"
2 books with covers of obviously different areas
6-inch square and 12-inch × 3-inch rectangle made from construction paper
1-inch tiles, container with at least 80

WARM-UP
Ask the child to multiply 926 by 9 in her journal. [8334] The check numbers are (8) × (0) = (0).

Ask her to write and say the multiples of 8. Then ask for 8s facts.

ACTIVITIES
Facts practice. Take out Quick Practice-8 and ask the child to do the facts in the **top** rectangle when you say start. Ask them to correct them. The answers are:

12	28	30	16	18	72	15	16	0	40
8	18	21	9	24	18	12	20	24	30
14	21	36	72	9	6	8	24	32	9
27	28	45	16	18	32	12	45	54	63
24	20	36	15	81	27	63	54	35	12

Tell her to practice correct strategies.

The term area. Ask the child, What does the word *area* mean to you? She may be familiar with phrases such as play area, work area, or parking area. Compare 2 areas by pointing to, for example, a small desk and a larger table and ask, Which has greater area? Is your play area greater or less than the hall area? Stress that area means how much flat space the object takes up.

Comparing areas. Show the two books whose covers are obviously different areas and ask which cover has the greater area, or takes up greater space. See below.

Comparing the area of two book covers. The cover on the left has the greater area.

Note: Squares are rectangles!

Show the child the 6 × 6 and 12 × 3 rectangles. Which rectangle do you think has the greater area? How can you find out? Give her time to think. If necessary, ask if she could use the tiles some way. [covering the rectangles]

Then give the rectangles and a container of tiles to the child to cover the figures, as shown on the next page.

Note: A child needs many experiences with grids before she is ready to "multiply length times width."

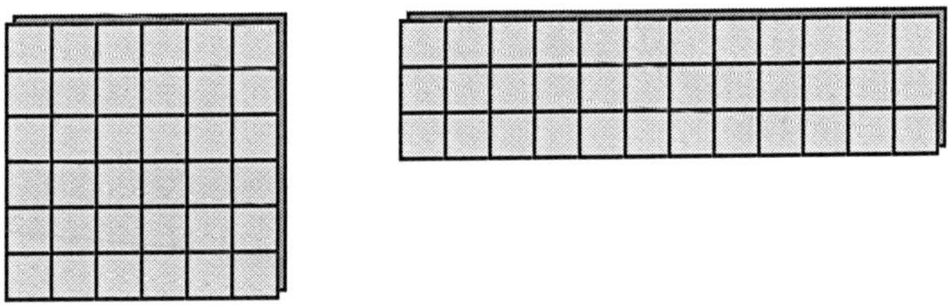

Comparing the *space* taken up by the two rectangles by covering both with tiles.

When she has finished, ask, How many tiles do you need to cover the figures? [36] Ask how she found her answer. Ignore counting by 1 strategies. Express delight if she suggests multiplying. So, which figure has the greater area? [the same]

Ask her to turn the 6 × 6 array into a 12 × 3 array and the 12 × 3 array into a 6 × 6 array. The figures below show the process.

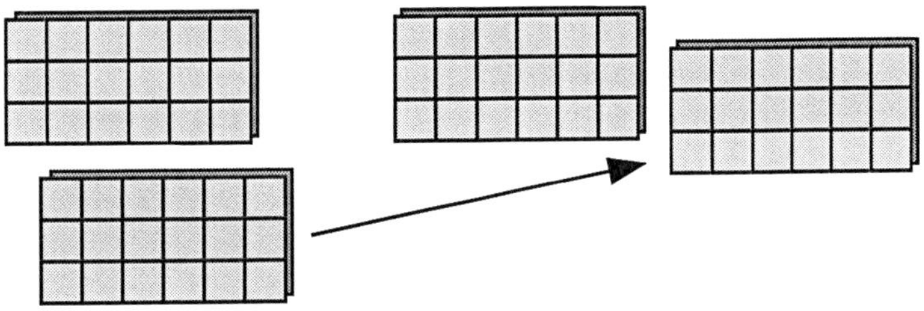

Showing the equivalence of the two rectangles.

What letter do we use to stand for perimeter? [P] Guess what letter is used for area? [A] Which measures the distance around a figure, perimeter or area? [perimeter] Which measures how much space is in a figure? [area]

Ask the child to write the equation for the area of the rectangles.

$$A = 36$$

Square inches. What did we use to measure the area? [a tile] What is its shape? [square] How long is a side? [1 inch] Tell the child we call it a *square inch.* Its abbreviation is *sq. in.* Write the units after the area equation.

$$A = 36 \text{ sq. in.}$$

Comparing perimeters. Since the areas are equal, do you think the perimeters are also equal? Ask the child to calculate the perimeters and write them.

The 6 × 6 rectangle: $P = 6 \times 4 = 24$; $(6 + 6 + 6 + 6)$

The 12 × 3 rectangle: $P = (12 + 3) \times 2 = 30$; $(12 + 3 + 12 + 3)$

Which has the greater perimeter? [the 12 × 3 rectangle]

Worksheet. For the worksheet, the child can use the grids to find the perimeters and areas. The solutions are:

1. $\mathbf{P = 16}$ **in.**	$\mathbf{A = 15}$ **sq. in.**
2. $\mathbf{P = 14}$ **in.**	$\mathbf{A = 10}$ **sq. in.**
3. $\mathbf{P = 14}$ **in.**	$\mathbf{A = 12}$ **sq. in.**

Lesson 76

Working With Square Inches

OBJECTIVE 1. To work with area by covering rectangles with square inches

MATERIALS Worksheet 72, "Quick Practice-8"
Worksheet 74, "Working With Square Inches"
One-inch tiles
Ruler
10 rectangles cut from 9-inch × 12-inch construction paper with the following dimensions in inches:

5 × 5	4 × 5	4 × 6	5 × 3	4 × 3
9 × 3	9 × 2	8 × 3	7 × 2	9 × 1

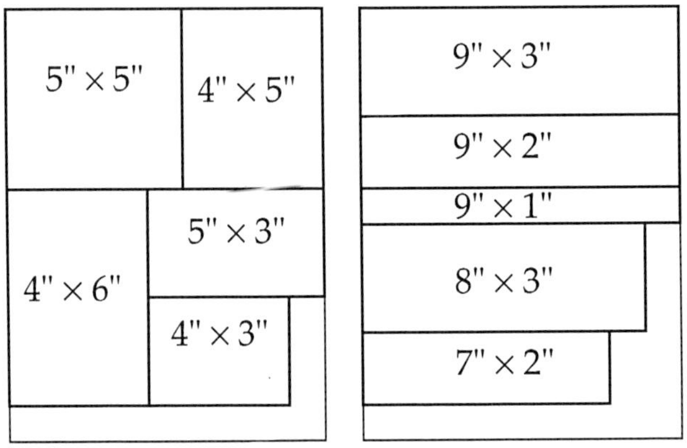

The layout for cutting out the rectangles from 9" × 12" construction paper.

WARM-UP Ask the child to say the multiples of 8. Then ask for 8s facts.

Ask the child to write in his journal the number of days in 8 years. [365 × 8 = 2,920, actually 2,922 because of the 2 leap years] The check numbers are (5) × (8) = (4).

ACTIVITIES ***Facts practice.*** Ask the child to take out his Quick Practice sheet and ask him to do the facts in the **bottom** rectangle when you say start. Ask him to correct them. The answers are:

18	32	12	45	20	0	6	24	18	20
54	63	21	45	16	15	16	8	72	30
24	14	12	15	18	72	9	24	9	40
12	9	18	36	24	28	54	8	27	21
36	28	12	63	35	81	30	16	27	32

Tell him to practice correct strategies.

Note: Laying out the tiles helps the child understand what area really means. Gradually, he comes to see the multiplication.

Finding the area. Mark each of the rectangles described above with a letter A to J. Set them out on the table

Explain to the child that he is to draw a sketch of each rectangle along with the letter in his journal. Next he writes the measurements in inches and finds the area in square inches. To find the area, he can use tiles or the ruler. The solutions are on the next page.

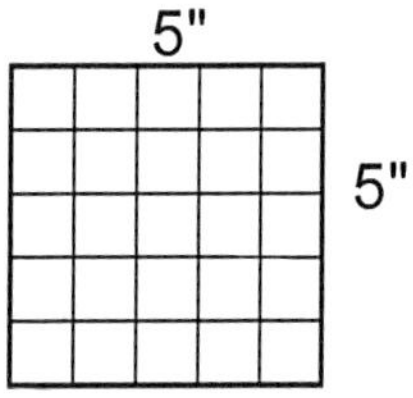

$A = 25$ sq. in.

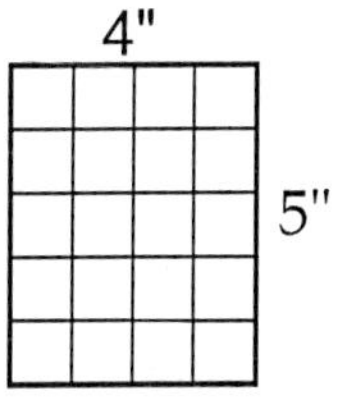

$A = 20$ sq. in.

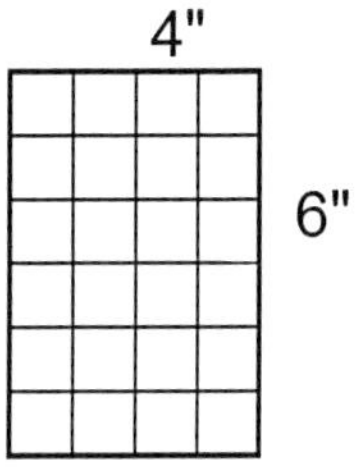

$A = 24$ sq. in.

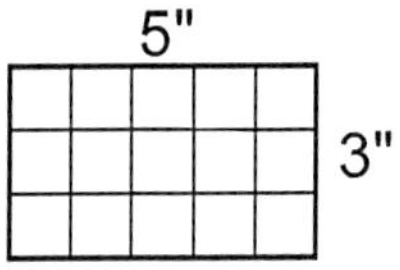

$A = 15$ sq. in.

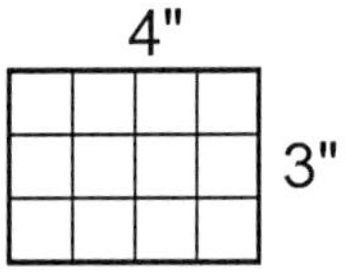

$A = 12$ sq. in.

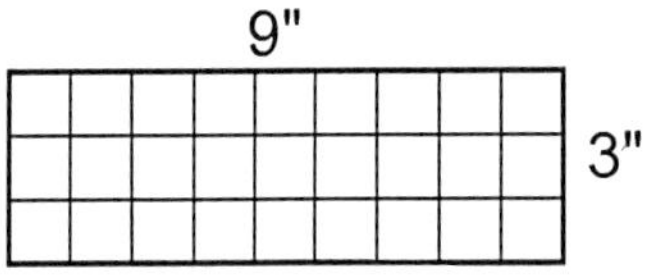

$A = 27$ sq. in.

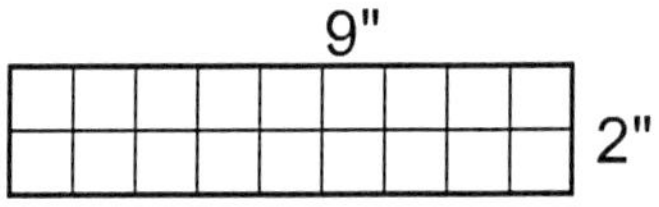

$A = 18$ sq. in.

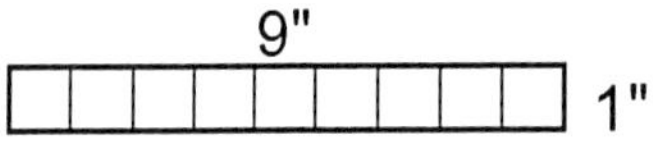

$A = 9$ sq. in.

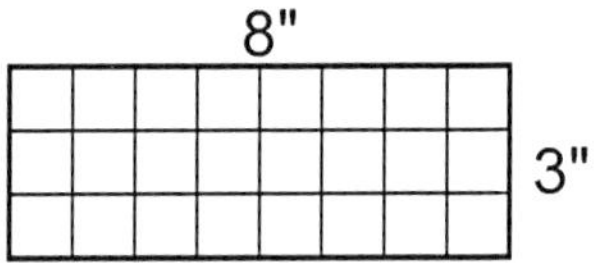

$A = 24$ sq. in.

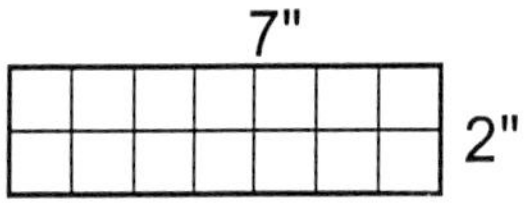

$A = 14$ sq. in.

Worksheet. The worksheet has three rectangles. The child is asked to draw in the square inches and find the areas. The figures and areas are given below.

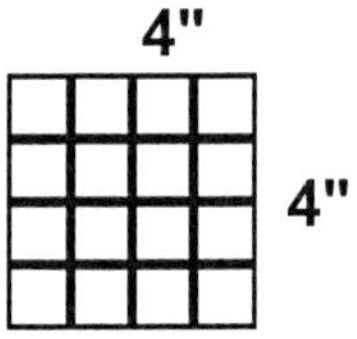

$A =$ **16 sq. in.**

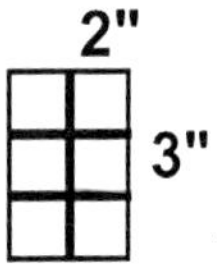

$A =$ **6 sq. in.**

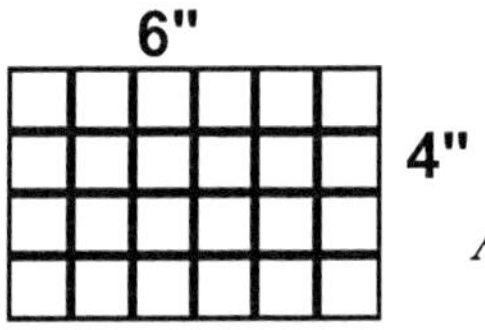

$A =$ **24 sq. in.**

Lesson 77

Working With Sixes

OBJECTIVES 1. To learn the 6s table
2. To learn the facts with 6 as the multiplier

MATERIALS Worksheet 75, "Quick Practice-9"
Abacus
Worksheet 54, "Multiplication Table"
Worksheet 76, "Working With Sixes"

WARM-UP Ask the child to multiply 9,534 by 8 in her journal. [76,272] The check numbers are (3) × (8) = (6).

Ask, What is 3 × 2 and then doubled? [12] What is 3 × 5 doubled? [30] What is 3 × 9 doubled? [54] What is 3 × 3 doubled? [18] What is 3 × 6 doubled? [36] What is 3 × 7 doubled? [42]

ACTIVITIES ***Facts practice.*** Take out Quick Practice-9 and ask the child to do the facts in the **top** rectangle when you say start. Ask her to correct them. The answers are:

63	54	18	72	8	30	0	27	28	24
24	63	72	36	8	24	20	45	27	18
28	40	16	15	45	15	12	12	81	6
16	18	21	9	54	35	9	9	20	30
21	36	18	12	32	24	14	12	32	16

Ask her to practice the ones that were incorrect.

Abacus practice. Enter 6 × 4 on an abacus, as shown below, and flash it for 2-3 seconds. What is the equation? [6 × 4 = 24] Ask, How did you figure it out? Repeat for other facts.

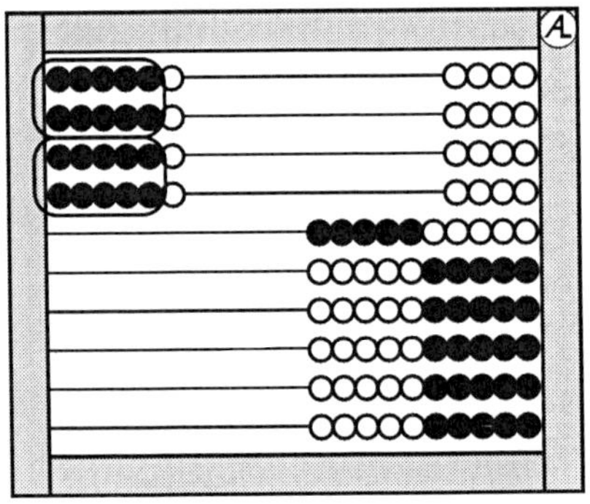

Seeing the 10s in 6 × 4.

6s multiples. Ask the child to write the multiples of 6 in 2 rows as shown below.

6	12	18	24	30
36	42	48	54	60

What patterns do you see with the 6s multiples? [All the numbers are even.] What is the difference between the numbers in each row? [30] What are the check numbers? [a repeating 6, 3, 0 pattern]

What is 6 × 4? [24] What is 6 × 7? [42] What is 6 × 8? [48] What is 6 × 7? [42] and so forth. Concentrate especially on the new facts, 6 × 6 to 6 × 8.

What is 6 × 2? [12] What is 6 × 4? [24] What is 6 × 6? [36] What is and 6 × 8? [48] **Ask what pattern she noticed. [The product ends in the same digit as the number being multiplied (for even numbers).]** Does it works when multiplying by odd numbers, such as 6 × 3 or 6 × 5? [no]

Comparing 6s to 3s. Ask the child to write the multiples of 3 in a row and the multiples of 6 directly below as shown.

3	6	9	12	15	18	21	24	27	30
6	12	18	24	30	36	42	48	54	60

What pattern do you see? [The 6s are double the 3s.]

Practice with written symbols. With the 6s written in 2 rows still visible, write 6 × 1 to 6 × 10 in order. Point to the various facts in random order and the child to say the answers.

6s on the multiplication table. Ask the child to write the 6s and multiplying by 6s facts on her multiplication table (Worksheet 54). See the table below. How many spaces are left to be filled? [4]

×	1	2	3	4	5	6	7	8	9	10
1	1	2	3	4	5	6	7	8	9	10
2	2	4	6	8	10	12	14	16	18	20
3	3	6	9	12	15	18	21	24	27	30
4	4	8	12	16	20	24	28	32	36	40
5	5	10	15	20	25	30	35	40	45	50
6	6	12	18	24	30	**36**	**42**	**48**	54	60
7	7	14	21	28	35	**42**			63	70
8	8	16	24	32	40	**48**			72	80
9	9	18	27	36	45	54	63	72	81	90
10	10	20	30	40	50	60	70	80	90	100

Entering 6s and the 6s as multipliers.

Note: The recent additions are shown in boldface.

Worksheet. The answer for the worksheet are below.

6	6	6	6	6	6	6	6	6	6
× 3	× 4	× 1	× 7	× 9	× 5	× 8	× 6	× 2	× 10
18	**24**	**6**	**42**	**54**	**30**	**48**	**36**	**12**	**60**
2	5	7	8	3	10	6	1	4	9
× 6	× 6	× 6	× 6	× 6	× 6	× 6	× 6	× 6	× 6
12	**30**	**42**	**48**	**18**	**60**	**36**	**6**	**24**	**54**

6 × 10 = **60**	7 × 6 = **42**	a. **42, less**
6 × 2 = **12**	10 × 6 = **60**	b. **$48,000,000**
6 × 5 = **30**	8 × 6 = **48**	c. **192 oz**
6 × 6 = **36**	3 × 6 = **18**	d. **48**
6 × 4 = **24**	9 × 6 = **54**	e. **2190**
6 × 1 = **6**	5 × 6 = **30**	
6 × 3 = **18**	2 × 6 = **12**	
6 × 7 = **42**	4 × 6 = **24**	
6 × 8 = **48**	1 × 6 = **6**	
6 × 9 = **54**	6 × 6 = **36**	

Lesson 78

Review

OBJECTIVE 1. To review and practice

MATERIALS Worksheet 77-A and 77-B, "Review" (2 versions)
Cards for playing games

ACTIVITIES ***Review worksheet.*** Give the child about 15 to 20 minutes to do the review sheet. The oral problems to be read are as follows:

$20 \times 4 =$ $\quad 70 \times 2 =$ $\quad 99 + 99 =$

Review Sheet 77-A.

1. Write only the answers to the oral questions. 80 140 198

4. Write only the answers. $2000 \times 7 =$ 14000 $\quad 156 - 49 =$ 107 $\quad 99 + 89 =$ 188

7. This line is an inch long. Draw a square inch.

9. What is the area of the figure below in square inches?

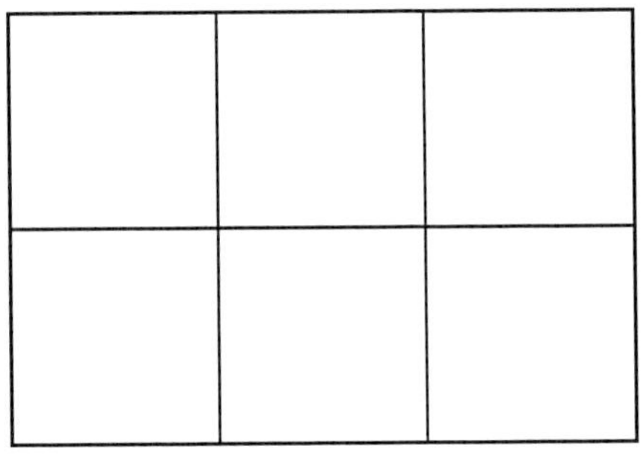

A = 3 × 2
A = 6 sq in.

8. A class spends 55 minutes every day in math class. How much time is that in a week (5 days)?

55
× 5
275 min

OR 55 min is 5 min less than 1 hr, so 5 times that 5 hr - 25 min = 4 hr 35 min

10. The population of North Dakota is 642,000. Write this number in words.

six hundred forty-two thousand

11. The population of Wyoming is four hundred ninety-three thousand seven hundred eighty-two. Write this number in digits.

493,782

12. Martin Luther King, Jr. was born in 1929 and died in 1968. How old was he when he died? Explain.

1968
-1929
39 years

13. A certain new car costs twenty-four thousand six hundred eighteen dollars. What do two new cars cost?

$24,618
× 2
$49,236

The oral problems to be read to the child are as follows:

$40 \times 3 =$ $\quad$ $70 \times 3 =$ $\quad$ $98 + 98 =$

Review Sheet 77-B.

1. Write only the answers to the oral questions. 120 210 196

4. Write only the answers. $2000 \times 9 =$ 18000 $163 - 56 =$ 107 $99 + 72 =$ 171

7. This line is an inch long. Draw a square inch.

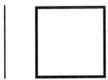

8. A grade class spends 65 minutes every day at lunch and recess. How much time is that in a week (5 days)?

$$\begin{array}{r} 65 \\ \times\ 5 \\ \hline 325 \end{array} \text{ min}$$

OR 65 min is 5 min more than 1 hr, so 5 times that 5 hr + 25 min = 5 hr 25 min

9. What is the area of the figure below in square inches?

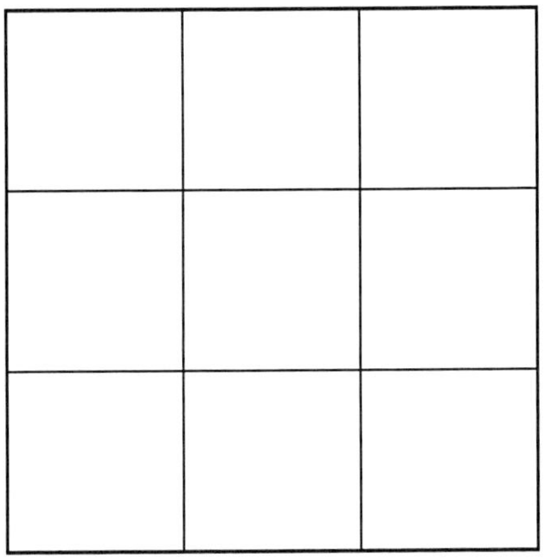

A = 3 × 3
A = 9 sq in.

10. The population of Alaska is 627,000. Write this number in words.

six hundred twenty-seven thousand

11. The population of Vermont is six hundred eight thousand eight hundred twenty-seven. Write this number in digits.

608,827

12. John F. Kennedy was born in 1917 and died in 1963. How long did he live?

$$\begin{array}{r} 1963 \\ -1917 \\ \hline 46 \end{array} \text{ years}$$

13. A certain new boat costs fifteen thousand seven hundred sixty-one dollars. What do 3 new boats cost?

$$\begin{array}{r} \$15,761 \\ \times\ 3 \\ \hline \$47,283 \end{array}$$

Games. Spend the remaining time playing games.

MULTIPLICATION. Show Your Product (P14) and Multiple Authors (P15) *(Math Card Games).*

Lesson 79

Working With Centimeters

OBJECTIVES 1. To become aware of the two measuring systems
2. To work with centimeters and square centimeters

MATERIALS Worksheet 75, "Quick Practice-9"
One-inch tiles
Meter stick
Ruler with inches and centimeters
Worksheets 78-1 & 78-2, "Working With Centimeters"

WARM-UP Ask the child to write in her journal the number of days in 6 years. [365 × 6 = 2,190] The check numbers are (5) × (6) = (3).

Ask the child to say the multiples of 8 to 80 and to write them with 5 numbers in two rows.

ACTIVITIES ***Facts practice.*** Ask the child to take out her Quick Practice sheet and ask her to do the facts in the **bottom** rectangle when you say start. Ask her to correct them. The answers are:

18	9	24	36	28	15	12	12	9	18
32	18	16	27	8	24	16	24	8	63
45	16	12	12	18	0	35	27	45	6
81	20	24	54	72	30	36	21	9	21
32	20	63	40	15	14	54	72	30	28

Tell her to practice correct strategies.

Reviewing centimeters. Ask the child to draw a line that is 1 inch long. Take out the ruler and ask her to measure to see how close she is. Then ask her to look at the centimeter side of the ruler and to draw a line that is 1 centimeter long.

Which is longer, an inch or a centimeter? [inch] What is 100 centimeters called? [meter]

Two systems. Explain that there are two measuring systems used in the world, one called *U.S. Customary* and the other called the *metric* system. Write these words on top of columns. Who uses the U.S. Customary system? [people in the United States] Who uses the metric system? [most of the rest of the world and scientists in the United States]

In which column does the inch belongs? [U.S. customary] centimeter? [metric] meter? [metric] ounces? [U.S. customary] and gallons? [U.S. customary] Encourage the child to think of other units and to place them in the appropriate column.

U.S. Customary	Metric
inch	centimeter
foot	meter
ounce	liter
gallon	
mile	kilometer
pound	
square inches	square centimeters

Area in square centimeters. Take out Worksheet 78-1. Ask the child to do the first four questions. Discuss her answers. The solutions are below.

6 cm
4½ cm
1 sq cm
3 sq cm

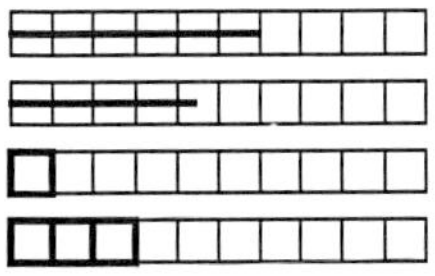

Next ask her to measure rectangle #5 [8 cm and 7 cm] and record the measurements. How many sides they need to measure? [2] Why? [The other 2 are the same.]

What does it mean to find the area? [to find the number of square centimeters that fill the space] How many square centimeters are in the first row? [8] How many rows are there? [7] How do we add the same number several times? [multiply] Show her the preferred way to write it.

$$A = 8 \times 7$$

$$A = 56 \text{ sq cm}$$

5. 8 cm

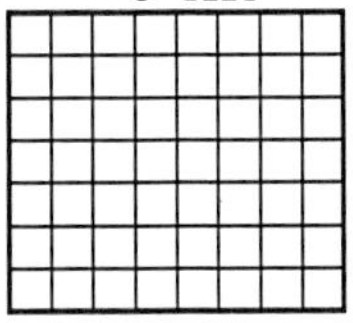

7 cm

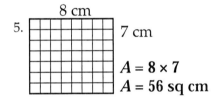

$A = 8 \times 7$
$A = 56$ sq cm

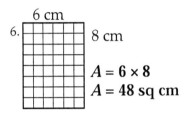

6. 6 cm

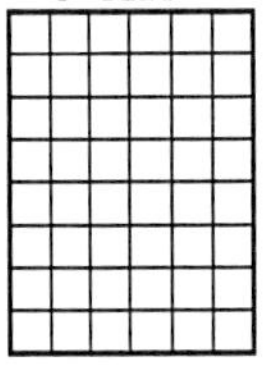

8 cm

$A = 6 \times 8$
$A = 48$ sq cm

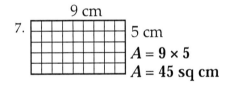

7. 9 cm

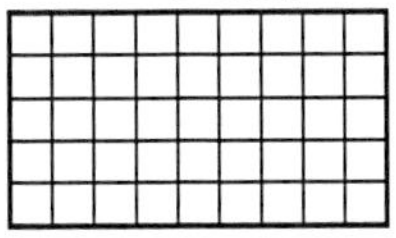

5 cm

$A = 9 \times 5$
$A = 45$ sq cm

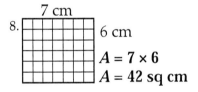

8. 7 cm

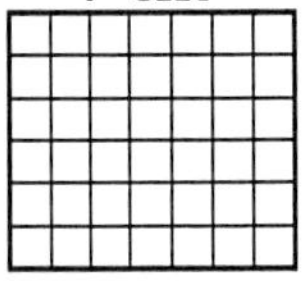

6 cm

$A = 7 \times 6$
$A = 42$ sq cm

Continue with the remaining rectangles. The answers are above.

Worksheet. Worksheet 78-2 also asks for area in square centimeters; however, the rectangles do not have the square centimeters shown. If the child experiences difficulty, help her to imagine the first row and then think about the number of rows. It might help if she draws the first row of squares. The solutions follow.

1. **$A = 9 \times 4$**
   $A = 36$ sq cm

2. **$A = 5 \times 10$**
   $A = 50$ sq cm

3. **$A = 10 \times 6$**
   $A = 60$ sq cm

4. **$A = 7 \times 7$**
   $A = 49$ sq cm

5. **$A = 6 \times 6$**
   $A = 36$ sq cm

Lesson 80

Finding Areas

OBJECTIVES
1. To use creativity in problem solving
2. To write equations for perimeter and area

MATERIALS Worksheet 79, "Quick Practice-10"
Worksheet 80-1 & 80-2, "Finding Areas"
Crayons or colored pencils

WARM-UP Ask the following orally for the child to do mentally, What is 128 + 28? [156] What is 100 – 54? [46] What is 1½ × 4? [6]

Ask the child to write the following problem in his journal: 8090 × 6 [48,540] The check numbers are (8) × (6) = (3).

Ask the child to say the multiples of 8 to 80 and to write them in two rows.

8	16	24	32	40
48	56	64	72	80

Then ask the child for various multiplication facts.

ACTIVITIES ***Facts practice.*** Take out Quick Practice-10 and ask the child to do the facts in the **top** rectangle when you say start. Ask him to correct them. The answers are:

45	**28**	**20**	**9**	**12**	**24**	**14**	**21**	**8**	**15**
72	**27**	**12**	**20**	**9**	**16**	**18**	**63**	**15**	**6**
30	**9**	**32**	**72**	**30**	**54**	**28**	**18**	**16**	**32**
16	**54**	**36**	**45**	**12**	**40**	**35**	**12**	**81**	**24**
0	**24**	**63**	**18**	**27**	**8**	**18**	**24**	**21**	**36**

Ask him to learn strategies and to practice the ones that were incorrect.

Finding the area of an irregular figure. Take out the worksheet. Ask the child to find the area two different ways using multiplication. Ask him to use different colors to show the different areas. Also ask him to write the equations and explain his work.

One solution is to find two rectangles shown below on the left and to find those areas and add them together.

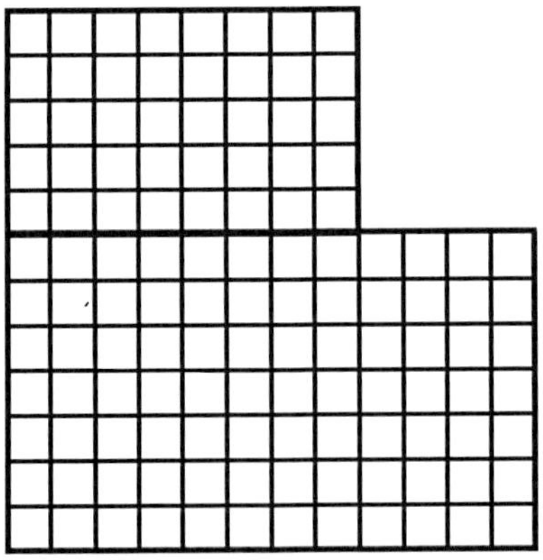

Separating the top from the bottom and adding them.

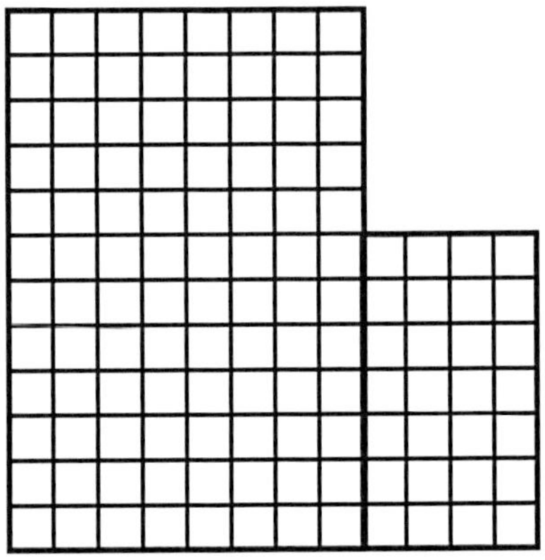

Separating the left from the right and adding them.

Note: You may prefer to use color names rather than *top* and *bottom.*

The calculations are

A of top = $8 \times 5 = 40$

A of bottom = $12 \times 7 = 84$

$A = 40 + 84 = 124$ sq. units

A second solution, shown at the bottom right on the previous page, is to break the figure vertically. The calculations are

A of left = $8 \times 12 = 96$

A of right = $4 \times 7 = 28$

$A = 96 + 28 = 124$ sq. units

Another solution, shown below, is to find the area of the whole rectangle and to subtract the missing part.

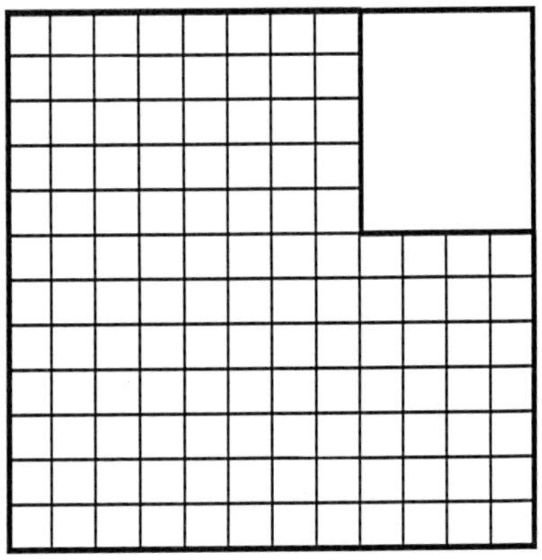

Seeing the whole rectangle and subtracting the missing portion.

A of whole rectangle = $12 \times 12 = 144$

A of missing part = $4 \times 5 = 20$

$A = 144 - 20 = 124$ sq. units

Still another solution is to break the figure into three rectangles. Ask the child which way he thinks is easiest and why. Ask if the answers are the same. [yes]

Worksheet. Give him the second worksheet where he is asked to find the area of a figure two ways and to find several other areas. Stress the importance of writing the equations.

The answers are given below.

1. $\mathbf{A = 5 \times 4 + 3 \times 2 + 5 \times 4}$
   $\mathbf{A = 20 + 6 + 20 = 46}$ **sq. units**
2. $\mathbf{A = 13 \times 4 - 3 \times 2}$
   $\mathbf{A = 52 - 6 = 46}$ **sq. units**
3. $\mathbf{A = 46}$ **sq. units**
4. $\mathbf{A = 38}$ **sq. units**
5. $\mathbf{A = 36}$ **sq. units**
6. $\mathbf{A = 38}$ **sq. units**

Lesson 81

Area Problems

OBJECTIVES
1. To solve area problems
2. To calculate cost

MATERIALS Worksheet 79, "Quick Practice-10"
Worksheet 81, "Area Problems"

WARM-UP Ask the following orally for the child to do mentally, 137 + 37? [174] What is 100 – 37? [63] What is 1½ × 6? [9]

Ask the child to write the multiples of 8 to 80. Then ask her for the 8s facts.

Then ask the child to mentally compute the following given orally: 25 + 85 [110], 99 + 43 [142], and 78 + 78. [156]

Write and ask the child to add mentally: 37 + 85 [122], 98 + 68 [166], and 87 + 53. [140]

What's my Rule? Tell the child you have a game for her that she played last year. Explain she is to say a number less than 25 and you will write that number and another number. After three such numbers, she can guess your rule.

For the first example, use the rule, + 3. So add 3 to the number the child gives. For example, she says 5, you write 5 in the left column and 8 in the right column; if she says 12, you write 12 and 15; if she says 20, you write 20 and 23. See below.

5	8	[+ 3]
12	15	
20	23	

5	10	[× 2]
23	46	
10	20	

If she does not discover the rule with 3 examples, continue with more examples.

For the second game, use multiply by 2. Ask the child to give you numbers; return the number times by 2. See the example above on the right.

ACTIVITIES ***Facts practice.*** Ask the child to take out her Quick Practice sheet and ask her to do the facts in the **bottom** rectangle when you say start. Ask her to correct them. The answers are:

20	**27**	**40**	**12**	**9**	**18**	**28**	**24**	**9**	**16**
54	**27**	**30**	**18**	**72**	**15**	**24**	**21**	**6**	**28**
24	**45**	**8**	**0**	**30**	**12**	**9**	**21**	**16**	**54**
36	**24**	**12**	**32**	**36**	**20**	**63**	**16**	**81**	**14**
18	**72**	**15**	**8**	**12**	**35**	**32**	**63**	**18**	**45**

Bulletin board problem 1. Part 1. Mr. Garcia is making a bulletin board with cork squares. Each square measures 1 foot on a side. He wants the board to be 12 feet long by 6 feet high. How many cork squares does he need? [72]

Review how much a foot is. [length of a ruler] Draw the figure as shown at the top of the next page. Ask the child to draw it in her journal. (She can let each square equal 1 square foot.)

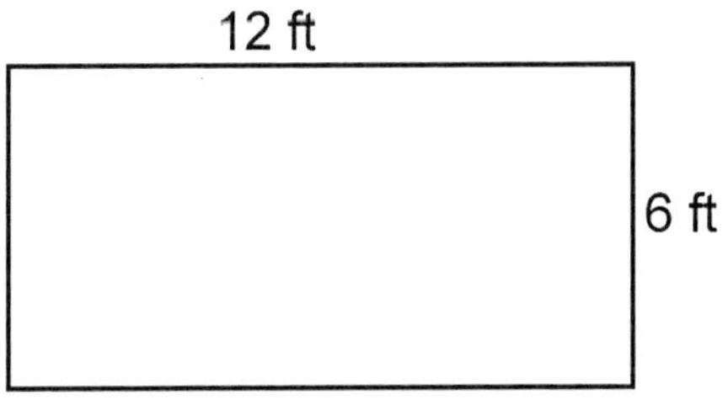

The bulletin board problem drawn on the board.

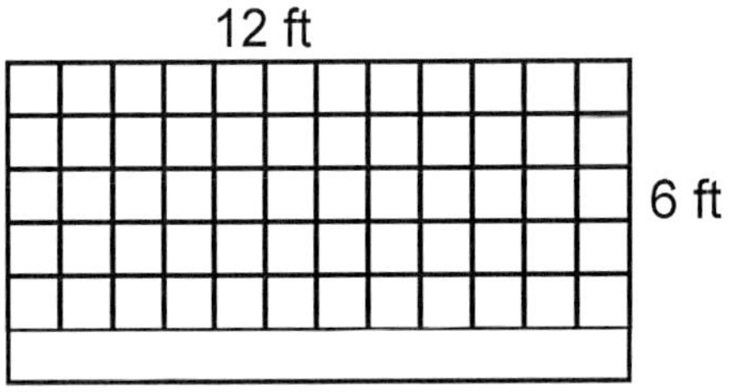

The bulletin board problem drawn in a math journal.

Ask her to find the number of cork squares needed. She may be able to solve it mentally, $10 \times 6 = 60$ and $2 \times 6 = 12$; $60 + 12 = 72$; or she may prefer to multiply on paper.

Part 2. Mr. Garcia needs to know what his bulletin board will cost. Each cork square costs $3. [$216]

Part 3. Mr. Garcia decided his bulletin board is too large. He gives half to another teacher. How many squares will his bulletin board have? [36 squares] What shape can his bulletin board be now? And what is the cost now? [rectangle, $108]

There are several solutions. Ask the child to share her method. Be sure all methods are discussed. Some arrangements are shown below.

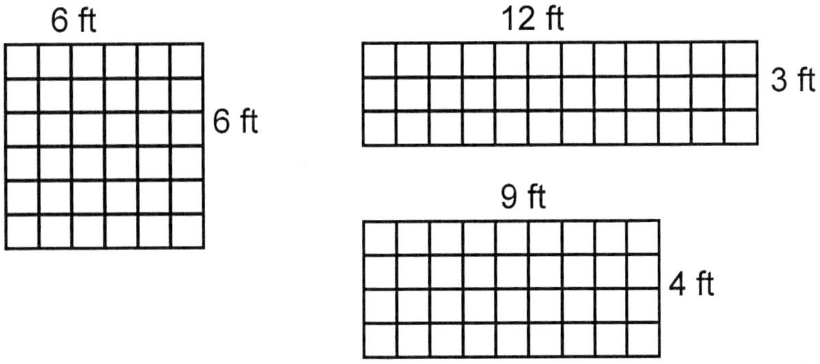

Solutions for using half of the cork squares to build the bulletin board.

Worksheet. The worksheet asks the child to find the area for building a patio, the cost of the patio, amount of fencing, the cost of the fencing, and the total cost. She also will be finding area of a rectangle measuring 2 inches by 2½ inches.

The answers are:

1. **$A = 16 \times 8 = 128$**
2. **$128 \times 6 = \$768$**
3. **$P = 48$ ft**
4. **$480**
5. **$1248**
6. **$A = 81 \times 100 = 8100$ sq ft**
7. **$A = 5$ sq in.**

Lesson 82

Working With Eights

OBJECTIVES
1. To learn the 8s table
2. To learn the facts with 8 as the multiplier

MATERIALS
Worksheet 82, "Quick Practice-11"
Abacus
Worksheet 54, "Multiplication Table"
Worksheet 83, "Working With Eights"

WARM-UP
Ask the child to multiply 9,534 by 6 in his journal. [57,204] The check numbers are (3) × (6) = (0).

Ask, What is 4 × 2 and then doubled? [16] What is 4 × 5 doubled? [40] What is 4 × 9 doubled? [72] What is 4 × 4 doubled? [32] What is 4 × 6 doubled? [48] What is 4 × 7 doubled? [56]

ACTIVITIES
Facts practice. Take out Quick Practice-11 and ask the child to do the facts in the **top** rectangle when you say start. Ask him to correct them. The answers are:

24	54	16	45	15
20	30	24	63	30
36	20	9	35	9
28	24	27	6	45
8	9	21	36	16

54	32	72	14	12
63	18	8	72	24
16	12	28	12	15
40	18	21	27	81
18	32	0	18	12

Ask him to learn strategies and to practice the ones that were incorrect.

Abacus practice. Enter 8 × 6 on an abacus, as shown at the left, and flash it for 2-3 seconds. What is the equation? [8 × 6 = 48] Ask, How did you figure it out? Discuss the several ways. Repeat for the other 8s facts.

8s multiples. Ask the child to write the multiples of 8 in 2 rows as shown below.

8	16	24	32	40
48	56	64	72	80

What patterns do you see with the 8s multiples? All the numbers are even. The ones are counting by 2s backwards (8, 6, 4, 2, 0). The check numbers decrease from 8, 7, . . . , 0, 8. The ones in each row decrease by 2 while the tens increase by 1. What number can be added to the first row to get the second row? [40]

What is 8 × 4? [32] What is 8 × 7? [56] What is 8 × 6? [48] What is 8 × 8? [64] Repeat for other 8s facts; concentrate especially on the 2 new facts, 8 × 7 and 8 × 8.

Comparing 8s to 4s. Ask the child to write the multiples of 4 in a row and the multiples of 8 directly below as shown.

4	8	12	16	20	24	28	32	36	40
8	16	24	32	40	48	56	64	72	80

Then ask what pattern he notices. [The 8s are double the 4s.]

Practice with written symbols. With the 8s written in 2 rows still visible, write 8 × 1 to 8 × 10 in order on the board. Point to the various facts in random order and ask the child to say the answers.

8s on the multiplication table. Ask the child to write the 8s and multiplying by 8s facts on his multiplication table (Worksheet 54). See the table below. How many spaces are left to be filled? [1]

×	1	2	3	4	5	6	7	8	9	10
1	1	2	3	4	5	6	7	8	9	10
2	2	4	6	8	10	12	14	16	18	20
3	3	6	9	12	15	18	21	24	27	30
4	4	8	12	16	20	24	28	32	36	40
5	5	10	15	20	25	30	35	40	45	50
6	6	12	18	24	30	36	42	48	54	60
7	7	14	21	28	35	42		**56**	63	70
8	8	16	24	32	40	48	**56**	**64**	72	80
9	9	18	27	36	45	54	63	72	81	90
10	10	20	30	40	50	60	70	80	90	100

Entering the 8s and the 8s as multipliers.

Note: The recent entries are shown in boldface.

Multiplication trick. Write 12 = 3 × _. Ask the child what goes in the blank.

$$12 = 3 \times 4$$

Ask if he notes what is special about it. [numbers in order 1 to 4]

Then tell him there is another one; write 56 = 7 × _. Ask what completes the equation

$$56 = 7 \times 8$$

He will see the pattern, the numbers 5 to 8.

Worksheet. The worksheet concentrates on 8s. The answers are as follows:

8	8	8	8	8	8	8	8	8	8
× 6	× 2	× 8	× 3	× 5	× 7	× 9	× 1	× 4	× 10
48	**16**	**64**	**24**	**40**	**56**	**72**	**8**	**32**	**80**

2	10	8	6	7	9	4	1	3	5
× 8	× 8	× 8	× 8	× 8	× 8	× 8	× 8	× 8	× 8
16	**80**	**64**	**48**	**56**	**72**	**32**	**8**	**24**	**40**

8 × 5 = **40**	2 × 8 = **16**	a. **56**
8 × 10 = **80**	5 × 8 = **40**	b. **$72**
8 × 8 = **64**	6 × 8 = **48**	c. **240**
8 × 1 = **8**	8 × 8 = **64**	d. **56**
8 × 6 = **48**	7 × 8 = **56**	e. **72**
8 × 3 = **24**	9 × 8 = **72**	f. **96 months**
8 × 7 = **56**	4 × 8 = **32**	
8 × 9 = **72**	10 × 8 = **80**	
8 × 2 = **16**	1 × 8 = **8**	
8 × 4 = **32**	3 × 8 = **24**	

Lesson 83

Multiplying Three Numbers

OBJECTIVES
1. To use the associative law of multiplication
2. To multiply by multiples of 10

MATERIALS
Worksheet 82, "Quick Practice-11"
Worksheet 84, "Multiplying Three Numbers"
Abacus

WARM-UP
Ask the child to solve the following in her journal: 8090 × 6 [48,540] The check numbers are (8) × (6) = (3).

Then ask the child to mentally compute the following given orally: 34 + 89 [123], 99 + 87 [186], and 67 + 66. [133]

Write on the board and ask her to find mentally: 69 + 74 [143], 98 + 91 [189], and 78 + 49. [127]

What's my Rule? Play 2 games of What's my Rule. Ask the child to say three numbers between 2 and 10; each time write 5 times the number. Examples are shown below on the left. For the second game, write one half the number. Examples are below on the right.

10	50	[× 5]
3	15	
7	35	

12	6	[÷ 2]
6	3	
7	$3\frac{1}{2}$	

ACTIVITIES
Facts practice. Ask the child to take out her Quick Practice sheet and ask her to do the facts in the **bottom** rectangle when you say start. Ask her to correct them. The answers are:

64	**24**	**72**	**54**	**35**	**63**	**42**	**18**	**56**	**36**
48	**36**	**15**	**40**	**18**	**24**	**16**	**81**	**56**	**36**
0	**30**	**32**	**20**	**14**	**8**	**16**	**45**	**40**	**21**
8	**72**	**30**	**21**	**24**	**27**	**28**	**45**	**48**	**12**
27	**63**	**32**	**16**	**20**	**28**	**24**	**54**	**18**	**12**

Reviewing terms. Review the term *product.* What do we call the answer when we add? [sum] When we multiply? [product]

Multiplying 3 numbers. Take out the worksheet and ask the child to do the first 3 rows, down to the solid line. The answers are below:

$5 \times 2 \times 4 = \mathbf{40}$	$4 \times 5 \times 3 = \mathbf{60}$	$2 \times 3 \times 6 = \mathbf{36}$	$3 \times 2 \times 10 = \mathbf{60}$
$4 \times 2 \times 5 = \mathbf{40}$	$4 \times 3 \times 5 = \mathbf{60}$	$6 \times 3 \times 2 = \mathbf{36}$	$10 \times 2 \times 3 = \mathbf{60}$
$5 \times 4 \times 2 = \mathbf{40}$	$5 \times 3 \times 4 = \mathbf{60}$	$2 \times 6 \times 3 = \mathbf{36}$	$3 \times 10 \times 2 = \mathbf{60}$

When she has completed it, ask, What patterns do you see? [The answers are the same in each column.] Why is that so? [The numbers are the same, but in a different order.]

Ask the child to show the equations in the first column with her abacus. She is to show: 5 × 2 four times, 4 × 2 five times, and 5 × 4 two times. See the figures at the top of the next page.

When we multiply numbers, does the order make any difference? [no]

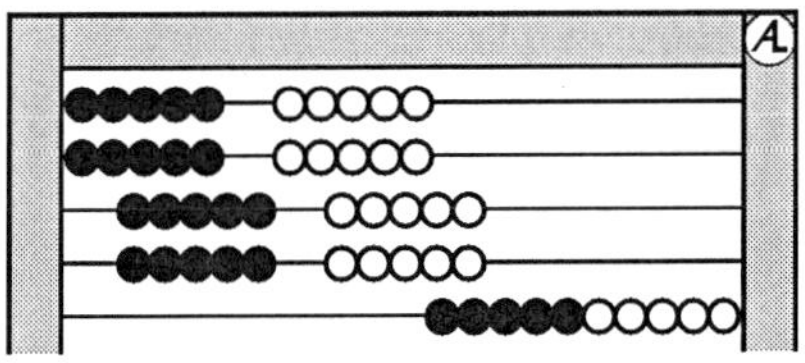

5 × 2 four times

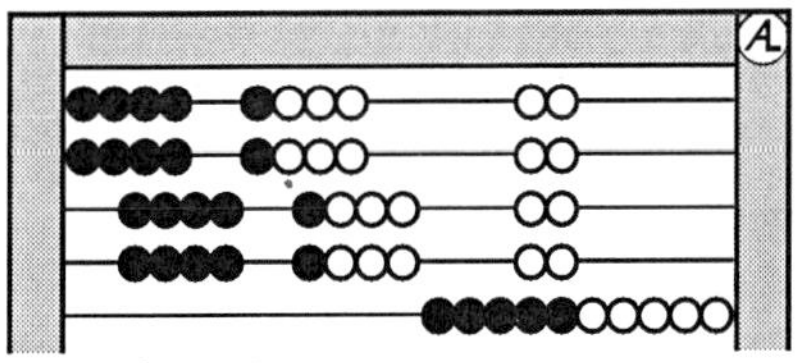

4 × 2 five times

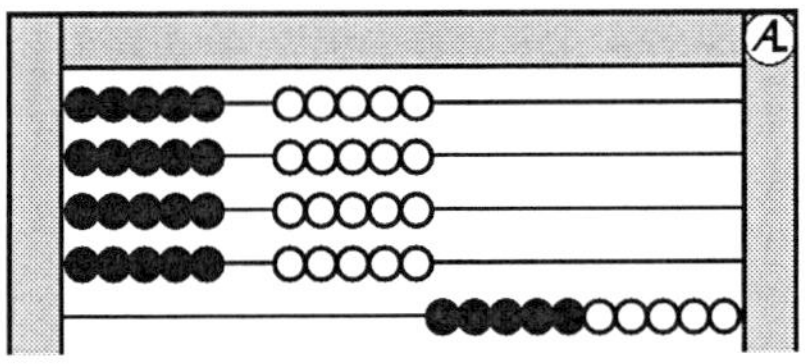

5 × 4 two times

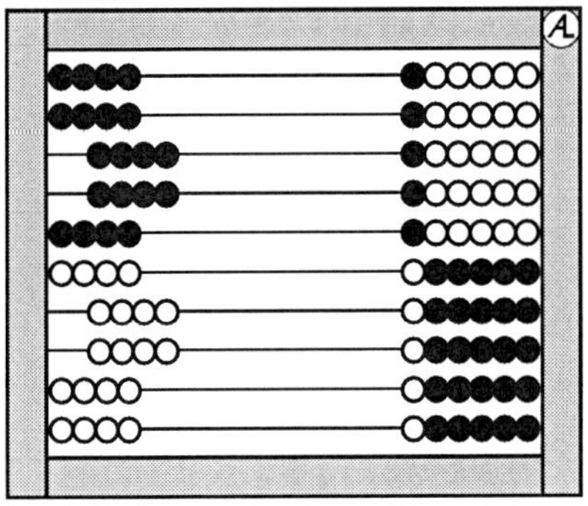

4 × 2 five times another way

Application. Write

$$2 \times 14 \times 5 = [140]$$

How could you do it in your head? Ask for additional solutions until she says to multiply the 2 × 5, getting 10 and multiplying that by 14 to get 140.

Repeat for 5 × 4 × 25. [500] Here the easiest way is to see 4 × 25 is 100–think of 4 quarters in a dollar–then multiply 5 × 100.

Repeat for $2 \times 5 \times 1\frac{1}{2}$ = [15] Here it is easier to multiply the 1½ by the even number, 2, to get 3 and then multiplying by 5, to get 15.

Multiplying by multiples of 10. Write

$$\begin{array}{r} 43 \\ \underline{\times 10} \end{array} \qquad\qquad \begin{array}{r} 43 \\ \underline{\times 20} \end{array}$$

[430, 860] Ask the child to discuss the solutions. Clearly, the second example is twice the first.

Finally, give her this example

$$\begin{array}{r} 404 \\ \underline{\times 30} \\ 12{,}120 \end{array} \qquad\qquad \begin{array}{r} 952 \\ \underline{\times 30} \\ 28{,}560 \end{array}$$

She can think of first multiplying by 3 and then by 10.

Worksheet. Ask the child to complete the worksheet, which has similar problems. The answers are as follows.

100	**40**	**0**	
170	**180**	**80**	
210	**480**	**260**	
600	**700**	**480**	
180	**0**	**4200**	
1360	**850**	**6450**	**1280**
1280	**4680**	**66,360**	**63,600**

Lesson 84

Review

OBJECTIVE 1. To review and practice

MATERIALS Worksheet 85-A and 85-B, "Review" (2 versions)
Cards for playing games

ACTIVITIES ***Review worksheet.*** Give the child about 15 to 20 minutes to do the review sheet. The oral problems to be read are as follows:

$30 \times 7 =$ $\quad$ $72 + 86 =$ $\quad$ $100 - 35 =$

Review Sheet 85-A.

1. Write only the answers to the oral questions. 210 158 65

4. Write only the answers. $2 \times 7 \times 5 =$ 70 $\quad 60 - 31 =$ 29 $\quad 101 + 67 =$ 168

7. Which rectangle below has the greater area? Explain your work.

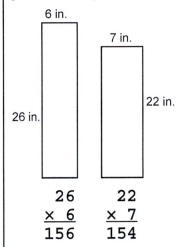

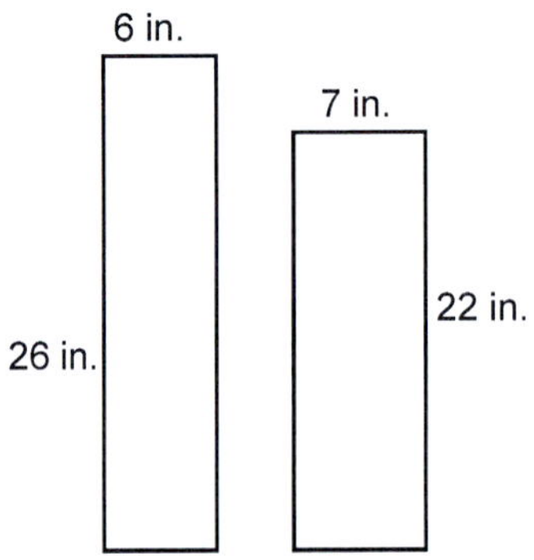

$$\begin{array}{r} 26 \\ \times\ 6 \\ \hline 156 \end{array} \qquad \begin{array}{r} 22 \\ \times\ 7 \\ \hline 154 \end{array}$$

The first rectangle.

11. Which rectangle has the greater perimeter? Show your work.

$P = (26 + 6) \times 2 = 64$

$P = (22 + 7) \times 2 = 58$

The first rectangle is greater.

13. What is the area of the room below in square feet?

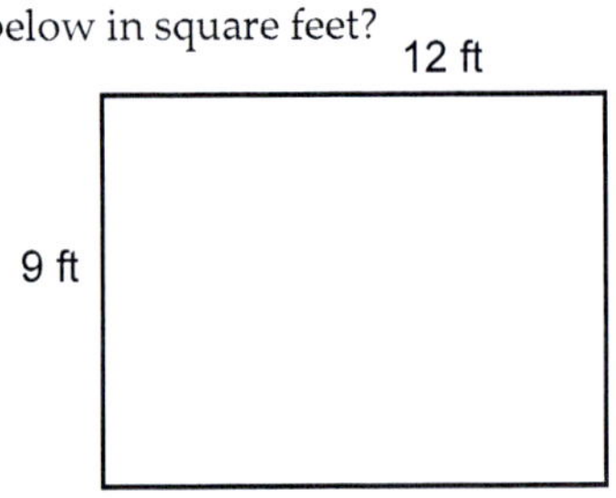

$$\begin{array}{r} 12 \\ \times\ 9 \\ \hline 108 \end{array}$$ sq ft

15. How many tiles are needed to cover the floor? Each tile is 1 square foot. What is the cost? Each tile costs $6.

108 tiles

$$\begin{array}{r} 108 \\ \times\ 6 \\ \hline \$648 \end{array}$$

17.

$$\begin{array}{r} 8{,}579 \\ +19{,}334 \\ \hline 27{,}913 \end{array} \qquad \begin{array}{r} 5{,}031 \\ -4{,}056 \\ \hline 975 \end{array} \qquad \begin{array}{r} 5{,}829 \\ \times\ 8 \\ \hline 46{,}632 \end{array} \qquad \begin{array}{r} 573 \\ \times\ 50 \\ \hline 28{,}650 \end{array}$$

The oral problems to be read to the children are as follows:

$30 \times 8 =$ $\quad 63 + 75 =$ $\quad 100 - 47 =$

Review Sheet 85-B.

1. Write only the answers to the oral questions. **240** **138** **53**

4. Write only the answers. $8 \times 2 \times 5 =$ **80** $\quad 80 - 19 =$ **61** $\quad 106 + 43 =$ **149**

7. Which rectangle below has the greater area? Explain your work.

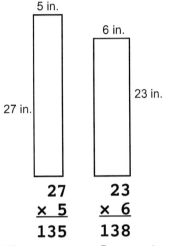

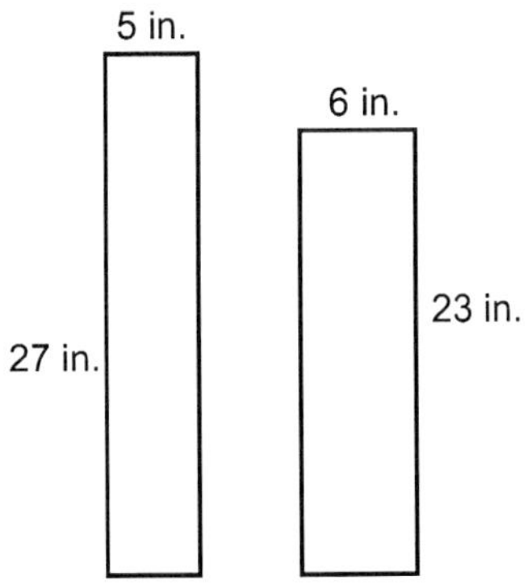

$$\begin{array}{r} 27 \\ \times\ 5 \\ \hline 135 \end{array} \qquad \begin{array}{r} 23 \\ \times\ 6 \\ \hline 138 \end{array}$$

The second rectangle.

11. Which rectangle has the greater perimeter? Show your work.

$P = (27 + 5) \times 2 = 64$

$P = (23 + 6) \times 2 = 58$

The 1st rectangle has greater perimeter.

13. What is the area of the room below in square feet?

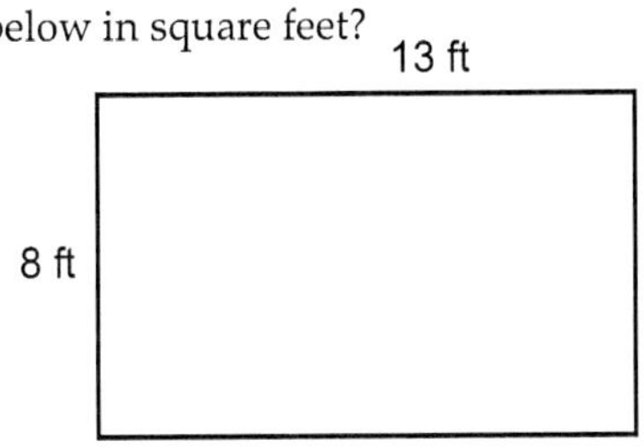

$$\begin{array}{r} 13 \\ \times\ 8 \\ \hline 104 \end{array}$$ **104 sq ft**

15. How many tiles are needed to cover the floor? Each tile is 1 square foot. What is the cost? Each tile costs $9.

104 tiles

$$\begin{array}{r} 104 \\ \times\ 9 \\ \hline \$936 \end{array}$$

17.

$$\begin{array}{r} 9{,}832 \\ +33{,}872 \\ \hline 43{,}704 \end{array} \qquad \begin{array}{r} 3{,}119 \\ -2{,}123 \\ \hline 996 \end{array} \qquad \begin{array}{r} 2{,}487 \\ \times\ 6 \\ \hline 14{,}922 \end{array} \qquad \begin{array}{r} 709 \\ \times\ 80 \\ \hline 56{,}720 \end{array}$$

Note: *Multiples Solitaire* was formerly called *Threesome Solitaire.*

Games. Spend the remaining time playing games.

MULTIPLICATION. Multiples Solitaire (*Math Card Games,* P19).

Lesson 85

Arrays of Cubes

OBJECTIVES 1. To find the number of cubes (volume) in an array
2. To draw arrays of cubes

MATERIALS Worksheet 86, "Quick Practice-12"
Worksheet 87, "Arrays of Cubes"
Cubes, either 1-inch cubes or 1 cm cubes

WARM-UP Ask the child to write in her journal the number of days in 8 years. [365 × 8 = 2,920 days. There will be 2 leap years, so the answer is 2,922 days.] The check numbers are (5) × (8) = (4).

Ask the child to recite and write the multiples of 7. Then ask for the 7s multiplication facts, such as, 7 × 3 = [21] and 7 × 6. = [42]

What's my Rule? Play 2 games of What's my Rule. Ask the child to say three numbers between 2 and 10; each time write 8 times the number. Examples are shown below on the left. For the second game, write 2 less than the number. Examples are on the right.

3	24	[× 8]
8	64	
5	40	

10	8	[−2]
6	4	
2	0	

ACTIVITIES ***Facts practice.*** Take out Quick Practice-12 and ask the child to do the facts in the **top** rectangle when you say start. Ask her to correct them. The answers are:

18	**40**	**45**	**63**	**24**	**72**	**32**	**8**	**81**	**45**
32	**36**	**8**	**16**	**54**	**16**	**36**	**30**	**12**	**20**
63	**24**	**27**	**72**	**35**	**0**	**27**	**56**	**24**	**21**
24	**48**	**56**	**14**	**28**	**40**	**18**	**21**	**48**	**12**
18	**20**	**15**	**30**	**64**	**42**	**16**	**54**	**28**	**36**

Building 3D arrays. Take out the cubes and ask the child to make an array of 2 by 3. Generally, the first number indicates how many in a row and the second number gives the number of rows. See the left figure below.

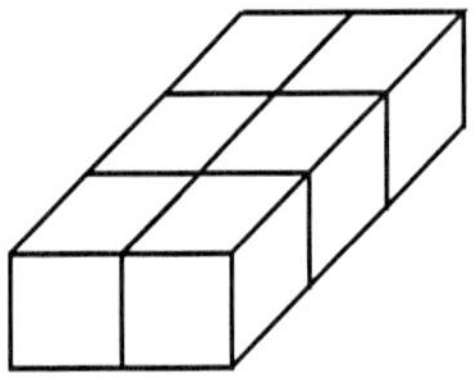

A 2 × 3 array.

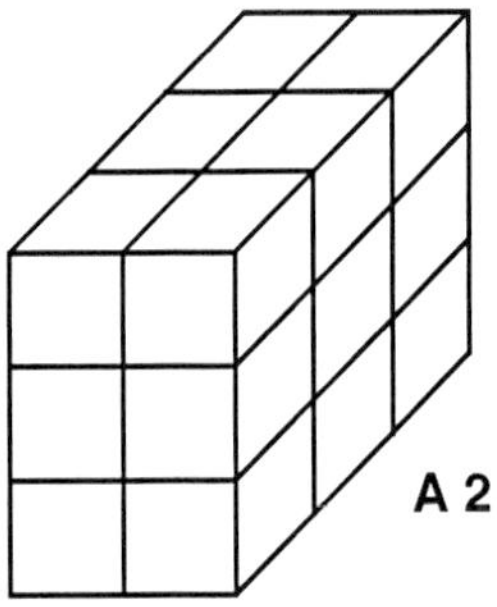

A 2 × 3 × 3 array.

Next ask her to change the array to 2 by 3 by 3. What do you think the second 3 means? [3 layers] See the figure on the right above.

How many cubes are in the array? [18 cubes] Ask her to write the equation:

$$2 \times 3 \times 3 = 18$$

Ask her to turn the array on the side as shown below. Then ask her to write the equation, observing the same order for the numbers. [$3 \times 3 \times 2 = 18$] See the left figure below.

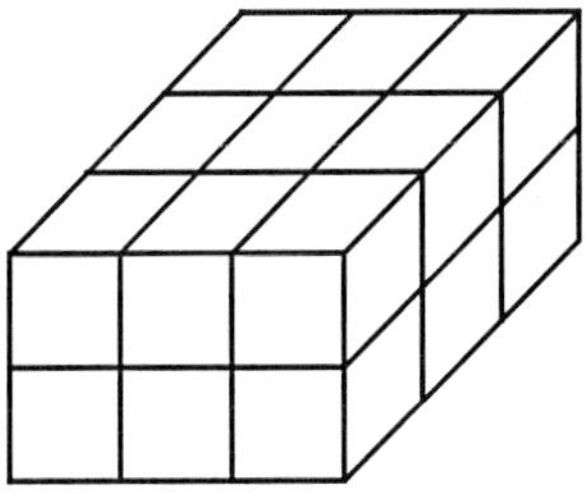

A 3 × 3 × 2 array.

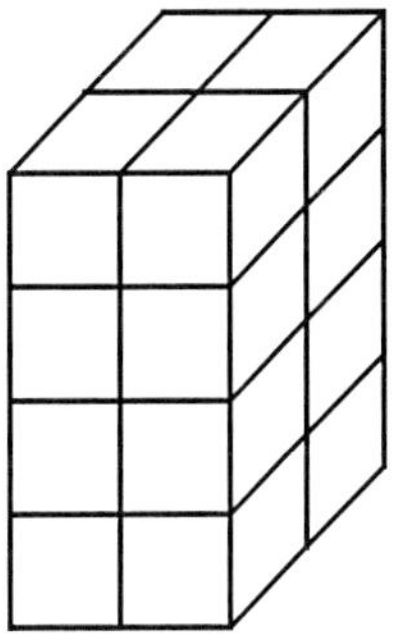

A 2 × 2 × 4 array.

Repeat for other arrays, such as, $2 \times 2 \times 4$. See the right figure above.

Worksheet. The worksheet has 6 figures for the child to write the array equations and find the total number of cubes. It also asks her to draw freehand several arrays. The answers are:

1.
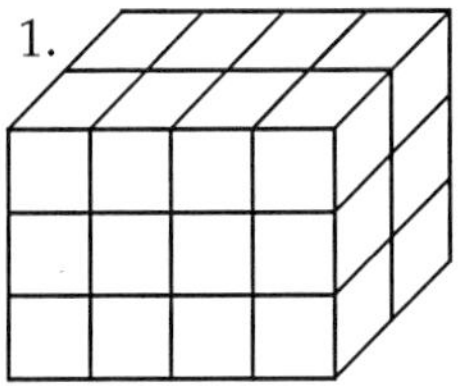

4 × 2 × 3 = 24 cubes

2.
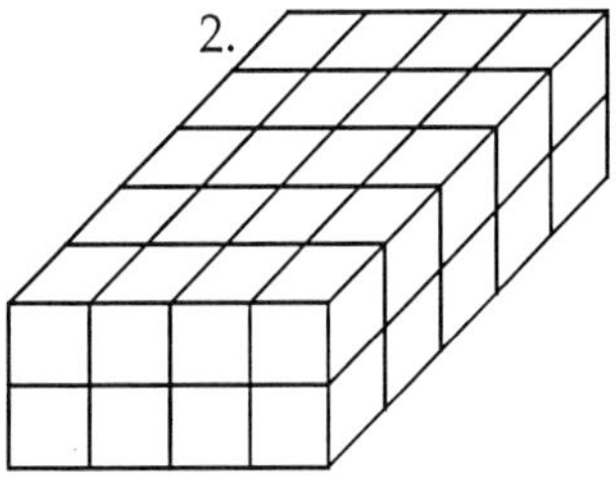

4 × 5 × 2 = 40 cubes

3.
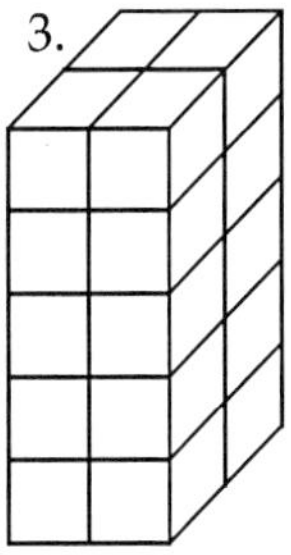

2 × 2 × 5 = 20 cubes

4.
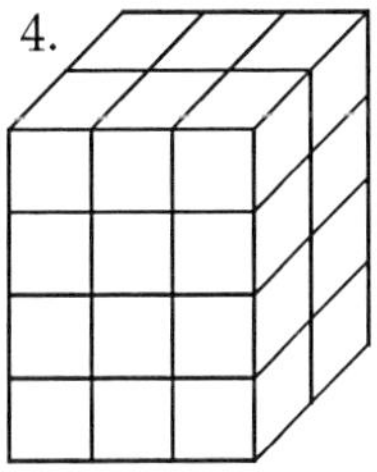

3 × 2 × 4 = 24 cubes

5.
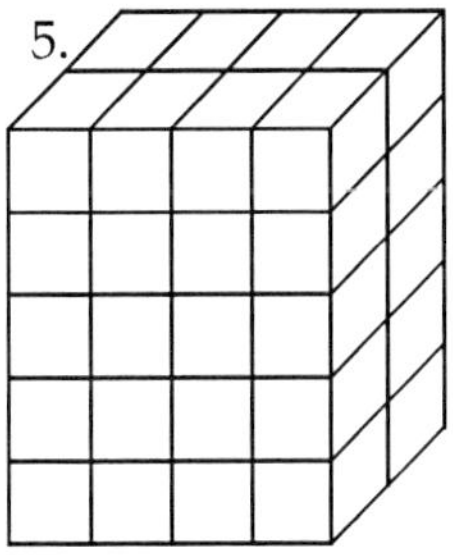

4 × 2 × 5 = 40 cubes

6.
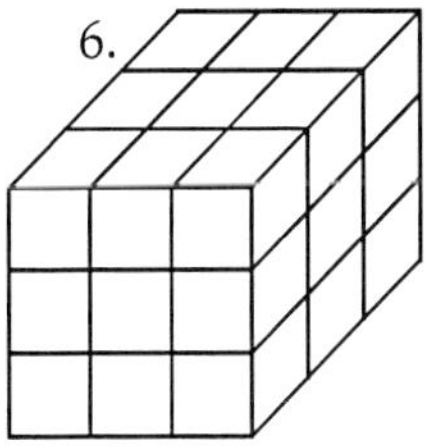

3 × 3 × 3 = 27 cubes

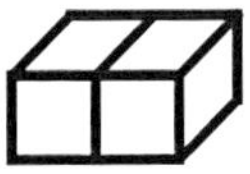

$2 \times 1 \times 1$

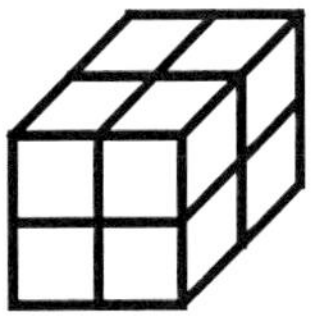

$2 \times 2 \times 2$

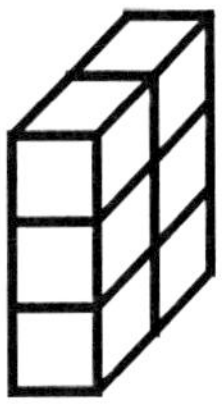

$1 \times 2 \times 3$

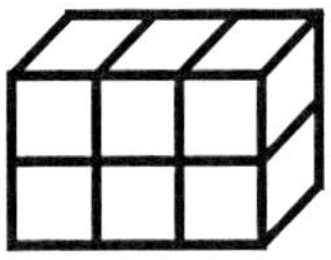

$3 \times 1 \times 2$

Lesson 86

Working With Sevens

OBJECTIVES
1. To learn the 7s table
2. To learn the facts with 7 as the multiplier

MATERIALS
Worksheet 86, "Quick Practice-12"
Abacus
Worksheet 54, "Multiplication Table"
Worksheet 88, "Working With Sevens"

WARM-UP
Ask the child to multiply 487 by 8 in his journal. [3896] The check numbers are (1) × (8) = (8).

What's my Rule? Play 2 games of What's my Rule. Ask the child to say three numbers between 2 and 10; each time write 3 less than the number; examples are below on the left. For the second game, write 5 less than the number; examples are on the right.

Note: This game employs basic algebraic thinking.

10	7	[−3]
3	0	
7	4	

3	−2	[−5]
6	1	
7	2	

ACTIVITIES
Facts practice. Ask the child to take out his Quick Practice sheet and ask him to do the facts in the **bottom** rectangle when you say start. Ask him to correct them. The answers are:

8	48	40	24	14	18	27	32	30	54
12	35	72	12	32	63	72	24	27	36
30	28	54	0	20	21	42	18	36	16
64	28	8	56	18	15	56	21	45	24
45	20	63	16	81	36	16	24	40	48

Ask him to learn strategies and to practice the ones that were incorrect.

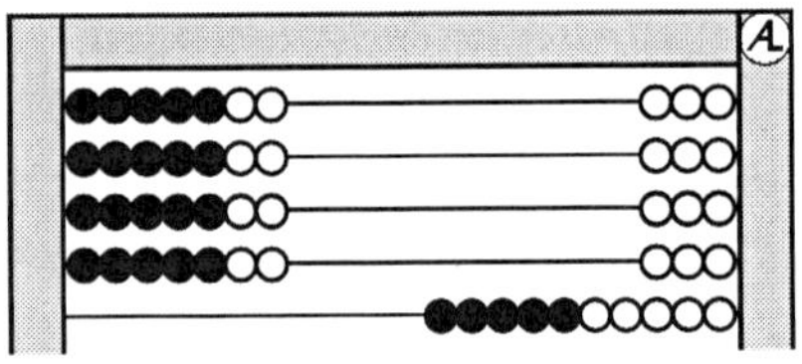

Abacus practice. Enter 7 × 4 on the abacus, as shown on the left. How did you figure it out? One way is to see 5 × 4 = 20 and add the remaining 8. Enter other 7s facts and flash the abacus for 2-3 seconds and ask for the equations.

7s multiples. Ask the child to write the multiples of 7 in three columns as shown below.

7	14	21
28	35	42
49	56	63
70		

Ask, What patterns do you see with the 7s multiples? Are the numbers even or odd? [both, they alternate] The ones increase by 1, starting at the right top and going down the columns going from right to left. (21, 42, 63, 14, and so forth). This means that no two 7 facts have the same number in the ones place. Also within each row, the tens increase by 1 (07, 14, 21, and 28, 35, 42).

Ask him to look at the multiples listing. Then ask, What is 7 × 3? [21], 7 × 6? [42], 7 × 9? [63], 7 × 7? [49], and so forth. Concentrate especially on the new fact, 7 × 7.

Ask, What happens when you multiply 7 by an even number? [even number] What happens when you multiply 7 by an odd number? [odd number]

Practice with written symbols. With the 7s written in three columns still visible, write 7 × 1 to 7 × 10 in order . Point to the various facts in random order and the child to say the answers.

7s on the multiplication table. Ask the child to write the 7s and multiplying by 7s facts on his multiplication table (Worksheet 54). See the table below.

×	1	2	3	4	5	6	7	8	9	10
1	1	2	3	4	5	6	7	8	9	10
2	2	4	6	8	10	12	14	16	18	20
3	3	6	9	12	15	18	21	24	27	30
4	4	8	12	16	20	24	28	32	36	40
5	5	10	15	20	25	30	35	40	45	50
6	6	12	18	24	30	36	42	48	54	60
7	7	14	21	28	35	42	**49**	56	63	70
8	8	16	24	32	40	48	56	64	72	80
9	9	18	27	36	45	54	63	72	81	90
10	10	20	30	40	50	60	70	80	90	100

Entering the 7s the 7s as multipliers, actually only 7 × 7.

Note: The last entry is shown in boldface.

Note: Tell him to keep his multiplication table for another activity.

Worksheet. The answers to the worksheet follow:

7	7	7	7	7	7	7	7	7	7
× 4	× 3	× 7	× 9	× 1	× 6	× 2	× 5	× 8	× 10
28	**21**	**49**	**63**	**7**	**42**	**14**	**35**	**56**	**70**
7	10	3	6	5	9	1	4	8	2
× 7	× 7	× 7	× 7	× 7	× 7	× 7	× 7	× 7	× 7
49	**70**	**21**	**42**	**35**	**63**	**7**	**28**	**56**	**14**

7 × 10 = **70**	2 × 7 = **14**	a. **42 days**
7 × 7 = **49**	9 × 7 = **63**	b. **$51**
7 × 9 = **63**	5 × 7 = **35**	c. **84 in.**
7 × 1 = **7**	10 × 7 = **70**	d. **420 min**
7 × 2 = **14**	1 × 7 = **7**	e. **168 hr**
7 × 4 = **28**	6 × 7 = **42**	
7 × 5 = **35**	7 × 7 = **49**	
7 × 3 = **21**	3 × 7 = **21**	
7 × 6 = **42**	4 × 7 = **28**	
7 × 8 = **56**	8 × 7 = **56**	

Lesson 87

Seeing Patterns

OBJECTIVE 1. To continue a pattern, both geometrically and numerically

MATERIALS Worksheet 89, "Quick Practice-13"
Worksheet 90, "Seeing Patterns"
Tiles, optional

WARM-UP Ask the child to write the following problem in her journal: 9999 × 7. [69,993] The check numbers are (0) × (7) = (0).

What's my Rule? Play 2 games of What's my Rule. Ask for 3 numbers between 2 and 10. For the first game write 2 times the number. Examples are shown below on the left. For the second game, double the number and add 1. Examples are on the right.

> **Note:** The second game will be harder for the child. It may be a challenge for the more advanced child.

8	16	[× 2]
3	6	
7	14	

8	17	[× 2 + 1]
3	7	
7	15	

ACTIVITIES ***Facts practice.*** Take out Quick Practice-13 and ask the child to do the facts in the **top** rectangle when you say start. Ask her to correct them. The answers are:

63	**28**	**35**	**30**	**28**	**21**	**20**	**48**	**72**	**40**
27	**56**	**30**	**12**	**54**	**36**	**72**	**16**	**42**	**14**
35	**32**	**63**	**24**	**42**	**56**	**24**	**0**	**45**	**40**
21	**20**	**25**	**32**	**7**	**24**	**49**	**64**	**7**	**18**
81	**14**	**54**	**27**	**16**	**36**	**48**	**36**	**45**	**18**

Extending pattern 1. Layout or draw the first three terms of the following pattern as shown below on the left.

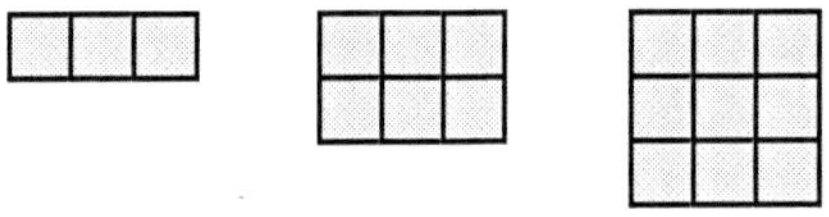

The first 3 terms of a pattern.

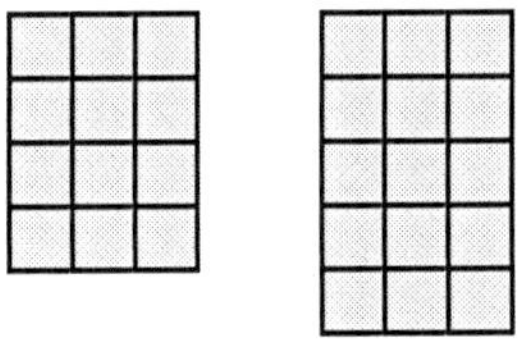

The next 2 terms of the pattern.

Ask the child to extend the pattern one more term and to explain it. [added 3 more, or built the next multiple of 3] See above on the right. Ask her to extend it one more term.

Next draw a table and explain that the information can be written in a table. Ask the child how many squares are in each term and write it in the appropriate space. See below.

Term	1	2	3	4	5
Number of squares	3	6	9	**12**	**15**

How many squares are in term 6? [18] Term 10? [30] Term 100? [300]

Extending pattern 2. Now draw the first three terms of the pattern as shown below. Ask the child to draw the next 2 terms and to complete the table, both shown below.

> **Note:** Tiles can be used rather than triangles.

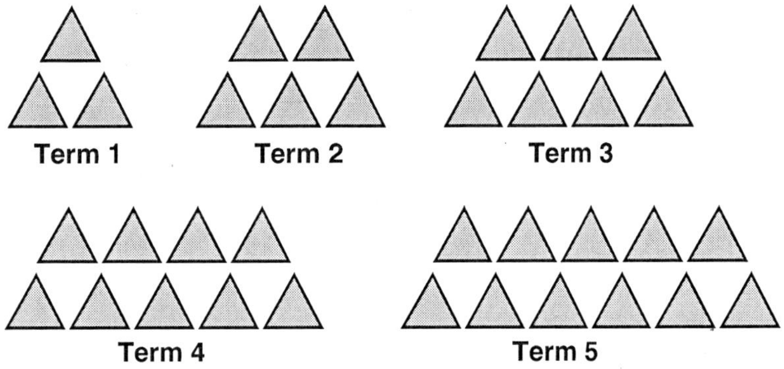

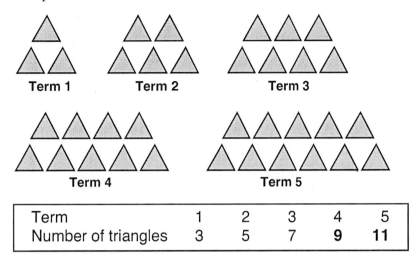

Term	1	2	3	4	5
Number of triangles	3	5	7	**9**	**11**

<u>How can you find the number of triangles without counting, by looking only at the term number?</u> [add the term number with the next higher term number, or multiply the term number by 2 and add 1]

<u>How many triangles are in term 6?</u> [13] <u>Term 10?</u> [21] <u>Term 100?</u> [201] <u>Term 200?</u> [401] <u>What term has 15 triangles?</u> [7] <u>What term has 41 triangles?</u> [20]

Worksheet. For her written work, the worksheet asks the child to extend 2 patterns 2 more terms and to complete the table. The answers are as follows.

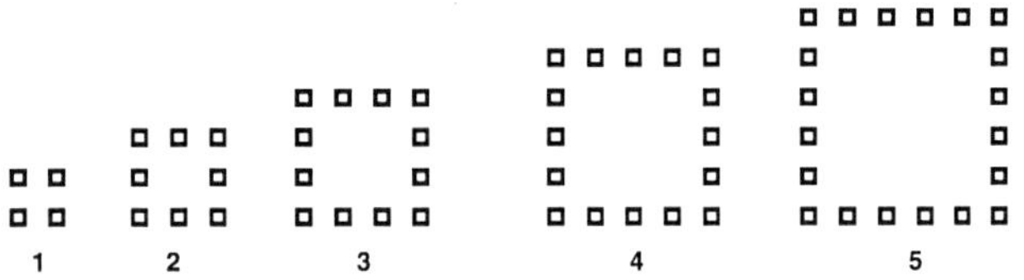

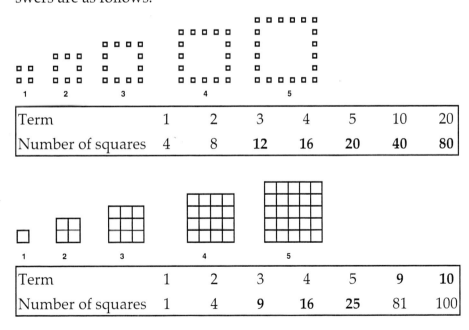

Term	1	2	3	4	5	10	20
Number of squares	4	8	**12**	**16**	**20**	**40**	**80**

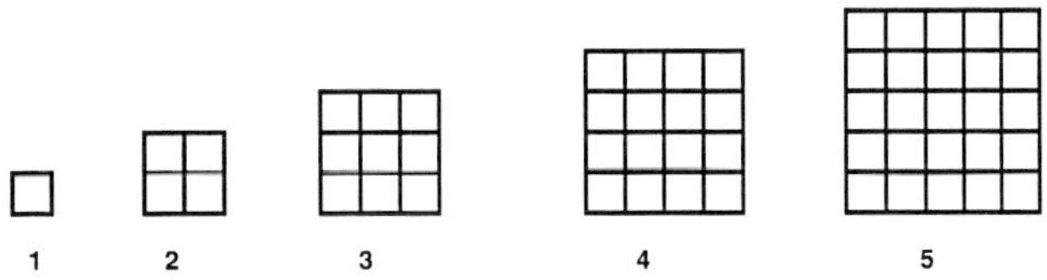

Term	1	2	3	4	5	**9**	**10**
Number of squares	1	4	**9**	**16**	**25**	81	100

Lesson 88

Patterns With Squares

OBJECTIVES 1. To discover some interesting patterns regarding squares
2. To learn the term *perfect square*

MATERIALS Worksheet 89, "Quick Practice-13"
Worksheet 91, "Patterns With Squares"
Tiles
Calculator

WARM-UP Ask the child to multiply 6,926 by 7 in his journal. [48,482] The check numbers are (5) × (7) = (8).

What's my Rule? Play 2 games of What's my Rule. Ask for 3 numbers between 2 and 10. For the first game write 2 times the number minus 1. Examples are below on the left. For the second game, double the number and add 2. Examples are on the right.

8	15	[× 2 − 1]
2	3	
7	13	

8	18	[× 2 + 2]
3	8	
5	12	

ACTIVITIES ***Facts practice.*** Ask the child to take out his Quick Practice sheet and ask him to do the facts in the **bottom** rectangle when you say start. Ask him to correct them. The answers are:

54	**27**	**42**	**16**	**28**	**45**	**16**	**24**	**63**	**35**
36	**35**	**24**	**48**	**20**	**30**	**18**	**7**	**48**	**24**
12	**64**	**81**	**49**	**32**	**20**	**18**	**40**	**14**	**28**
45	**0**	**40**	**56**	**27**	**32**	**36**	**72**	**42**	**21**
54	**14**	**36**	**72**	**30**	**63**	**25**	**7**	**56**	**21**

Sum of consecutive odds. Take out the worksheet. Ask the child to extend the first pattern and fill in the table. See below.

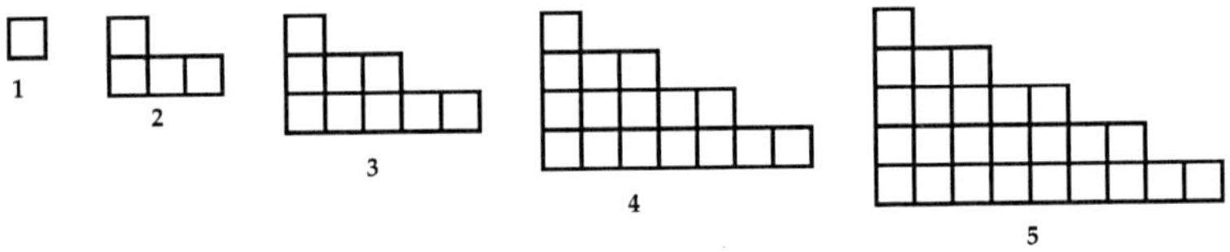

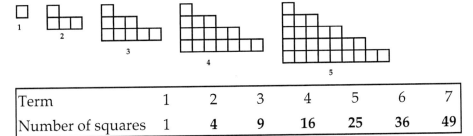

Term	1	2	3	4	5	6	7
Number of squares	1	**4**	**9**	**16**	**25**	**36**	**49**

The pattern and table from the worksheet.

<u>What is the pattern?</u> [consecutive odd numbers] <u>Do you notice anything is special about the number of squares.?</u> [They are the term squared!] If desired, remind him of the squares in the previous lesson.

Ask him to transform a row of consecutive odd numbers into squares. Choose the 4th or higher term and ask him to first make the pattern and then to rearrange it into a square. It is best done in an organized manner as shown on the next page for 5 rows.

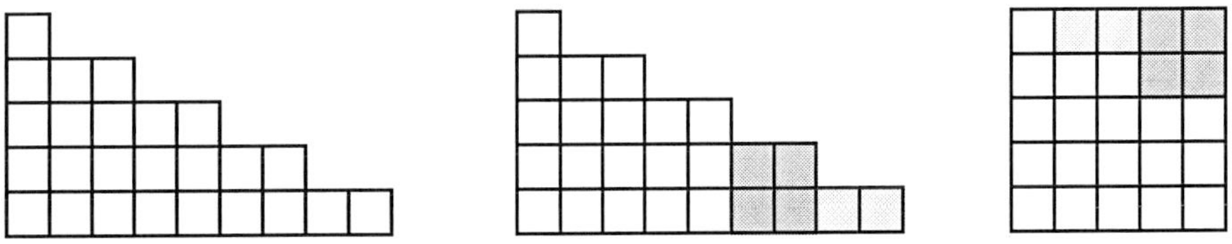

Rearranging 5 rows of consecutive odd squares into 5 squared.

Note: The phrase *perfect square* is a mathematical term.

Perfect squares. Explain that a number that can be made into a square using that number of tiles is called a *perfect square.* Is 9 a perfect square? [yes] Is 16? [yes] Is 10? [no] Is 1? [yes] Is 2? [no] Ask him to name some other perfect squares.

Perfect squares on the calculator. Tell the child there is a way to use a calculator to check if a number is a perfect square: enter the number and press the square root symbol (a check mark with a line). If there are no digits after the decimal point, it is a perfect square; if there are digits after the decimal point, it is not.

For example, to see if 9 is a perfect square,

Press 9 then [√] (3 will show, indicating a perfect square.)

Next ask them to see if 10 is a perfect square.

Press 10 then [√] (3.162276 will show, not a perfect square.)

Ask him to try it with 81 [9, yes], 100 [10, yes], 1000 [no], 1936 [44, yes], and other numbers of his choosing.

Ask him to add the odd numbers from 1 to 15 on the calculator. [64] Is it a perfect square? [yes, 8 squared] Ask him to add the odd numbers from 1 to any odd number of their choosing and to check if it is a perfect square. [yes]

Consecutive numbers. Ask the child to draw 2 more terms for the second pattern on the worksheet and to complete the table, both of which are shown below.

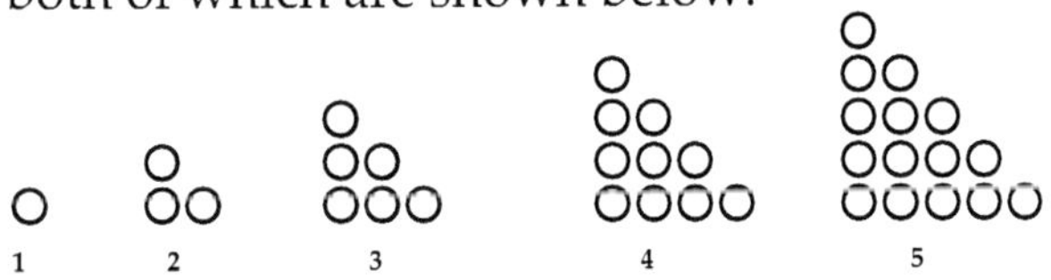

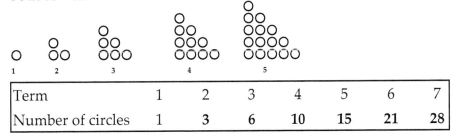

Term	1	2	3	4	5	6	7
Number of circles	1	**3**	**6**	**10**	**15**	**21**	**28**

When he has finished, ask Are any of the numbers squares? [only 1] Then ask him to add any 2 consecutive numbers in the table. Are they squares? [yes] For example 1 + 3 [4], 3 + 6 [9], or 21 + 28. [49] Ask him to look at the pattern and figure out why that is so. [The circles fit together to create a square.]

Worksheet. Ask him to complete the worksheet. Answers are:

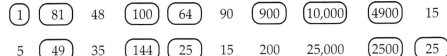

(1)	(81)	48	(100)	(64)	90	(900)	(10,000)	(4900)	15
5	(49)	35	(144)	(25)	15	200	25,000	(2500)	(25)

Lesson 89

A Squares Pattern

OBJECTIVE 1. To learn the pattern that $n^2 = (n-1)(n+1) + 1$

MATERIALS Worksheet 92, "Quick Practice-14"
Worksheet 93, "A Squares Pattern"
1-inch tiles
Calculator

> **Note:** Worksheets 92 and 96 will be used again. You may want to use sheet protectors.

WARM-UP Ask the child to write the following problem in her journal: 8,097 × 7 [56,679] The check numbers are (6) × (7) = (6).

Continue my Rule. The game of Continue my Rule is similar to What's my Rule. However once the child has discovered the rule, she continues it with several more cases. The rule here is the number multiplied by itself.

3	5	1	10	7	8
9	25	1	[100]	[49]	[64]

ACTIVITIES ***Facts practice.*** Take out Quick Practice-14 and ask the child to do the facts in the **top** rectangle when you say start. Ask her to correct them. The answers are:

> **Note:** The answers are also found on Worksheet 92-Key.

54	**49**	**72**	**24**	**48**	**45**	**28**	**64**	**45**	**36**
12	**30**	**42**	**9**	**48**	**27**	**63**	**72**	**28**	**24**
18	**24**	**15**	**36**	**20**	**81**	**40**	**32**	**27**	**32**
40	**18**	**21**	**20**	**15**	**21**	**35**	**18**	**56**	**24**
36	**54**	**30**	**35**	**42**	**56**	**63**	**12**	**25**	**16**

Finding the pattern. Write the following:

2 + 2 = [4]	4 + 4 = [8]	5 + 5 = [10]
3 + 1 = [4]	5 + 3 = [8]	6 + 4 = [10]

<u>What is the pattern in each set of equations?</u> [The sum is the same. One addend is 1 less and the other addend is 1 more.]

Now write

2 × 2 = [4]	5 × 5 = [25]	6 × 6 = [36]	7 × 7 = [49]
3 × 1 = [3]	6 × 4 = [24]	5 × 7 = [35]	6 × 8 = [48]

<u>Is this pattern the same as the one above?</u> [One number is one more and one is 1 less in the second row, but the product is always 1 less.]

Understanding the pattern. Give about 25 tiles to the child. First ask her to make a 4 × 4 with the tiles. Then ask her change it into 3 × 5. See the figures below.

Then ask her to start with a 5 × 5 and build 4 × 6. See the figures on the next page.

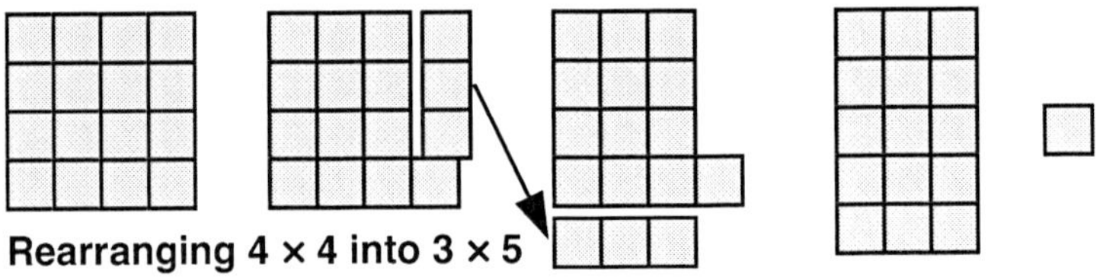

Rearranging 4 × 4 into 3 × 5 with 1 left over.

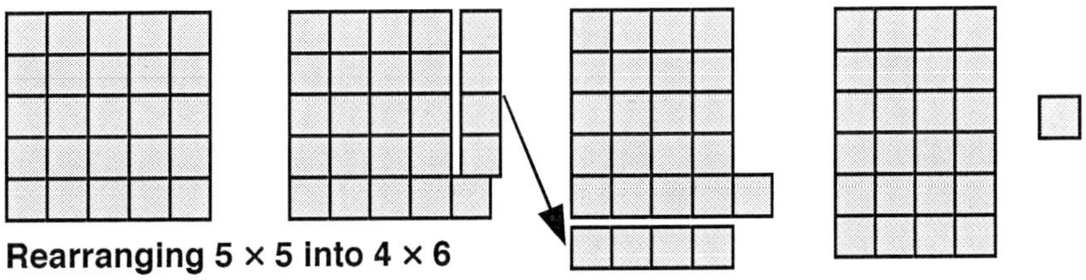

Rearranging 5 × 5 into 4 × 6 with 1 left over.

Squaring on a calculator. Tell the child that multiplying a number by itself is called *squaring* the number. Ask what is 3 squared [9], 5 squared [25], 9 squared. [81]

Take out the calculator. Tell the child there is a special way of squaring a number on a calculator. The procedure is as follows to square 6.

Press 6 [×] [=] (The answer 36 will show.) Or

Press 6 [×] [M+] (The answer 36 will be added to memory.)

Ask her to find 29 squared [841], 345 squared [119,025], and 689. [474,721]

Applying the new strategy. What is 10 × 10? [100] What 2 numbers multiplied together will give 99? [9 × 11] What is 20 × 20? [400] What two numbers multiplied together will give 399? [19 × 21] Repeat for 30 × 30. [900: 29 × 31 = 899]

Worksheet. The worksheet explains how to write 2 as an exponent. A calculator will be needed. The solutions are below:

$5^2 = \mathbf{25}$ $4^2 = \mathbf{16}$ $2^2 = \mathbf{4}$

$1^2 = \mathbf{1}$ $8^2 = \mathbf{64}$ $10^2 = \mathbf{100}$

$3^2 = \mathbf{9}$ $9^2 = \mathbf{81}$ $7^2 = \mathbf{49}$

$6^2 = \mathbf{36}$ $0^2 = \mathbf{0}$ $8^2 = \mathbf{64}$

3 × 3 = **9** 9 × 9 = **81**
4 × 2 = **8** **10 × 8 = 80**

4 × 4 = **16** 8 × 8 = **64**
3 × 5 = 15 **7 × 9 = 63**

×	1	2	3	4	5	6	7	8	9	10
1			3							
2		4		8						
3	3		9		15					
4		8		16		24				
5			15		25		35			
6				24		36		48		
7					35		49		63	
8						48		64		80
9							63		81	
10								80		

17 × 17 = **289** 15 × 15 = **225** 11 × 11 = **121** 13 × 13 = **169**
18 × 16 = **288** 16 × 14 = **224** 12 × 10 = **120** 14 × 12 = **168**

20 × 20 = **400** 29 × 29 = **841** 30 × 30 = **900** 40 × 40 = **1600**
19 × 21 = **399** 28 × 30 = **840** 29 × 31 = **899** 39 × 41 = **1599**

Lesson 90

Review

OBJECTIVE 1. To review and practice

MATERIALS Worksheet 94-A and 94-B, "Review" (2 versions)
Cards for playing games

ACTIVITIES ***Review worksheet.*** Give the child about 15 to 20 minutes to do the review sheet. The oral problems to be read are as follows:

3 squared = 83 + 87 = 87 – 38 =

Review Sheet 94-A.

1. Write only the answers to the oral questions. **9** **170** **49**

4. Write only the answers. 9^2 = **81** 200 – 41 = **159** 146 + 39 = **185**

7. Find the number of cubes in the array. Write the equation.

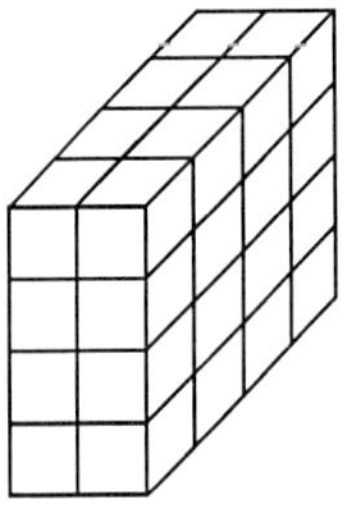

2 × 4 × 4 =

32 cubes

8. Use a pattern to find the products.

7 × 7 = **49**
8 × 6 = **48**

27 × 27 = 729
28 × 26 = **728**

9. Draw the fourth term and complete the table.

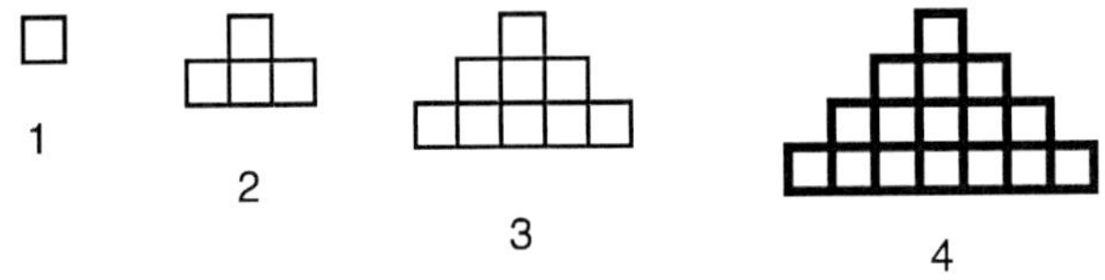

Term	1	2	3	4	5	6	10
Number of squares	1	4	**9**	**16**	**25**	**36**	**100**

16. How many hours are in a week?

$$\begin{array}{r} 24 \\ \times 7 \\ \hline 168 \end{array}$$

17. Jamie has 9 quarters, 8 dimes, and 12 pennies. How much money is that?

9 quarters = 9 × .25 = $2.25
8 dimes = 8 × .10 = $0.80
12 pennies = 12 × .01 = $0.12
$3.17

19.

$$\begin{array}{r} 3,916 \\ 3,682 \\ +\ 3,405 \\ \hline 11,003 \end{array} \qquad \begin{array}{r} 4,000 \\ -\ 2,098 \\ \hline 1,902 \end{array} \qquad \begin{array}{r} 4,157 \\ \times 7 \\ \hline 29,099 \end{array} \qquad \begin{array}{r} 623 \\ \times 20 \\ \hline 12,460 \end{array}$$

The oral problems to be read to the child are as follows:

$30 \times 8 =$ $\quad$ $63 + 75 =$ $\quad$ $100 - 47 =$

Review Sheet 94-B.

1. Write only the answers to the oral questions. **240** **138** **53**

4. Write only the answers. $8^2 =$ **64** $\quad$ $200 - 29 =$ **171** $\quad$ $153 + 28 =$ **181**

7. Find the number of cubes in the array. Write the equation.

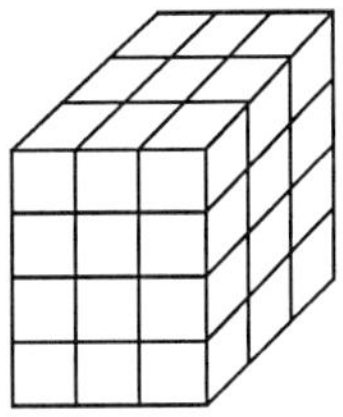

3 × 3 × 4 =

36 cubes

8. Use a pattern to find the products.

$7 \times 7 =$ **49**

$6 \times 8 =$ **48**

$28 \times 28 = 784$

$29 \times 27 =$ **783**

9. Draw the fourth term and complete the table.

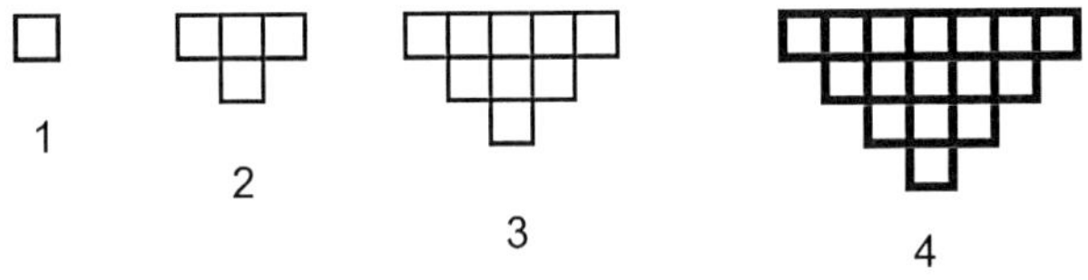

Term	1	2	3	4	5	6	10
Number of squares	1	4	**9**	**16**	**25**	**36**	**100**

16. How many minutes are in a day?

```
  24
× 60
1440
```

17. Jordan has 7 quarters, 9 dimes, and 13 pennies. How much money is that?

```
7 quarters = 7 × .25    = $1.75
9 dimes = 9 × .10       = $0.90
13 pennies = 13 × .01   = $0.13
                          $2.78
```

19.

$$\begin{array}{r} 5{,}829 \\ 6{,}868 \\ +\ 5{,}309 \\ \hline \mathbf{18{,}006} \end{array} \qquad \begin{array}{r} 6{,}000 \\ -\ 3{,}076 \\ \hline \mathbf{2{,}924} \end{array} \qquad \begin{array}{r} 6{,}482 \\ \times\ 7 \\ \hline \mathbf{45{,}374} \end{array} \qquad \begin{array}{r} 752 \\ \times\ 20 \\ \hline \mathbf{15{,}040} \end{array}$$

Games. Spend the remaining time playing games.

MULTIPLICATION. Square Memory (P21) and Crazy Squares (P23) *(Math Card Games).*

Lesson 91

Continuing the Pattern

OBJECTIVE 1. To analyze and continue numerical patterns

MATERIALS Worksheet 92, "Quick Practice-14"
Worksheet 95, "Continuing the Pattern"
Calculator

WARM-UP Ask the child to write the following problem in her journal: 2423 – 549. [1874] The check numbers are (0) + (2) = (2).

Tell the child that the opposite of squaring a number is find its *square root.* The square root of 9 is 3 because 3 × 3 = 9 and the square root of 36 is 6. What is the square root of 100? [10] What is the square root of 4? [2] What is the square root of 49? [7]

ACTIVITIES ***Facts practice.*** Ask the child to take out her Quick Practice sheet and ask her to do the facts in the **bottom** rectangle when you say start. Ask her to correct them. The answers are:

72	**24**	**64**	**54**	**12**	**24**	**49**	**48**	**28**	**63**
18	**54**	**20**	**56**	**35**	**42**	**27**	**20**	**24**	**32**
15	**36**	**30**	**12**	**28**	**32**	**36**	**42**	**15**	**36**
45	**16**	**21**	**40**	**35**	**9**	**18**	**48**	**56**	**25**
18	**63**	**40**	**21**	**45**	**72**	**27**	**30**	**81**	**24**

Multiples patterns. Write the following pattern and ask the child for the next 3 terms.

2 4 6 8 _ _ _

This pattern is obvious with the solution given below.

2 4 6 8 10 12 14

To help her analyze other patterns ask her to find the differences between the terms. How much is the difference between the first and second terms? [2] Write it as shown below between the terms. How much is the difference between the second and third terms? [2] How much is the difference between the third and fourth terms? [2]. Continue with all the terms.

2 4 6 8 10 12 14
 2 2 2

Next write the following pattern.

4 14 24 34 _ _ _

Ask her to find the differences between the terms and to write three more terms. See below.

4 14 24 34 44 54 64
 10 10 10

Decreasing patterns. Write the following pattern

15 13 11 9 __ __ __

For these differences, tell the child to write -2 (negative 2). The pattern then becomes

15 13 11 9 7 5 3
-2 -2 -2

Doubling and halving patterns. Next give the child the following doubling, or multiplying by 2, pattern.

5 10 20 40 __ __ __

Here the difference between the terms does not help much, although it is interesting because it is the same as the pattern.

5 10 20 40 80 160 __
5 10 20

Now challenge her with this halving pattern.

16 8 4 2 __ __ __

Here also the differences repeat the patterns. This pattern also reviews halves and fourths. It will be

16 8 4 2 1 $\frac{1}{2}$ $\frac{1}{4}$

Square pattern. The following pattern should be familiar

1 4 9 16 __ __ __

The differences should also be familiar.

1 4 9 16 25 36 49
3 5 7

Worksheet. The worksheet requires finding the next 2 terms for the patterns given. The child is to make up her own pattern for the last 2 problems for you to find and continue. The answers are as follows.

1.	3	6	9	12	**15**	**18**
2.	2	4	8	16	**32**	**64**
3.	1	4	9	16	**25**	**36**
4.	6	11	16	21	**26**	**31**
5.	1	3	6	10	**15**	**21**
6.	11	22	33	44	**55**	**66**
7.	20	19	18	17	**16**	**15**
8.	25	50	75	100	**125**	**150**
9.	2	12	22	32	**42**	**52**
10.	1	10	100	1,000	**10,000**	**100,000**
11.	$\frac{1}{2}$	1	$1\frac{1}{2}$	2	$\mathbf{2\frac{1}{2}}$	**3**
12.	16	8	4	2	**1**	$\mathbf{\frac{1}{2}}$
13.	$\frac{1}{2}$	$1\frac{1}{2}$	$2\frac{1}{2}$	$3\frac{1}{2}$	$\mathbf{4\frac{1}{2}}$	$\mathbf{5\frac{1}{2}}$

Lesson 92

The Distributive Property

OBJECTIVES
1. To introduce estimating
2. To realize the calculator problem of finding a + b × c
3. To use the distributive law in problem solving

MATERIALS
Worksheet 96, "Quick Practice-15"
Worksheet 97, "The Distributive Property"
Calculator

Note: Worksheet 96 is used again.

WARM-UP
Ask the child to write the following problem in his journal: 8034 – 644. [7390] The check numbers are (5) + (1) = (6). (Subtraction check numbers: add the subtrahend and remainder check numbers.)

Note: The term *distributive property* replaces the older term *distributive law*. A property emphasizes a quality, not a rule.

Remind him that the square root of 9 is 3 because 3 × 3 = 9. What is the square root of 64? [8] What is the square root of 4? [2] What is the square root of 25? [5] What is the square root of 81? [9]

ACTIVITIES
Facts practice. Take out Quick Practice-15 and ask the child to do the facts in the **first** rectangle when you say start. Ask him to correct them. The answers are:

30	16	63	54	28
18	30	24	35	15
20	36	21	35	64
18	24	36	25	48
20	32	27	45	40
56	18	54	32	27
12	49	9	42	63
24	81	12	72	48
24	15	56	28	40
45	21	36	72	42

Note: Estimating is best taught in context, not as an isolated skill.

Problem 1. Ask the child to solve the following problem two ways on the calculator. Diane bought 1 toy for $4.25 and 3 toys for $5.95. [$22.10] Write the amounts on the board as follows:

1	$4.25
3	$5.95

Tell him before he presses any keys he needs to think about what the answer should be. Will the answer be around 1, 22, or 30? [22] Ask him to explain. Emphasize thinking of just the whole dollars, 4 + 6 × 3; 6 is used because $5.95 is closer to 6 than to 5.

Now ask him to use multiplication to find the answer on the calculator. If he proceeds with

$$4.25 + 5.95 \times 3$$

the calculator will show 30.60 (or 43.1375 if the 3 is pressed before the 5.95), obviously an error. Help him to realize that the calculator is adding 4.25 and 5.95 and then multiplying by 3. Tell him that more expensive calculators, such as scientific calculators, will give the right answer because they will multiply first and then add.

How can you get the right answer? One way is to multiply 5.95 × 3 and then add the 4.25. However, a better solution is use the memory feature. Be sure memory is cleared; then add 4.25 to memory with [M+]. Next multiply 5.95 by 3 and add it to memory with [M+]. To find the sum, press [MR]. (Memory Return). The procedure follows:

Press 4.25 then [M+] (4.25 will show)

Press 5.95 [×] 3 [M+] (17.85 will show)

Press [MR] (The answer 22.1 will show)

If necessary, explain that the calculator thinks the 0 after 1 is extra, but it needs to be written when it is money, $22.10.

Problem 2. Dean bought 6 toys on Monday, each costing $3.79. On Tuesday he bought 7 more at the same price. What was the total cost? [$49.27] First ask the child to solve it on paper.

One solution is to add the 6 and 7, getting 13, and to multiply it by 3.79.

$3.79	$3.79	$ 22.74
× 6	× 7	+ 26.53
$22.74	$26.53	$ 49.27

Could you multiply by 10 and by 3 and get the same answer? [yes] The calculations are shown below.

$3.79	$3.79	$37.90
× 10	× 3	+ 11.37
$37.90	$11.37	$49.27

Now ask him to solve it on the calculator 2 different ways.

$$3.79 \times 6 + 3.79 \times 7 = [\$49.27]$$

and

$$3.79 \times 13 = [\$49.27]$$

Worksheet. The worksheet has three problems similar to the one above to be solved in two ways. A calculator is not necessary. The solutions are shown below.

1. 13	11	104	24
× 8	× 8	+ 88	× 8
104	88	192	192
2. 13	13	65	13
× 5	× 4	+ 52	× 9
65	52	117	117
3. 27	27	162	27
× 6	× 4	+ 108	× 10
162	108	270	270

Lesson 93

Square Inches in a Square Foot

OBJECTIVES

1. To find the number of square inches in a square foot
2. To multiply 2-digit numbers by 2-digit numbers, informally, that is, without an algorithm (procedure)

MATERIALS

Worksheet 96, "Quick Practice-15"
Worksheet 98, "Square Inches in a Square Foot"
About 2 hundred 1-inch tiles
Four 12-inch rulers

WARM-UP

Ask the child to write the following problem in her journal: 12,345 – 6,789. [5556] The check numbers are (3) + (3) = (6).

What is 8 × 10? [80] What is 18 × 10? [180] What is 20 × 10? [200] What is 21 × 10? [210] What is 30 × 10? [300] What is 33 × 10? [330]

ACTIVITIES

Facts practice. Ask the child to take out her Quick Practice sheet and ask her to do the facts in the **second** rectangle when you say start. Ask her to correct them. The answers are:

25	54	28	72	20	81	36	20	35	24
28	56	24	48	15	45	48	45	32	36
49	42	12	56	27	18	24	9	30	64
36	15	63	40	27	42	18	24	21	72
16	18	40	35	30	32	63	12	21	54

Square foot. Show the child a square inch and ask How long is each side? [1 inch] What is the perimeter? [4 inches] What is its area? [1 square inch] What is a square inch? [a square that is 1-inch on each side]

Show her a 12-inch ruler. How long is it? [1 foot, or 12 inches] How many inches in a foot? [12] What would a square foot look like? [a square 1 foot on each side] Build a square foot with 4 rulers as shown below. What is the size of the square? [square foot]

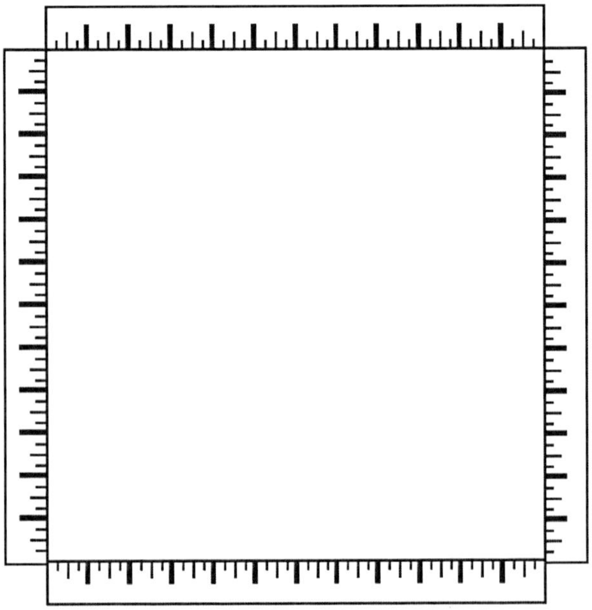

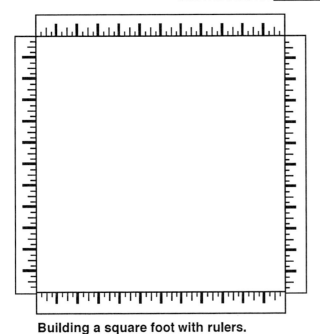

Building a square foot with rulers.

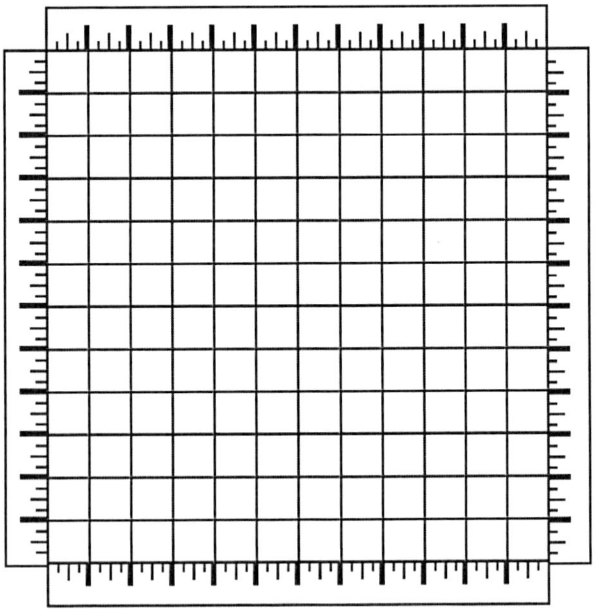

Filling a square foot with square inches (1" tiles).

How many square inches are there in a square foot? Suggest filling the area with tiles as shown above on the right. How many tiles will fit in a row? [12] How many rows can be make? [12]

Worksheet. Next ask the child to draw the square foot on the worksheet. Explain that because 12 tiles will not fit on the paper, the squares are made smaller, but each square represents 1 inch. Then ask her to find the total number [144] and to write out her thinking. Ask her to explain her method. Two ways are shown below.

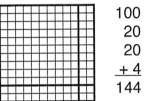

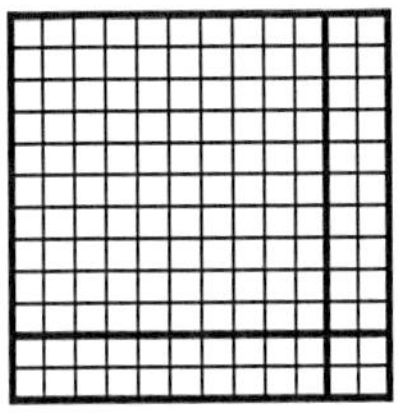

100
20
20
+ 4
144

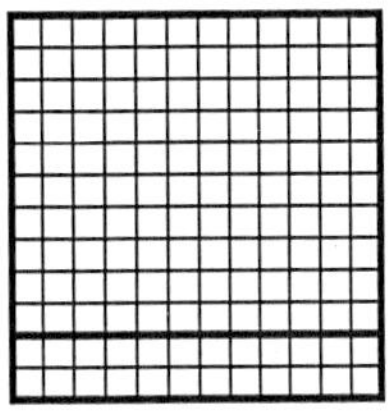

12
× 10
120

12
× 2
24

120
+ 24
144

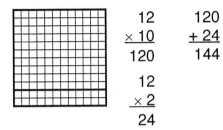

Two ways of finding 12 × 12. [144]

Discuss in detail the second solution shown since it is similar to the common algorithm. She knows 12 × 10 [120] and she can easily calculate 12 × 2. [24] The 2 sums are then added together.

Another example. Modify the 12 × 12 arrangement to show 14 × 13. Ask the child to find the number of square inches in the new rectangle. [182] She can show her work in her journal. Two solutions are below.

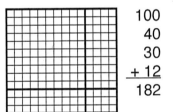

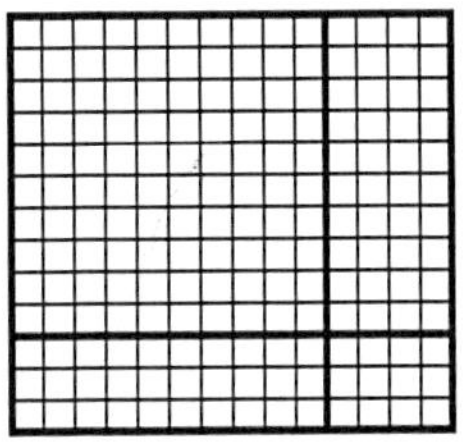

100
40
30
+ 12
182

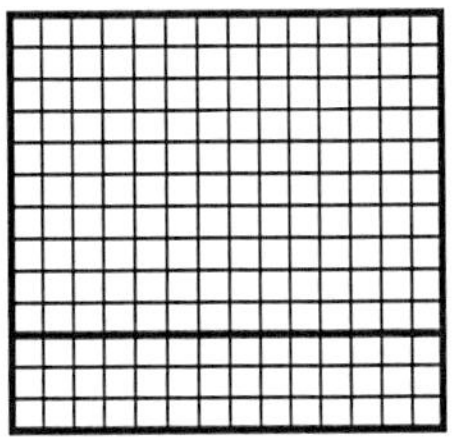

14
× 10
140

14
× 3
42

140
+ 42
182

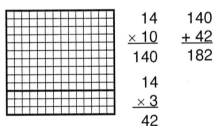

Two ways of finding 14 × 13. [182]

Worksheet. The worksheet has two problems similar to these. Some solutions are shown below.

Note: Encourage the child to solve these problems in a way that makes sense to her.

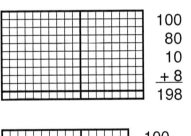

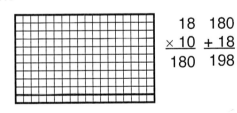

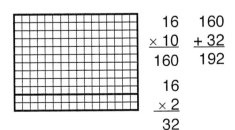

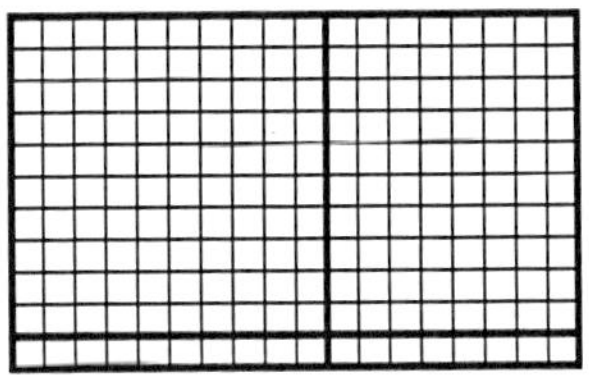

100
80
10
+ 8
198

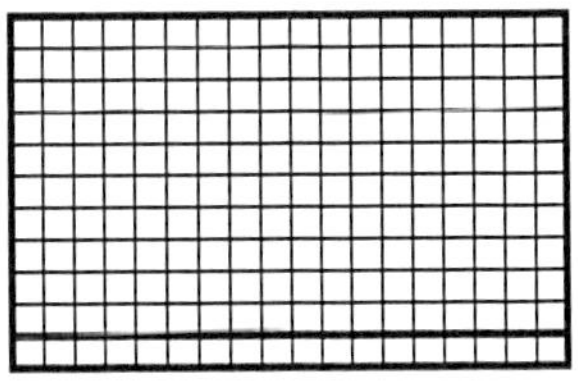

18
× 10
180

180
+ 18
198

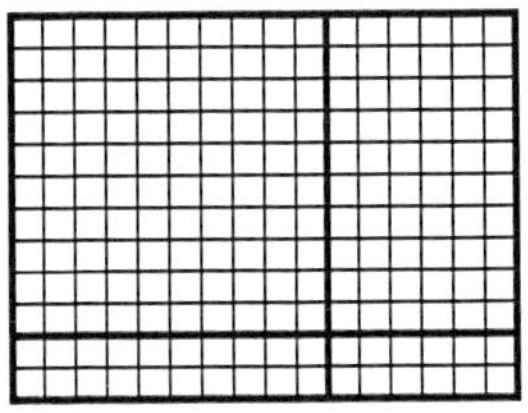

100
60
20
+ 12
192

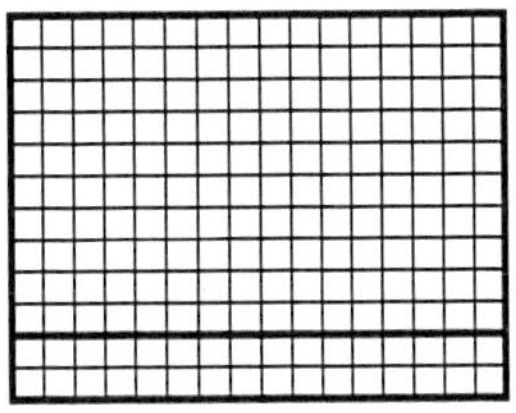

16
× 10
160

16
× 2
32

160
+ 32
192

Lesson 94

Multiplying by Two Digits

OBJECTIVE 1. To work with multiplying by 2 digits

MATERIALS Worksheet 92, "Quick Practice-14" or Worksheet 96, "Quick Practice-15"
Worksheet 99, "Multiplying by Two Digits"

WARM-UP Ask the child to write the following problem in his journal: 1085 × 5. [5425] The check numbers are (5) × (5) = (7).

What is 3 squared? [9] What does it mean to square a number? [multiply the number by itself] What is 6 squared? [36] What is 7 squared? [49] What is 9 squared? [81] What is 20 squared? [400]

ACTIVITIES ***Facts practice.*** For multiplication practice, alternate Worksheet 92 (answers in Lessons 89 & 91) and Worksheet 96 (answers in Lessons 92-93).

Multiplying by multiples of 10. Ask the child to write the following numbers in a column in his journal. Then ask him to multiply them by 10, 2, 20, 3, and 30. See below.

Number	Number × 10	Number × 2	Number × 20	Number × 3	Number × 30
8	80	16	160	24	240
12	120	24	240	36	360
40	400	80	800	120	1200

How can you easily multiply by 10? [annex a zero] How can you easily multiply by 20? [multiply by 2 and annex a zero]

Note: When multiplying by ten, we *annex a zero*. We do not "add" a zero.

Problem 1. Ask the child to solve the following problem in his journal: How many hours are in 3 days? [72 hr] How many hours are in the month of April? (30 days) [720 hr] How many hours in the month of March? [(31 days) [744 hr] Ask him to explain his solutions.

Summarize by saying that he added the number of hours in 30 days plus the number of hours in 1 day to get the number of hours in 31 days.

$$\begin{array}{r} 24 \\ \times\ 3 \\ \hline 72 \end{array} \text{ hr in 3 days} \qquad \begin{array}{r} 24 \\ \times\ 30 \\ \hline 720 \end{array} \text{ hr in April} \qquad \begin{array}{r} 720 \\ +\ 24 \\ \hline 744 \end{array} \text{ hr in March}$$

Problem 2. Next give the child a money problem. A store owner decided to give 20 people a treat, which costs 39¢ each. What will be the total cost? [$7.80] Ask the child to explain his work in detail.

$$\begin{array}{r} \$0.39 \\ \times\ 20 \\ \hline \$7.80 \end{array} \text{ for 20 people}$$

Change the problem. The owner and the store clerk also wanted treats. Now what is the total cost? [$8.58]

$0.39	$0.39	$7.80
× 20	× 2	+ $0.78
$7.80 for 20 people	$0.78 for 2	$8.58 for 22

Problem 3. The third problem does not split the tens and ones. In the town of Madison, the children go to school 175 days each year. How many days does a person go to school from first grade to graduating from high school? [2100]

If necessary, help the child figure out how many years of school there are until graduation. [12] Then ask him to figure out the answers. Suggest that if it makes sense to him, he write the multiplication using the algorithm shown.

175	175	1750
× 10	× 2	+ 350
1750 in 10 yr	350 in 2 yr	2100 in 12 yr

Of course, the split does not have to be 10 and 2, other numbers also work, for example, 6 and 6.

175	175	1050
× 6	× 6	+ 1050
1050 in 6 yr	1050 in 6 yr	2100 in 12 yr

Worksheet. The worksheet has three problems similar to those above to be solved without the multiplication algorithm. It also has a table for multiplying by multiples of 10. The answers are:

1. **$7.50** **$2.25** **$9.75**
2. **324**
3. **1196 hr**

Number	N × 10	N × 2	N × 20	N × 3	N × 30
9	**90**	**18**	**180**	**27**	**270**
11	**110**	**22**	**220**	**33**	**330**
21	**210**	**42**	**420**	**63**	**630**
30	**300**	**60**	**600**	**90**	**900**
25	**250**	**50**	**500**	**75**	**750**
100	**1000**	**200**	**2000**	**300**	**3000**
401	**4010**	**802**	**8020**	**1203**	**12,030**

Lesson 95

The Multiplication Algorithm

OBJECTIVE 1. To introduce the multiplication algorithm

MATERIALS Worksheet 92, "Quick Practice-14" or Worksheet 96, "Quick Practice-15"
Worksheet 100, "The Multiplication Algorithm"

WARM-UP What is 5 × 10? [50] What is 50 × 10? [500] What is 15 × 10? [150] What is 24 × 10? [240] What is 61 × 10? [610] What is 95 × 10? [950]

ACTIVITIES ***Facts practice.*** For multiplication practice alternate, Worksheet 92 (answers in Lessons 89 & 91) and Worksheet 96 (answers in Lessons 92-93).

Multiplying by multiples of 10. Write the following.

$$\begin{array}{r} 58 \\ \underline{\times 3} \end{array}$$

Ask the child to multiply it. Then write 58 × 30 and ask them to explain their procedure.

$$\begin{array}{r} {}^{2} \\ 58 \\ \underline{\times 3} \\ 174 \end{array} \qquad \begin{array}{r} {}^{2} \\ 58 \\ \underline{\times 30} \\ 1740 \end{array}$$

Why do you write the 4 in the tens place (for the second problem)? [The product is 10 times greater than multiplying by 3.] How is multiplying a number by 3 like multiplying it by 30? [same multiplying, but a 0 is needed]

Ask her to practice with 37 × 60 and 148 × 70.

$$\begin{array}{r} {}^{4} \\ 37 \\ \underline{\times 60} \\ 2220 \end{array} \qquad \begin{array}{r} {}^{3\,5} \\ 148 \\ \underline{\times 70} \\ 10360 \end{array}$$

Multiplying by two digits. Write the following and ask the child to find the products.

$$\begin{array}{r} {}^{2} \\ 53 \\ \underline{\times 9} \\ 477 \end{array} \qquad \begin{array}{r} 53 \\ \underline{\times 20} \\ 1060 \end{array}$$

Now how can we find 53 × 29? [Add 53 × 9 and 53 × 20]

$$\begin{array}{r} 53 \\ \underline{\times 29} \\ 477 \\ \underline{1060} \\ 1537 \end{array}$$

Ask her to try the following.

$$\begin{array}{r} 62 \\ \underline{\times 70} \end{array} \qquad \begin{array}{r} 62 \\ \underline{\times 3} \end{array} \qquad \begin{array}{r} 62 \\ \underline{\times 73} \end{array}$$

The solution is on the next page.

62	62	62
× 3	× 70	× 73
186	4340	186
		4340
		4526

The algorithm. If the child is ready for the next step, continue as follows. Otherwise, ask her to do the first rows on the worksheet. Those answers are at the bottom of this page.

Do you think you are ready to multiply in one step? Write the following and explain that you multiply by 4 in the first row below the line and multiply by 30 in the second row.

```
  1
  1̶
  24 (6)
× 34 (7)
  96
 720
 816 (6)
```

For writing the carries on top, one way is to cross out the row before starting the next multiplier. If the child is accurate without writing them, do not insist that she write them.

Check numbers. Encourage the child to use check numbers for checking her work, as shown above. If there is an error, use check numbers for the partial products.

Worksheet. The answers to the multiplication algorithm practice are as follows.

49	49	49	78	78	78
× 4	× 50	× 54	× 3	× 60	× 63
196	**2450**	**196**	**234**	**4680**	**234**
		2450			**4680**
		2646			**4914**

61	61	61	85	85	85
× 2	× 80	× 82	× 5	× 90	× 95
122	**4880**	**122**	**425**	**7650**	**425**
		4880			**7650**
		5002			**8075**

53	24	62	80	46	17
× 32	× 14	× 26	× 96	× 43	× 35
106	**96**	**372**	**480**	**138**	**85**
1590	**240**	**1240**	**7200**	**1840**	**510**
1696	**336**	**1612**	**7680**	**1978**	**595**

191	802	246	357	182	107
× 29	× 32	× 26	× 77	× 66	× 29
1719	**1604**	**1476**	**2499**	**1092**	**963**
3820	**24060**	**4920**	**24990**	**10920**	**2140**
5539	**25664**	**6396**	**27489**	**12012**	**3103**

Lesson 96

Review

OBJECTIVE 1. To review and practice

MATERIALS Worksheet 101-A and 101-B, "Review" (2 versions)
Cards for playing games

ACTIVITIES ***Review worksheet.*** Give the child about 15 to 20 minutes to do the review sheet. The oral problems to be read are as follows:

8 squared = $8\frac{1}{2} + 4\frac{1}{2} =$ 87 – 29 =

Review Sheet 101-A.

1. Write only the answers to the oral questions. **64** **13** **58**

4. Write only the answers. $3^2 \times 4 =$ **36** 158 – 39 = **119** 209 + 25 = **234**

7. Find the number of squares in the rectangle.

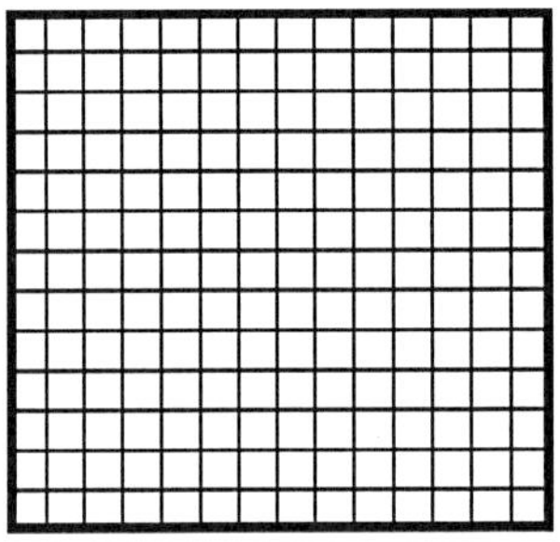

14
× 13
42
140
182

9. Ms. Garcia is buying 24 stamps, each costing 33¢. She also is buying 2 stamps for 20¢ each. What is the total cost?

24
× $.33
72
720
$7.92
+.40
$8.32

11. Find the number of bars in the array.

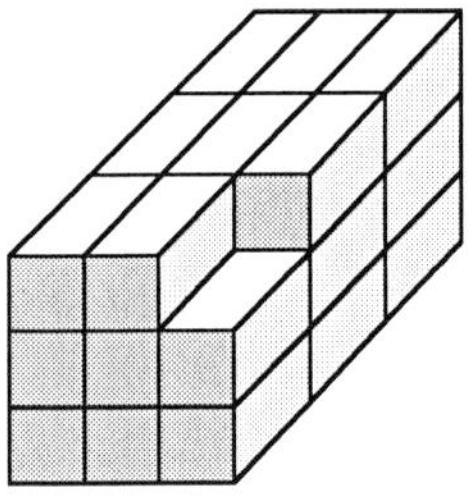

3 × 3 × 3 – 1 = 26

12. Write the following number:
3 million 270 thousand.

3,270,000

13.

78	78	78
× 5	× 60	× 65
390	**4680**	**390**
		4680
		5070

16.

808,791	560,567
+ 352,240	– 63,945
1,161,031	**496,622**

The oral problems to be read to the child are as follows:

9 squared = $7\frac{1}{2} + 8\frac{1}{2} =$ $93 - 18 =$

Review Sheet 101-B.

1. Write only the answers to the oral questions. 81 16 75

4. Write only the answers. $4^2 \times 2 =$ 32 $147 - 29 =$ 118 $207 + 46 =$ 253

7. Find the number of squares in the rectangle.

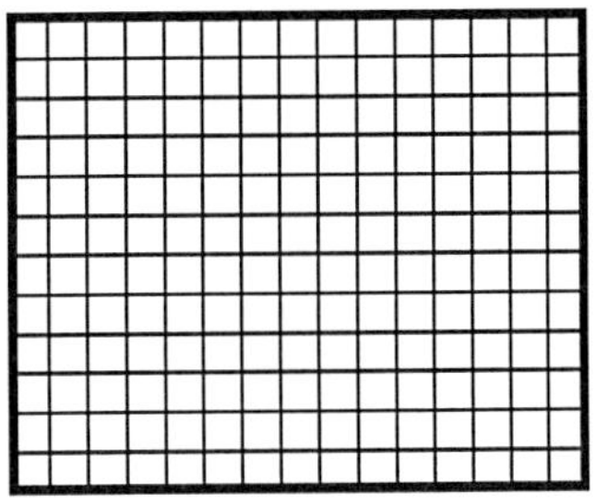

```
   15
 × 12
   30
  150
  180
```

9. Mr. Johnson is buying 24 stamps each costing 34¢. He also is buying 2 stamps for 20¢ each. What is the total cost?

```
    24
 ×$.34
    96
   720
 $8.16
  +.40
 $8.56
```

11. Find the number of bars in the array.

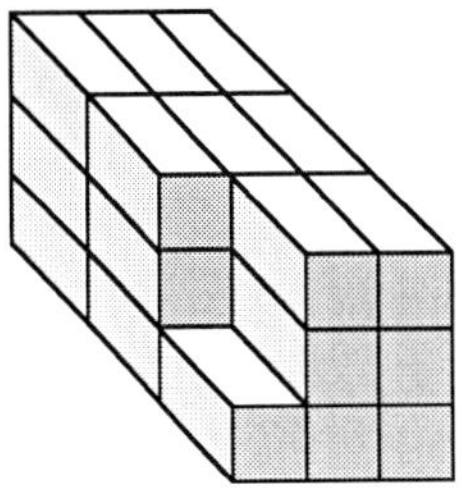

3 × 3 × 3 − 2 = 25

12. Write the following number: 2 million 750 thousand.

2,750,000

13.

```
  57      57      57
 × 3    × 70    × 73
 171    3990     171
                3990
                4161
```

16.

```
   871,569     486,912
 + 352,240    − 87,374
 1,223,809     399,538
```

Games. Spend the remaining time playing games.

MULTIPLICATION. Square Memory (P21) and Crazy Squares (P23) *(Math Card Games).*

LOGIC. Math Puzzles (in *Worksheets* immediately before the math journal). Solutions are in Appendix pg. 4-5).

Lesson 97

Problem Solving Using a Table

OBJECTIVE 1. To use a table for solving problems

MATERIALS Worksheet 92, "Quick Practice-14" or Worksheet 96, "Quick Practice-15"
Worksheet 102, "Problem Solving Using a Table"

WARM-UP Ask the child to solve the following problem in her journal: 24 × 12. [288] The check numbers are (6) × (3) = (0).

What number is 2 more than 6? [8] What number is 2 more than 12? [14] What number is 5 more than 45? [50] What number is 12 more than 6? [18] What number is 6 more than 10? [16] What number is 100 more than 6? [106]

ACTIVITIES ***Facts practice.*** For multiplication practice, alternate Worksheet 92 (answers in Lessons 89 & 91) and Worksheet 96 (answers in Lessons 92-93).

Problem 1. Give the child this problem. A family ordered milk and pieces of pizza at a restaurant for a total of 13 items. The milk cost $1 for each glass and the pizza cost $3 for each piece. The total bill was $29. How many glasses of milk and how many pieces of pizza were ordered? [5 glasses of milk and 8 pieces of pizzas.]

Let her think about the problem for a few minutes. Then explain that one way of solving it is to use a table. See the table below. To help remember the costs, write them above the words.

Milk	Pizza	($1) Cost of Milk	($3) Cost of Pizza	($29) Total Cost

Suggest she start with 2 as the number of orders of milk. Ask her to fill in the other numbers. How many pizzas will that be? [11, 13 – 2] What will the milk cost? [$2] What will the pizza cost? [$33, 3 × 11] See the first line below.

Note: Learning to set up such tables is a skill also needed for using spreadsheets on computers.

Ask the child to copy the table into her journal and to solve the problem. Several other guesses and the solution are shown below.

Milk	Pizza	($1) Cost of Milk	($3) Cost of Pizza	($29) Total Cost
2	11	2	33	35
6	7	6	21	27
7	6	7	18	25
8	5	8	15	23
5	**8**	**5**	**24**	**29**

Then ask her how many guesses they needed. Ask her to share her work. Do you see any patterns? [When the number of milk orders increase in order, the number of pizza orders decrease. See the table on the next page.

Milk	Pizza	($1) Cost of Milk	($3) Cost of Pizza	($29) Total Cost
1	12	1	36	37
2	11	2	33	35
3	10	3	30	33
4	9	4	27	31
5	**8**	**5**	**24**	**29**

Problem 2. Give the child the following problem. Bobby has 10 coins, all nickels and dimes. The total value of the coins is 85¢. How many coins are nickels and how many are dimes? [3 nickels and 7 dimes] See the table below for some possible values and the solution.

Nickels	Dimes	Value of N	Value of D	(85¢) Total Value
6	4	30¢	40¢	70¢
7	3	35¢	30¢	65¢
5	5	25¢	50¢	75¢
3	**7**	**15¢**	**70¢**	**85¢**

Be sure to write several lines, even if the child gives the correct answer on the first attempt, so she may see a pattern develop.

Worksheet. The worksheet has three similar problems. The tables, some values, and solutions are shown below.

1.

Tri	Square	TP for Tri	TP for Sq	(49) Total TP
7	8	21	32	53
8	7	24	28	52
9	6	27	24	51
10	5	30	20	50
11	**4**	**33**	**16**	**49**

2.

Pennies	Nickels	Value of P	Value of N	(82¢) Total Value
5	7	5¢	35¢	40¢
6	8	6¢	40¢	46¢
12	**14**	**12¢**	**70¢**	**82¢**

3.

Blue	Red	($.30) Cost of Blue	($.25) Cost of Red	($4.00) Total Cost
2	4	$0.60	$1.00	$1.60
4	8	$1.20	$2.00	$3.20
3	6	$0.90	$1.50	$2.40
6	12	$1.80	$3.00	$4.80
5	**10**	**$1.50**	**$2.50**	**$4.00**

Lesson 98

Times Greater

OBJECTIVE 1. To solve problems involving *times greater*

MATERIALS Worksheet 92, "Quick Practice-14" or 96, "Quick Practice-15"
Worksheet 103, "Times Greater"

WARM-UP Ask the child to write the following problem in his journal: 382 × 28. [10,696] The check numbers are (4) × (1) = (4).

Ask the child to compute mentally. What is 56 + 80? [136] What is 56 + 88? [144] What is 107 + 50? [157] What is 107 + 57? [164] What is 85 + 95? [180] What is 146 + 39? [185]

ACTIVITIES ***Facts practice.*** For multiplication practice, alternate Worksheet 92 (answers in Lessons 89 & 91) and Worksheet 96 (answers in Lessons 92-93).

Problem 1. Give the child the following problem. On Tuesday Kevin walked 16 minutes. The following day he walked twice as long. How long did he walk? [32 minutes]

Ask him to explain how he solved the problem.

Problem 2. Keesha rode her bicycle 17 blocks on Friday. The next day she bicycled 3 times as far. How far did she bike on Saturday? [51 blocks]

Ask the child to explain how he solved the problem. The following diagram might help him.

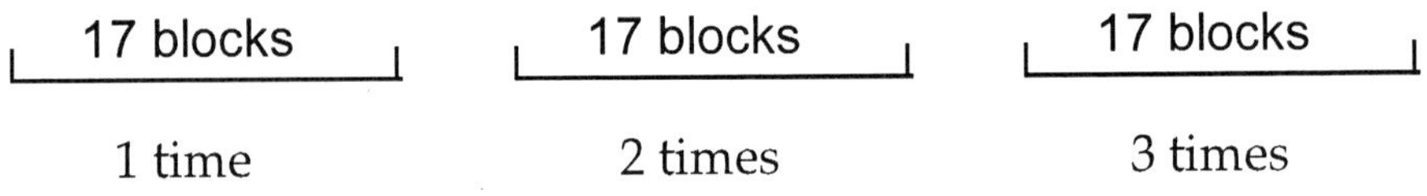

Problem 3. A certain used car costs $3999. A new car costs five times as much. What does the new car cost? [$19,995]

After the child has solved the problem, discuss what 5 times greater means. Then discuss the multiplication; ask if he used any shortcuts. If not, explain that he could do it by multiplying 4 thousand by 5, getting 20 thousand. Then the 5 needs to be subtracted from the 20 thousand because 3999 × 5 is 5 less than 4000.

Problem 4. Draw the rectangle shown below on the left; include the dimensions. Then ask the child to find the area.

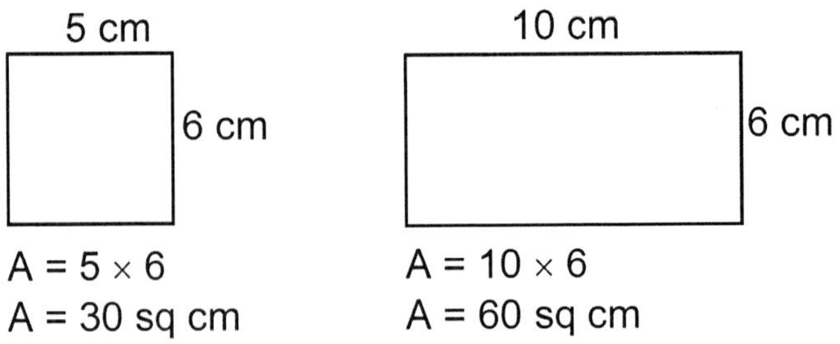

Comparing the area when the width of a rectangle is doubled.

Note: The terms, *width, length, height,* and *depth,* do not have precise meanings.

Then ask him to draw a rectangle twice as long; that is, the width is doubled. See the previous page on the right. What happens to the area? [It is twice as much.]

What would happen if the rectangle were three times as long? [The area would triple, be three times greater.]

What happens to the area if the width is back to 5 centimeters, but the height is doubled? Draw the rectangle below. Ask him for the dimensions.

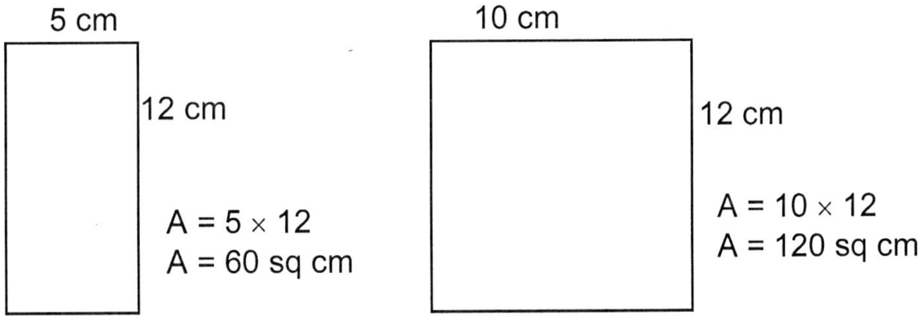

What happens to the area if both the length and height are doubled? [The area is 4 times greater.] See the figure above on the right.

Worksheet. The worksheet has several problems, including an area problem and problems using miles per hour. It also has more practice in 2-digit multiplying. The answers are below.

1. **32 yrs old** 2. **240**
3. **9 mph** 4. **72 mph**
5. **576 mph** 6. **45 mph**
7. **A = 5 × 4 = 20 sq cm**
8. **A = 10 × 4 = 40 sq cm, 2 times as much**

$$\begin{array}{r} 59 \\ \times\ 47 \\ \hline \mathbf{2773} \end{array} \quad \begin{array}{r} 25 \\ \times\ 98 \\ \hline \mathbf{2450} \end{array} \quad \begin{array}{r} 37 \\ \times\ 55 \\ \hline \mathbf{2035} \end{array} \quad \begin{array}{r} 62 \\ \times\ 34 \\ \hline \mathbf{2108} \end{array} \quad \begin{array}{r} 13 \\ \times\ 66 \\ \hline \mathbf{858} \end{array}$$

Lesson 99

Combination Problems

OBJECTIVES 1. To solve combination problems
2. To use a calculator for division

MATERIALS Worksheet 104, Quick Practice-Subtraction
Worksheet 105, Combinations
24 tiles

WARM-UP Ask the child to solve the following problem in her journal: 678 × 73. [49,494] The check numbers are (3) × (1) = (3).

Ask the child, Two times what is 12? [6] Two times what is 8? [4] Two times what is 18? [9] Two times what is 16? [8] Two times what is 14? [7] Two times what is 2? [1] Two times what is 8? [4]

ACTIVITIES ***Facts practice.*** Take out Quick Practice-Subtraction and ask the child to do the facts in the **top** rectangle when you say start. Ask her to correct them. The answers are:

5	7	2	7	4	2	5	8	9	9
3	9	8	4	6	8	9	3	4	4
3	8	3	6	7	6	5	4	2	6
9	7	7	2	6	7	2	8	9	2
4	8	6	5	9	8	3	5	3	5

Combinations with 3. Ask the child to read the first problem on her worksheet. Discuss what it means, if necessary.

1. Use the three digits 1, 2, and 3 to write all the possible 3-digit numbers using each digit once. Stress the importance of being organized.

123	**231**	**312**
132	**213**	**321**

Which number is the greatest? [**321**]

Which number is the least? [**123**]

Combinations with 4. Ask the child to read the second problem on her worksheet.

2. Use the digits 1, 2, 3, and 4 to write all possible 4-digit numbers using each digit once. Ask her to guess how many combinations she could make with 4 digits. [24] Then ask her to find them all.

1234	**2134**	**3124**	**4123**
1243	**2143**	**3142**	**4132**
1324	**2314**	**3214**	**4213**
1342	**2341**	**3241**	**4231**
1423	**2413**	**3412**	**4312**
1432	**2431**	**3421**	**4321**

Which number is the greatest? [**4321**]

Which number is the least? [**1234**]

Pattern. Ask the child to write her results in a table.

Number objects	1	2	3	4	5
Combinations	1	2	6	24	(120)

Ask her to guess how many combinations she can make with 5 objects.

If she is interested in the pattern, help her to see the combinations are each number multiplied up to the combination. For example, $24 = 2 \times 3 \times 4$. Why does that happen?

Worksheet. Ask the child to find the combinations for the four letters *O, P, S,* and *T.* The combinations are given below.

OPST	POST	SOPT	TOPS
OPTS	POTS	SOTP	TOSP
OSPT	PSOT	SPOT	TPOS
OSTP	PSTO	SPTO	TPSO
OTPS	PTOS	STOP	TSOP
OTSP	PTSO	STPO	TSPO

Six of these are words.

There are also five multiplication problems for practice.

37	44	89	54	86
× 51	× 66	× 64	× 36	× 61
1887	**2904**	**5696**	**1944**	**5246**

Lesson 100

Beginning Division

OBJECTIVES
1. To experience the partition model of division
2. To use a calculator for division

MATERIALS
Worksheet 104, "Quick Practice-Subtraction"
Worksheet 106, "Beginning Division"
18 tiles

WARM-UP
Ask the child to solve the following problem in his journal: 4098 × 54. [221,292] The check numbers are (3) × (0) = (0).

Ask, Ten times what is 20? [2] Ten times what is 80? [8] Ten times what is 100? [10] Ten times what is 110? [11] Ten times what is 140? [14] Ten times what is 200? [20] Ten times what is 10? [1]

ACTIVITIES
Facts practice. Ask the child to take out his Quick Practice-Subtraction sheet and ask him to do the facts in the **bottom** rectangle when you say start. Ask him to correct them. The answers are:

2	8	3	5	5	7	4	8	9	7
4	8	8	2	8	4	9	5	2	5
2	9	6	7	8	2	6	7	7	9
3	7	6	4	9	3	4	8	6	5
3	6	2	9	3	3	6	4	9	5

Note: This lesson uses the partition model of division, which answers the question of how many in a group.

Problem 1. Show the child the 18 tiles. How can we divide these 18 tiles equally among three people. [Take 3 tiles at a time and distribute 1 to each pile.] How many does each person get? [6] (Most likely, the word *divide* will be in the child's vocabulary; so using it here extends its meaning to the mathematical sense.)

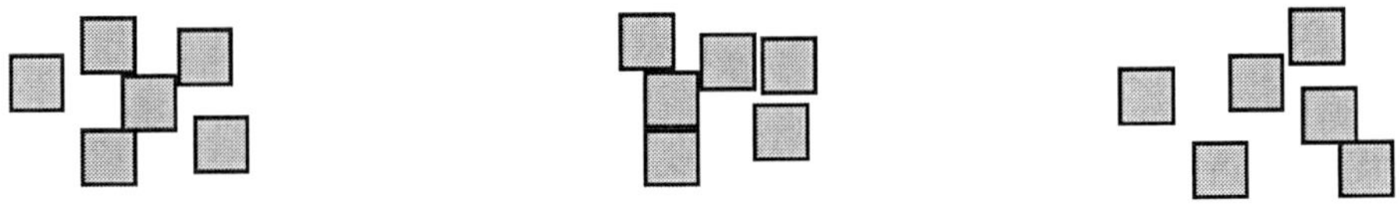

The 18 tiles split among the 3 people.

Next draw a part-whole circle set with three parts as shown below on the left. What do we write in the large circle: what is the whole? [18] Write it in the whole circle. Why do we need 3 part circles? [We divided the tiles into 3 parts.] Point to the first small circle. How much is the part? [6] Repeat for the other part circles. Emphasize that all the parts are equal. See the figure below on the right.

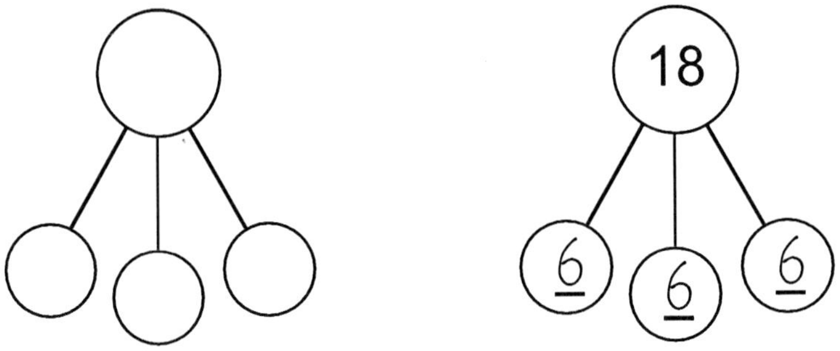

Representing the division problem dividing 18 into 3 groups with wholes and parts.

Ask the child to write the equation. He may write a multiplication equation: $3 \times 6 = 18$ or $6 \times 3 = 18$.

Or he may remember writing division equations from last year and write: $18 \div 3 = 6$.

If not, write it and read it as, 18 divided by 3 equals 6.

Problem 2. Suppose we divided 28 tiles among 4 people. How many parts do we need? [4] You could use dots to show distributing the tiles by placing dots in each part, counting by 4s. See the figures below. Then write the numbers in the parts.

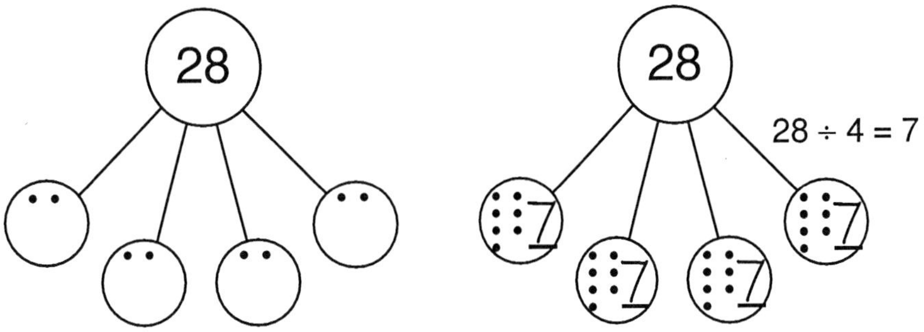

Using dots to divide 28 by 4.

Ask the child to write the equation. [$28 \div 4 = 7$]

Problem 3. Four scouts are having a pizza party. They have 2 pizzas cut into eighths. How many pieces of pizza does each scout get? Ask the child to explain how many pieces there. [Each pizza has 8 pieces, so 16.] How many parts do we need? [4] The solution and equation are shown below.

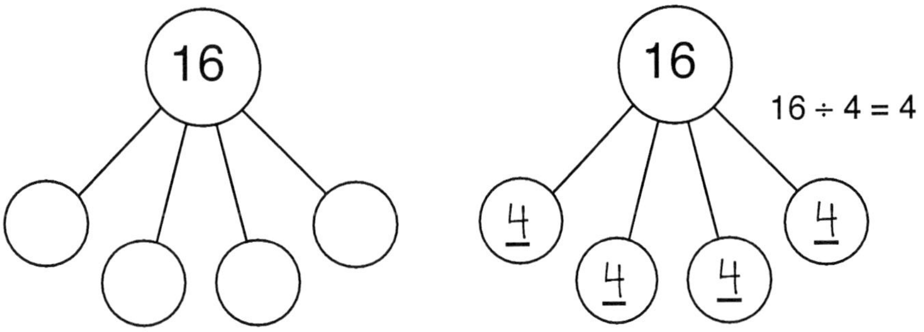

The scouts pizza problem.

Worksheet. The worksheet has four similar problems. The child is to complete the circles, write the numbers and equations. The solutions are as follows:

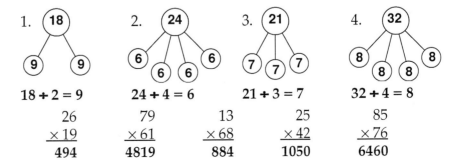

1. 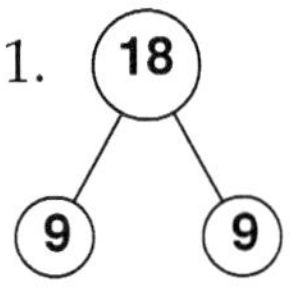

$18 \div 2 = 9$

2. 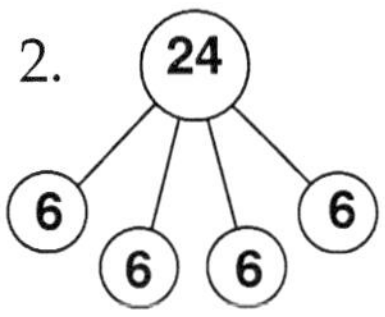

$24 \div 4 = 6$

3. 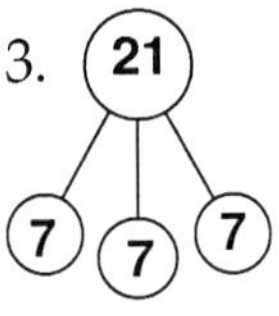

$21 \div 3 = 7$

4. 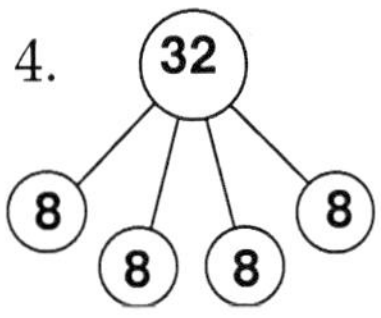

$32 \div 4 = 8$

$\begin{array}{r} 26 \\ \times 19 \\ \hline \mathbf{494} \end{array}$ $\begin{array}{r} 79 \\ \times 61 \\ \hline \mathbf{4819} \end{array}$ $\begin{array}{r} 13 \\ \times 68 \\ \hline \mathbf{884} \end{array}$ $\begin{array}{r} 25 \\ \times 42 \\ \hline \mathbf{1050} \end{array}$ $\begin{array}{r} 85 \\ \times 76 \\ \hline \mathbf{6460} \end{array}$

Lesson 101

Operations With Parts and Wholes

OBJECTIVES

1. To review the basic operations of arithmetic through the part-whole model
2. To solve partition problems using the part-whole model

MATERIALS

Worksheet 92, 96, or 104, "Quick Practice"
Worksheet 107, "Operations With Parts and Wholes"

WARM-UP

Ask the child to solve the following problem in her journal: 5827 × 9 [52,443] The check numbers are (4) × (0) = (0).

Ask the child 10 times what is 60? [6] Ten times what is 80? [8] Ten times what is 30? [3] Ten times what is 80? [8] Ten times what is 10? [1] Ten times what is 100? [10] Ten times what is 110? [11]

ACTIVITIES

Facts practice. For multiplication, use Worksheet 92 (answers in Lessons 89 & 91) and Worksheet 96 (answers in Lessons 92-93). For subtraction, use Worksheet 104 (answers in Lessons 99-100).

Basic operations. Tell the child that addition is one of the operations of arithmetic, or mathematics. Can you name other operations? [subtraction, multiplication, and division]

Draw a part-whole circle set and write 19 and 7 in the parts as shown below on the left.

> **Note:** Research shows that the part-whole model helps a child solve problems better.

What operation do we need to find the missing amount? [addition] What is the whole? [26] Write it in the whole circle. Ask the child to write the equation. [19 + 7 = 26]

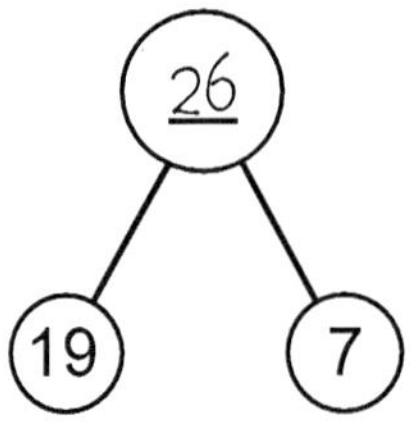

19 + 7 = 26

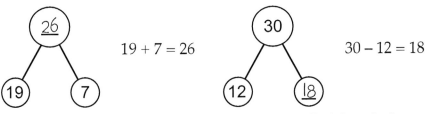

Adding to find the whole.

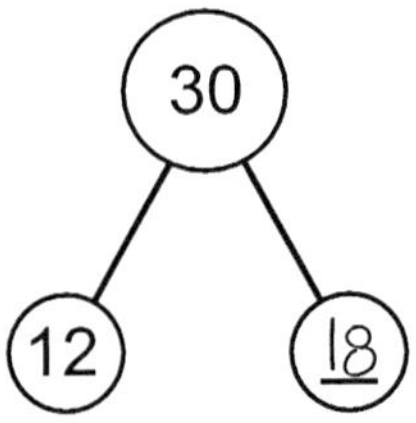

30 – 12 = 18

Subtracting to find the missing part.

Next draw another part-whole circle set and write 30 in the whole circle and 12 in a part circle as shown above on the right. What operation do we need to find the missing amount? [subtraction (addition by going up is also correct)] What is the missing part? [18] Write it in the part circle. Ask her to write the equation. [30 – 12 = 18, or 12 + 18 = 30]

Next draw a part-whole circle set with three parts; write 4, 5, and 6 in the parts as shown below on the left. What operation do we need to find the missing amount? [addition] What is the whole? [15] Write it in the whole circle. Ask her to write the equation.

Draw another part-whole circle set with 3 parts. Write 5 in all the parts as shown on the next page on the right. What operation do we need to find the whole? [multiplication] What is the whole? [15] Write it in the whole circle. Discuss why addition was used in one case and multiplication in the other. Stress that the parts must be the same for multiplying. Ask her to write the equation. [5 × 3 = 15]

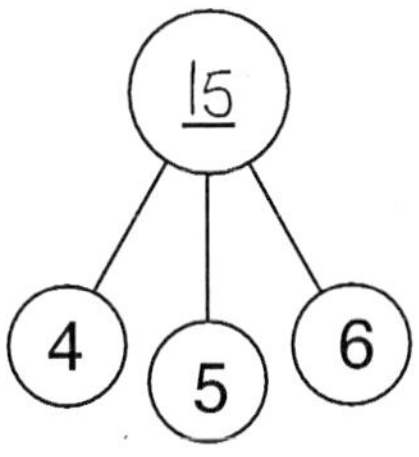

$4 + 5 + 6 = 15$

Adding to find the whole.

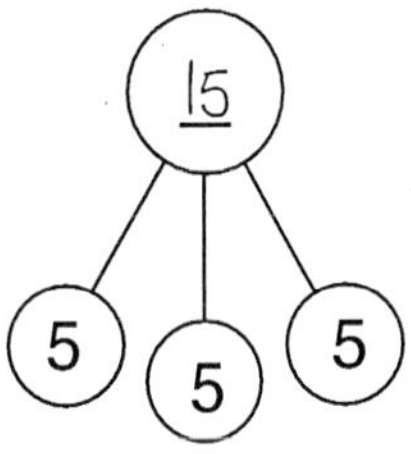

$5 \times 3 = 15$

Multiplying to find the whole.

Use a part-whole circle set with 4 parts and write 20 in the whole. What operation do we need to find the missing amount? [division] Ask her to write the equation. [$20 \div 4 = 5$]

Problem 1. Ask the child to listen to the following problem and to draw part-whole circles. Jo bought three stamps, each costing 33¢. What was the total cost? See below on the left.

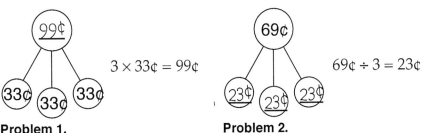

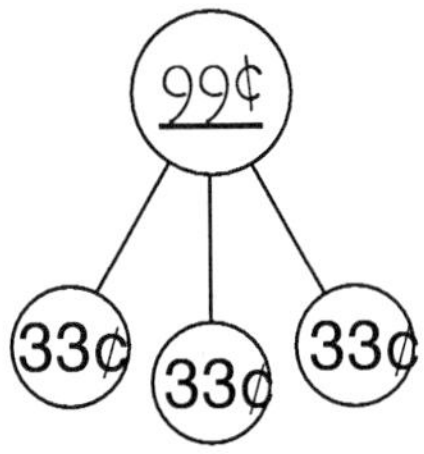

$3 \times 33¢ = 99¢$

Problem 1.

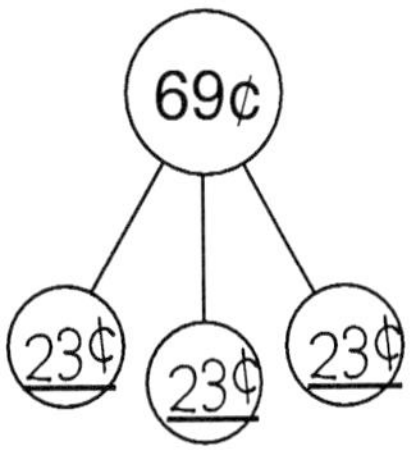

$69¢ \div 3 = 23¢$

Problem 2.

Problem 2. Ask the child to solve the next problem. Three erasers cost 69¢. What did each 1 cost? See the above figure on the right.

Worksheet. The child completes 8 part-whole sets and writes the equations and solves two problems with part-whole sets. She also practices 2-digit multiplication. Answers are:

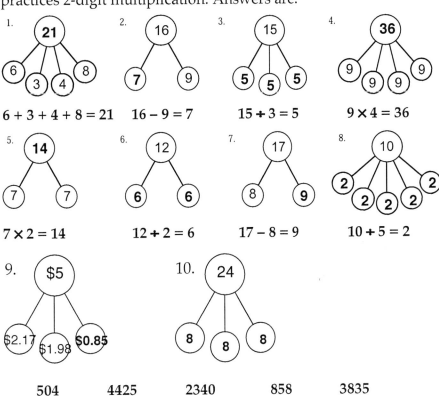

1.
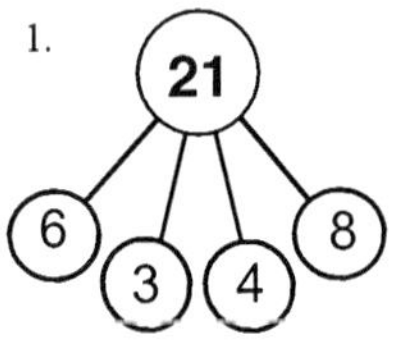

$6 + 3 + 4 + 8 = 21$

2.
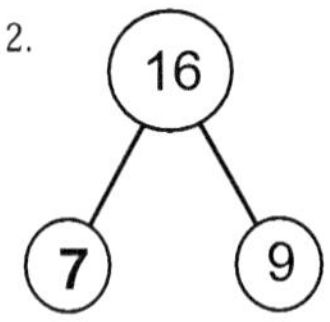

$16 - 9 = 7$

3.
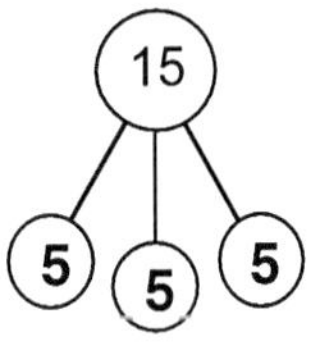

$15 \div 3 = 5$

4.
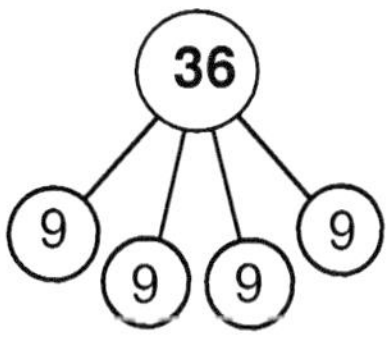

$9 \times 4 = 36$

5.
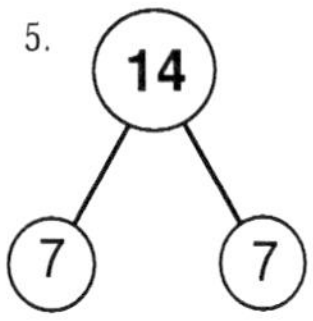

$7 \times 2 = 14$

6.
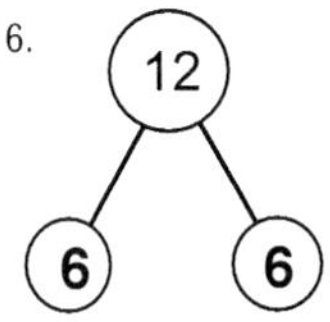

$12 \div 2 = 6$

7.
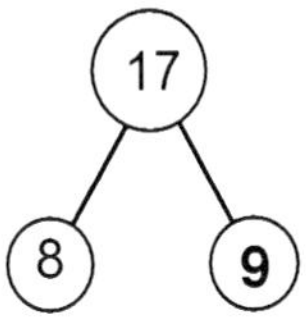

$17 - 8 = 9$

8.
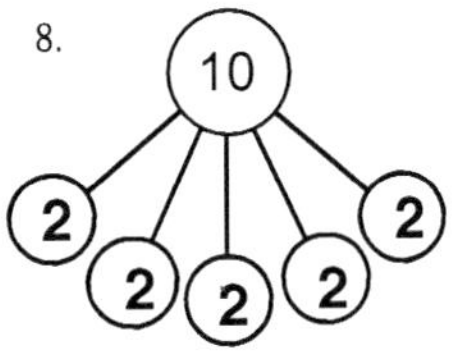

$10 \div 5 = 2$

9.
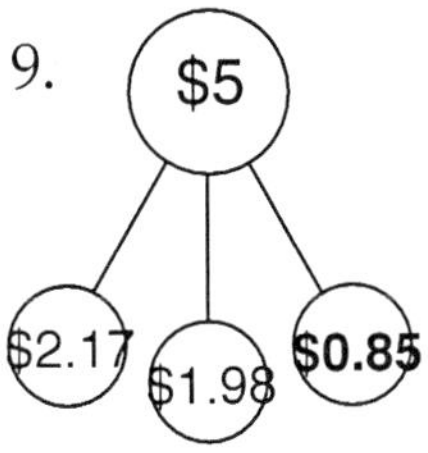

10.
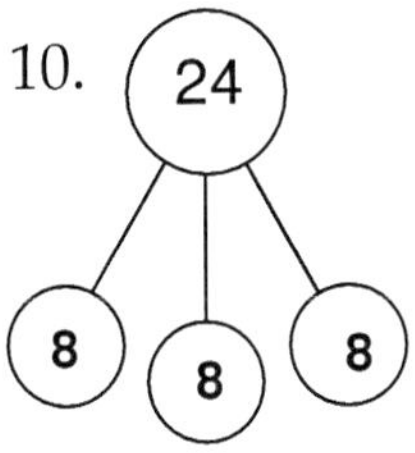

504 4425 2340 858 3835

Lesson 102

Review

OBJECTIVE 1. To review and practice

MATERIALS Worksheet 108-A and 108-B, "Review" (2 versions)
Cards for playing games

ACTIVITIES ***Review worksheet.*** Give the child about 15 to 20 minutes to do the review sheet. The oral problems to be read are as follows:

10 squared = 2 – 1½ = ½ of 18 =

Review Sheet 108-A.

1. Write only the answers to the oral questions. 100 $\frac{1}{2}$ 9

4. Write only the answers. 8^2 = 64 61 – 42 = 19 83 + 89 = 172

7. A group of 3 children have 15 sheets of special paper to share. How many sheets does each one get? Complete the part-whole circle sets. Then write the equation.

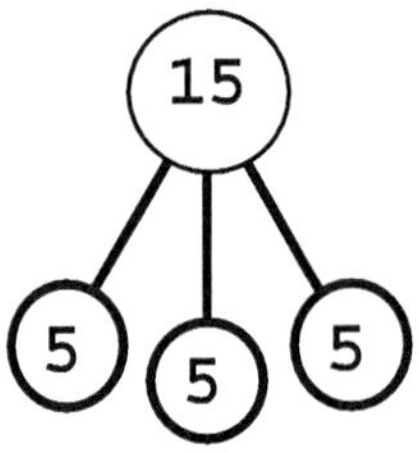

15 ÷ 3 = 5

9. Wen bought 2 identical parts, each costing $25.49. How much change did he get back from $60?

```
 $25.49     $60.00
 ×    2   –  50.98
 $50.98      $9.02
```

12. Draw a rectangle with the horizontal side 2 times wider than the one below. What is the area of each rectangle.

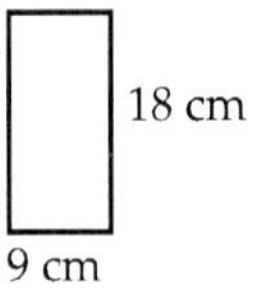


```
   18
 ×  9
  162 sq cm
```

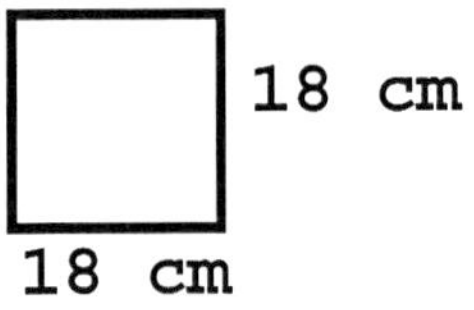


```
   18
 × 18
  144
  180
  324 sq cm
```

13. Write the largest number you can using all of these digits: 2, 7, 1, 4, 9. 97,421

14. Wendy is buying 17 favors. Each one cost 29¢. What is her total cost?

```
 $0.29
 ×  17
   203
   290
 $4.93
```

15. Write all the 3-digit numbers possible using 7, 8, and 9 for each number.

789
798
879
897
978
987

The oral problems to be read to the child are as follows:

9 squared = 1 – ½ = ½ of 16 =

Review Sheet 108-B.

1. Write only the answers to the oral questions. 81 $\frac{1}{2}$ 8

4. Write only the answers. 6^2 = 36 93 – 39 = 54 76 + 89 = 165

7. A group of 3 children have 18 dollars to share. How much does each one get? Complete the part-whole circle sets. Then write the equation.

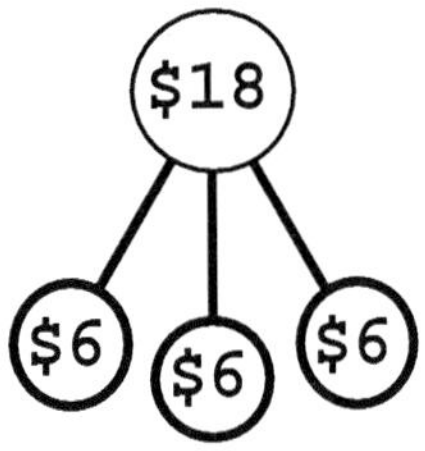

$18 ÷ 3 = $6

9. Wendi bought 2 identical gifts, costing $17.53 each. How much change did she get back from $50?

```
 $17.53        $50.00
 ×    2      -  35.06
 $35.06        $14.94
```

12. Draw a rectangle with the vertical side 2 times higher than the one below. What is the area of each rectangle?

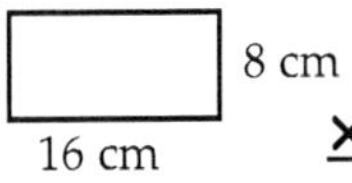


```
  16
 × 8
 128 sq cm
```

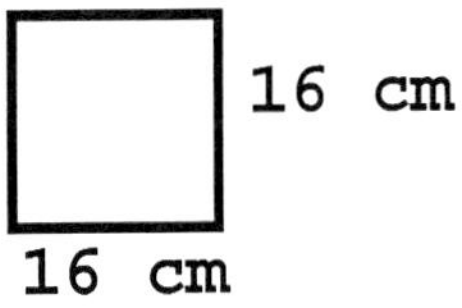


```
   16
 × 16
   96
  160
  256 sq cm
```

13. Write the smallest number you can using all of these digits: 2, 7, 1, 4, 9. 12,479

14. William is buying 19 favors. Each one cost 39¢. What is his total cost?

```
  $0.39
  ×  19
    351
    390
  $7.41
```

15. Write all the 3-digit numbers possible using 1, 2, and 5 for each number.

125
152
215
251
512
521

Games. Spend the remaining time playing games.

MULTIPLICATION. Square Memory (P21) and Crazy Squares (P23) *(Math Card Games).*

LOGIC. Math Puzzles (in *Worksheets* immediately before the math journal). Solutions are in Appendix pg. 4-5.

Lesson 103

Division: Number in a Group

OBJECTIVES

1. To solve division problems involving finding the number in a group and the remainder
2. To learn the term *remainder*

MATERIALS

Worksheet 92, 96, or 104, "Quick Practice"
Worksheet 109, "Division: Number in a Group"
Tiles

WARM-UP

Ask the child, Five times what is 10? [2] Five times what is 50? [10] Remind her of the relationship of these numbers to a clock. Five times what is 15? [3] Five times what is 35? [7] Five times what is 45? [9] Five times what is 25? [5] Five times what is 40? [8] Ten times what is 80? [8]

ACTIVITIES

Facts practice. For multiplication, use Worksheet 92 (answers in Lessons 89 & 91) and Worksheet 96 (answers in Lessons 92-93). For subtraction, use Worksheet 104 (answers in Lessons 99-100).

Problem 1. Ask the child to solve the following problem in any way she can. Jason is throwing food to the 5 ducks on a pond. He has 22 pellets of food. How many pellets can he feed each duck? [4 with 2 left over]

Note: What to do with remainders after dividing depends upon the context. So, it is important for the child to solve problems with remainders.

Ask her to explain her method. Then discuss the various additional solutions, 1 of which is shown below. Explain that the amount left over, that cannot be divided, is called a *remainder.* If the equation is not already written, do so as follows: $22 \div 5 = 4$ r2.

Note: These problems use the partition form of division; that is, finding how many in a group.

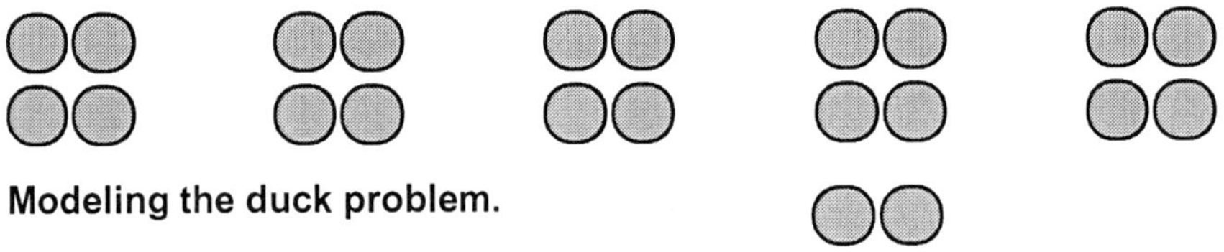

Modeling the duck problem.

Then ask her to draw the part-whole circles. The remainder can be written in an extra circle, as shown below.

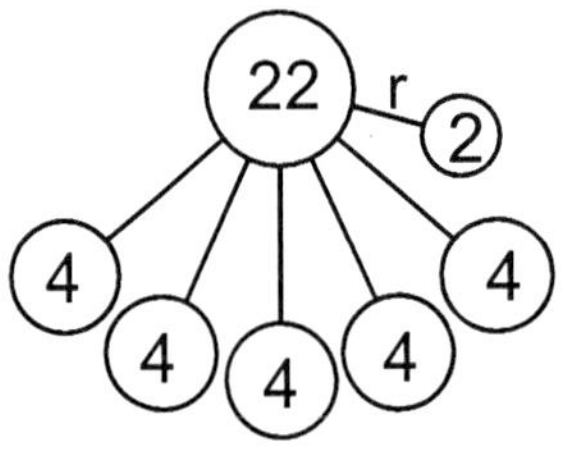

Part-whole circles for the duck problem.

Problem 2. This problem requires splitting a dollar. Four children earn \$5. How much does each one get? [\$1.25]

Giving the child enough time, along with the expectation that she can do it, will help her to solve the problem. She will have to think about what to do with the extra dollar. If she thinks about 4 quarters in a dollar, the problem is solved.

The equation is \$5.00 ÷ 4 = \$1.25. See the figure on the next page.

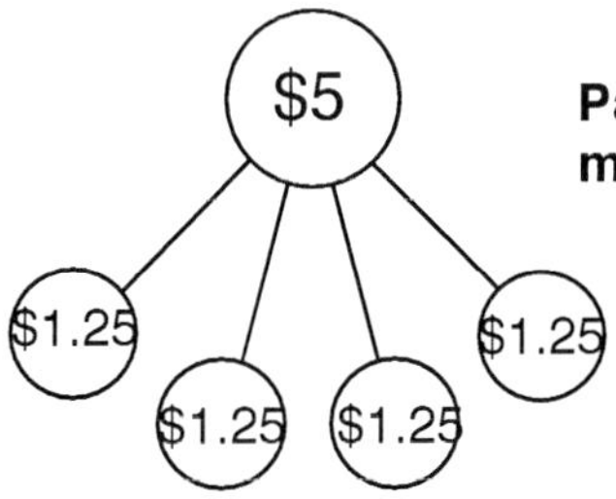

Part-whole circles for the money problem.

Problem 3. The remainder also needs to be divided in the next problem. Six children are splitting a pizza that is cut into 9 pieces. How many pieces does each one get?

Encourage the child to draw a diagram to show her solution. To draw a circle into ninths, first suggest she draw it into thirds and then divide each third into ninths as shown below on the left.

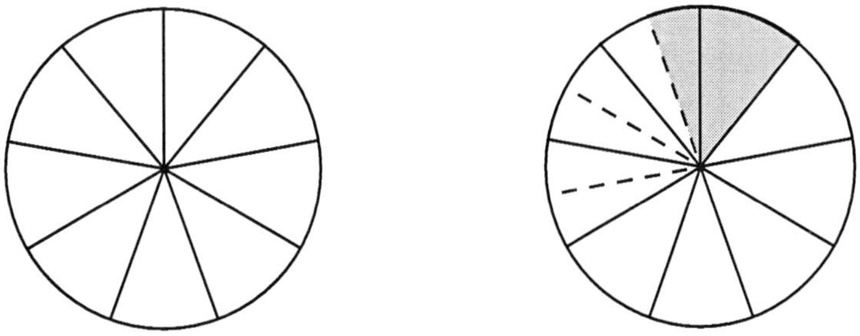

The problem of dividing a pizza with 9 pieces among 6 children. The amount for one child is shown shaded.

Give the child time to work the problem. Then discuss the solution. If each child gets 1 piece, there will be 3 left over, which can be split into half. See the figure above on the right. The equation is $9 \div 6 = 1\,{}^{1}/_{2}$. Here the part-whole circles are a bit cumbersome.

Problem 4. In this problem, the remainder is the answer. Zoe is decorating cookies, which must be all alike, with chocolate chunks. She has 10 cookies and 53 chunks. She can eat any chunks left over. How many does she eat? [3]

The equation is $53 \div 10 = 5$ r3

Worksheet. The worksheet has three similar problems and five multiplication problems. The answers follow.

1. **7, 3, 38 ÷ 5 = 7 r3**
2. **$5.50, $11 ÷ 2 = $5.50**
3. **1 1/2, 6 ÷ 4 = 1½**

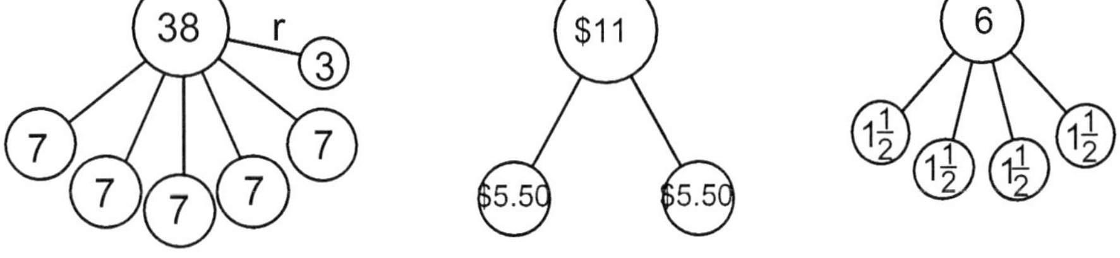

2982 2496 3024 2028 7425

Lesson 104

Division: Number of Groups

OBJECTIVE

1. To solve division problems involving finding the number of groups and the remainder

MATERIALS

Worksheet 92, 96, or 104, "Quick Practice"
Worksheet 110, "Division: Number of Groups"
Tiles

WARM-UP

Challenge the child to solve the following problem in his journal: 10,324 × 12 [123,888] The check numbers are (1) × (3) = (3).

Although the child has not multiplied 5-digit numbers previously, he has learned the mathematics to do so. He should not be led to expect to be shown the way down every mathematical path. He needs to be expected to venture out on his own.

Ask the child, What comes next, 12, 14, 16? [18] What comes next, 15, 20, 25? [30] What comes next, 12, 15, 18? [21] What comes next, 100, 95, 90? [85] What comes next, 12, 18, 24? [30] What comes next, 1, 4, 9? [16]

ACTIVITIES

Facts practice. For multiplication, use Worksheet 92 (answers in Lessons 89 & 91) and Worksheet 96 (answers in Lessons 92-93). For subtraction, use Worksheet 104 (answers in Lessons 99-100).

Measurement division. The child needs to experience this second type of division, referred to as *measurement,* where he is looking for the number of groups, not the number in a group. Do not bother teaching the distinction between the two types. For your reference, the chart below clarifies the differences.

	Partition	Measurement
Known	How many groups [5]	Number in a group [4]
Find	How many in a group [4]	How many groups [5]
Group	Into 5s	By 4s
Related to	Subtraction & multiplication	Inverse of multiplication
Equation*	_ × 4 = 20	5 × _ =
*Number in a group × number of groups: 4 × 5 = 20		

Comparing partition and measurement types of division, using 5 groups of 4.

Note: It is important that remainders be a part of division from the start because they are an integral part of division.

Wheel problem. Make the tiles available for the first problem. Mr. McCoy is assembling little toy cars by adding 4 wheels to each one. How many cars can he assemble with 20 wheels? A figure is shown below. Ask the child to explain how he found his answer and ask him to write the equation. [20 ÷ 4 = 5]

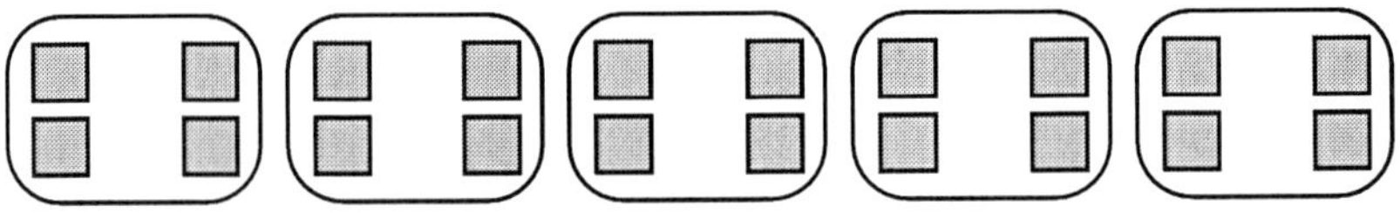

Finding the number of groups. [4]

Modeling on the abacus. The abacus demonstrates the two types of division. In the duck problem from the previous page, a partition problem, we are looking for the number in a group [4], represented by the number of beads on a wire. See the left figure below. In the wheel problem from the previous page, a measurement problem, we are looking for the number of groups, represented by the number of wires. See the right figure below.

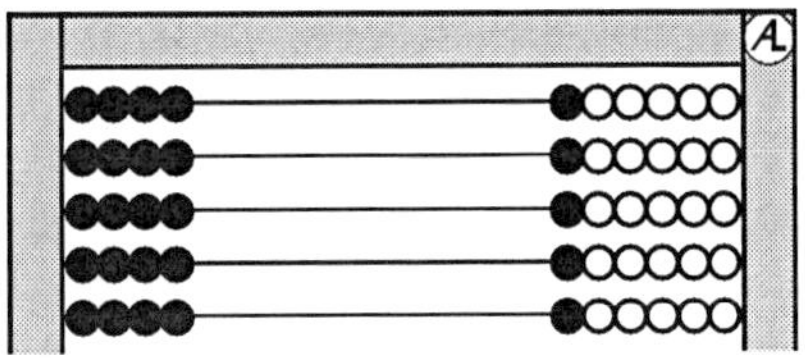

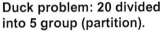

Duck problem: 20 divided into 5 group (partition).

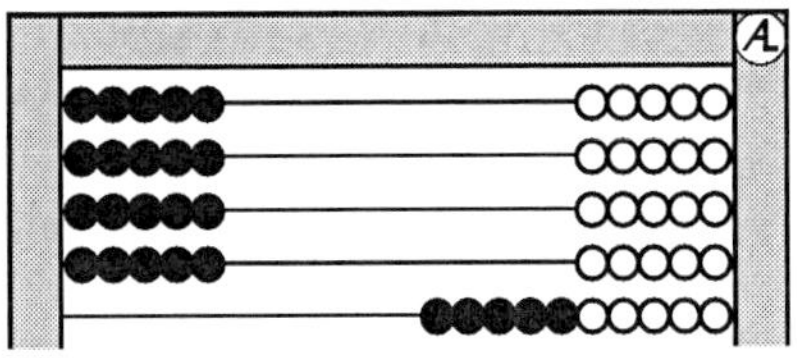

Wheel problem: 20 divided into groups of 4 (measurement)

Field trip problem. The following problem is a classic one, in which the final answer must be rounded up. An older child often is so engrossed in the calculations that he ignores the context and gives the wrong answer. A younger child tends to model the problem and find the correct solution.

<u>Thirteen children are going on an field trip together. Three children can ride in a car. How many cars are needed?</u> [5 cars] Ask the child to solve it any way that makes sense to him. Then ask him to explain his reasoning. It can be solved with tiles or on the abacus or by skip counting.

Ask for the equation. [$13 \div 3 = 4$ r1] Be sure the child realizes the answer to the problem is not the same as the answer to the equation.

Planting problem. Next give the child this problem, where the remainder must be rounded down. <u>Mrs. Lee is planting roses, exactly 10 in a row. She bought a box of 42 rose bushes. How many rows can she plant?</u> [4 rows]

Be sure the child understands that the answer is 4, and not 1 more, as it was in the previous problem. He may dislike the idea that 2 rose bushes must be "thrown away."

Cookie problem. A fractional answer makes the most sense for this next problem. <u>Scott has 13 cookies that he wants to divide evenly among 4 people. How many does each person get?</u> [3¼]

As before, discuss the child's answers. This time a remainder makes no sense–there won't be any cookies left. The equation is $13 \div 4 = 3\frac{1}{4}$.

Worksheet. The worksheet has similar problems. The solutions are below.

1. **10 teams, $62 \div 6 = 10$ r2**
2. **4 tables, $14 \div 4 = 3$ r2**
3. **6 packages, $55 \div 10 = 5$ r5**
4. **8 dollars, $33 \div 4 = 8$ r1**

Lesson 105

Parts and Wholes With Number of Groups

OBJECTIVE 1. To solve division problems involving finding the number of groups and the remainder

MATERIALS Worksheet 92, 96, or 104, "Quick Practice"
Worksheet 111, "Parts and Wholes With Number of Groups"
Tiles

WARM-UP Ask the child to solve the following problem in her journal: 817 × 23. [18,791] The check numbers are (7) × (5) = (8).

Ask the child What comes next, 60, 50, 40? [30] What comes next, 50, 45, 40? [35] What comes next, 12, 10, 8? [6] What comes next, 80, 72, 64? [56] What comes next, 70, 63, 56? [49]

ACTIVITIES ***Facts practice.*** For multiplication, use Worksheet 92 (answers in Lessons 89 & 91) and Worksheet 96 (answers in Lessons 92-93). For subtraction, use Worksheet 104 (answers in Lessons 99-100).

Problem 1. Give this problem to the child for her to discuss from the standpoint of parts and wholes. I have 12 tiles and I want to put them into groups of 4. How many groups can I make? [3] Draw a whole circle. What is the whole? [12] Write it in the whole circle as shown. How many in a group? [4] Draw an unconnected part circle as shown below on the left.

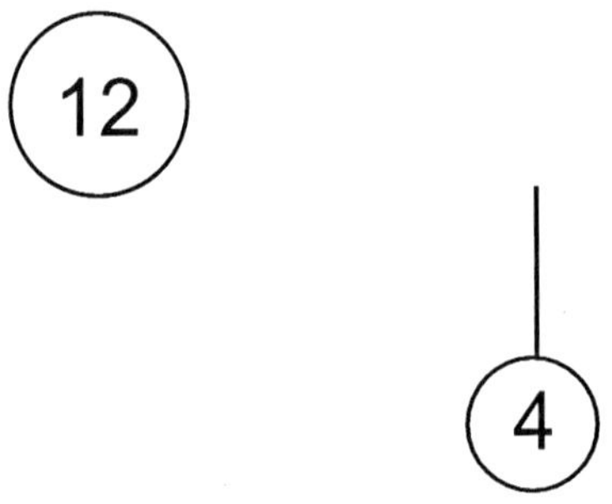

The *whole* circle and one *part* circle unconnected for problem 1.

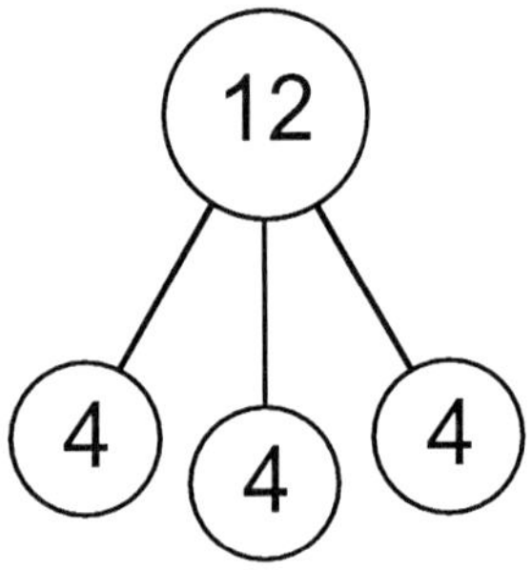

The part-whole circle showing the correct number of parts.

How many parts circles do I need? [3] Draw connected parts circles as she counts by 4s: 4, 8, 12. How many part circles do we have? [3] See the completed part-whole circles on the right above.

Ask her to write the equation. [12 ÷ 4 = 3] Remind her that a few days ago, she looked at the part-whole circles and wrote the equation a different way. What was that? [12 ÷ 3 = 4]

Problem 2. Sam is attending camp for 28 days. How many weeks is that? [4] Again draw the whole circle and unconnected part as shown on the next page.

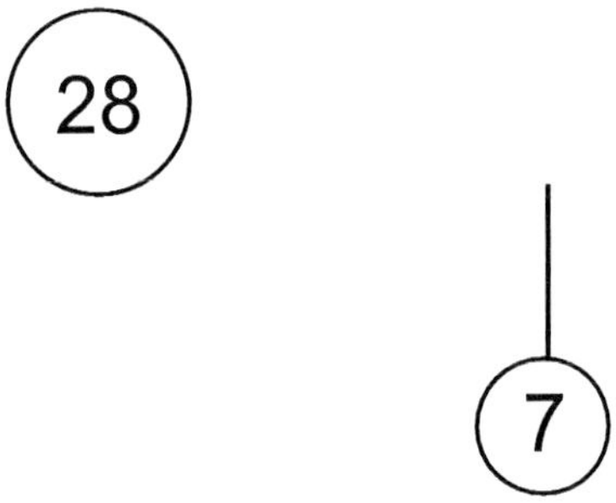

The *whole* circle and one *part* circle unconnected for problem 2.

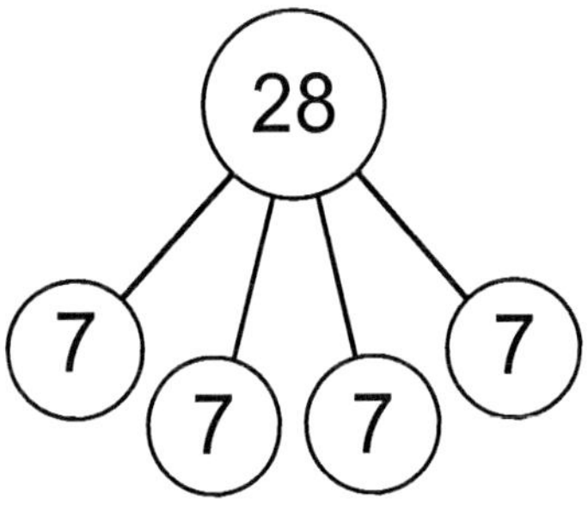

The part-whole circle showing the correct number of parts.

What is the whole? [28] Write it in the whole circle. How many in a group? [7, the number of days in a week] How many weeks in 28 days? [4] Ask for the equation. [28 ÷ 7 = 4]

Problem 3. It takes Sarah 20 minutes to walk around a park. How many times can she walk around it in 40 minutes? [2]

Draw the circles as above. Ask the child to complete it and write the equation. [There will 2 parts, 40 ÷ 20 = 2]

Worksheet. The worksheet has 6 part-whole circles to complete and some multiplication practice. The solutions are below.

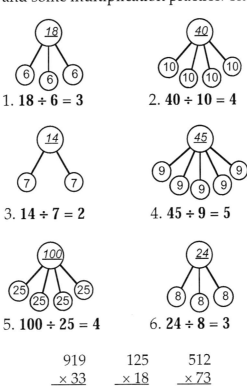

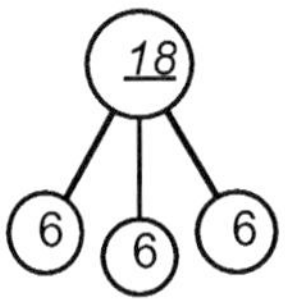

1. **18 ÷ 6 = 3**

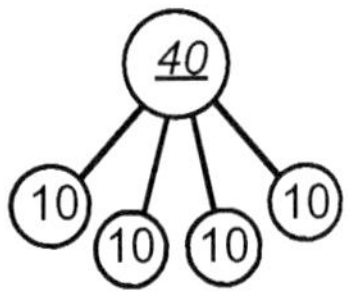

2. **40 ÷ 10 = 4**

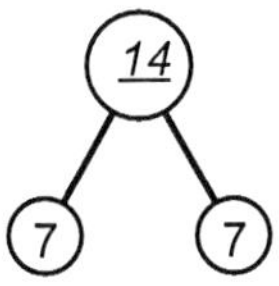

3. **14 ÷ 7 = 2**

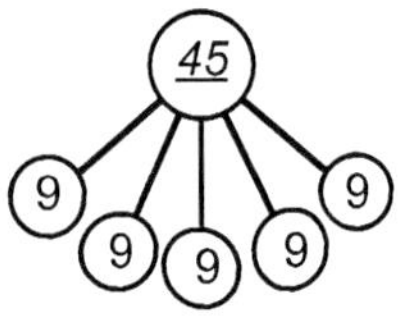

4. **45 ÷ 9 = 5**

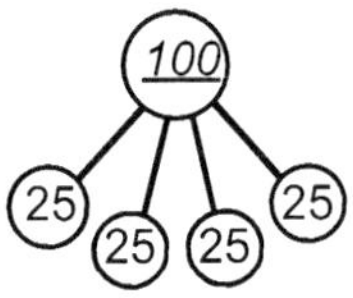

5. **100 ÷ 25 = 4**

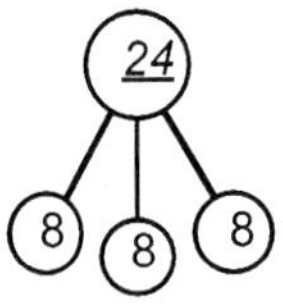

6. **24 ÷ 8 = 3**

919	125	512
× 33	× 18	× 73
30,327	**2,250**	**37,376**

Lesson 106

Problems Using Part Whole

OBJECTIVES
1. To solve different types of problems
2. To relate the problems to part whole
3. To learn the term *quotient*

MATERIALS Worksheet 92, 96, or 104, "Quick Practice"
Worksheet 112, "Problems Using Part Whole"

WARM-UP Ask the child to solve the following problem in his journal: 6423 × 35. [224,805] The check numbers are (6) × (8) = (3). Ask him to read the answer. [224 thousand 8 hundred 5]

Ask the child, How many are inches are in a foot? [12] How many ounces are in a quart? [32] How many hours are in a day? [24] How many minutes are in an hour? [60] How many degrees are in a circle? [360]

ACTIVITIES ***Facts practice.*** For multiplication, use Worksheet 92 (answers in Lessons 89 & 91) and Worksheet 96 (answers in Lessons 92-93). For subtraction, use Worksheet 104 (answers in Lessons 99-100).

The term quotient. Write

$$8 + 4 = 12$$

Point to the 12 and ask what name we call the number we get when we add 2 numbers. [sum or total]

Change the "+" sign to "–"

$$8 - 4 = 4$$

Point to the second 4 and ask what we call the number we get after subtracting. [remainder or difference]

Change the "–" sign to "×"

$$8 \times 4 = 32$$

Point to the 32 and ask what we call the number we get after multiplying. [product]

Change the "×" sign to "÷"

$$8 \div 4 = 2$$

Point to the 2 and tell the child the number we get when we divide 2 numbers is called the *quotient*. What is the quotient of 10 ÷ 2? [5] What is the quotient of 6 ÷ 2? [3] What is the quotient of 8 ÷ 2? [4] What is the quotient of 4 ÷ 2? [2]

Worksheet. The child is to work problems on the worksheet, complete the part-whole circles, and write the equations. Emphasize in writing the equations that he writes the operation he *needs* to find the solution.

Problem 1. Barry, Carrie, and Larry each have 18 dollars. How much do they have together? [$54]

Problem 2. Barry wants to spend his $18 on books. Each book costs $6. How many books can he buy? [3]

1.

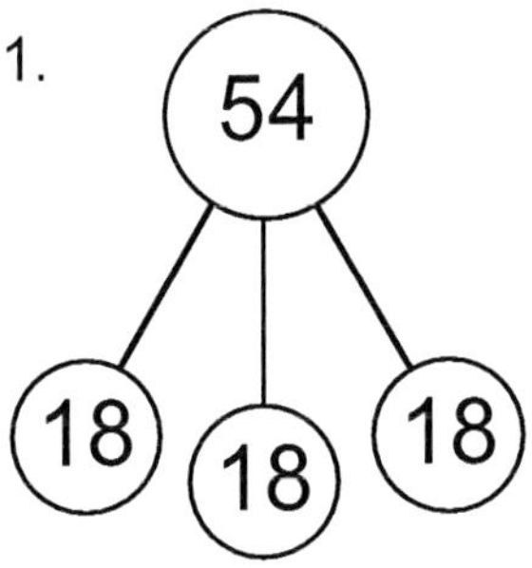

18 × 3 = 54

2.

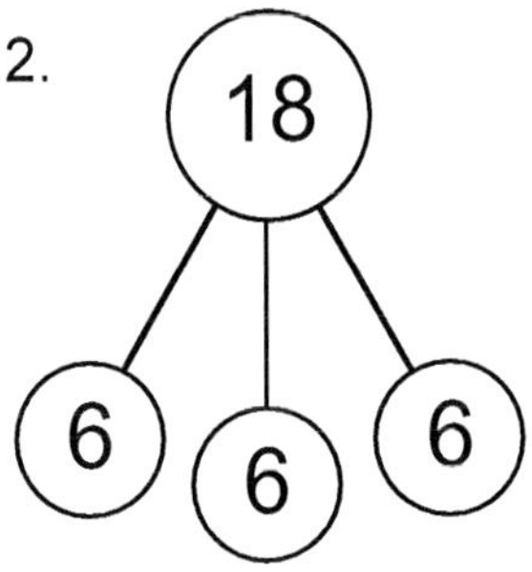

18 ÷ 6 = 3

Problem 3. Carrie earned $24.67. How much money does she have now? [$42.67]

Problem 4. Carrie spent $16. 32 for 4 plants. How much did each plant cost? [$4.08]

3.

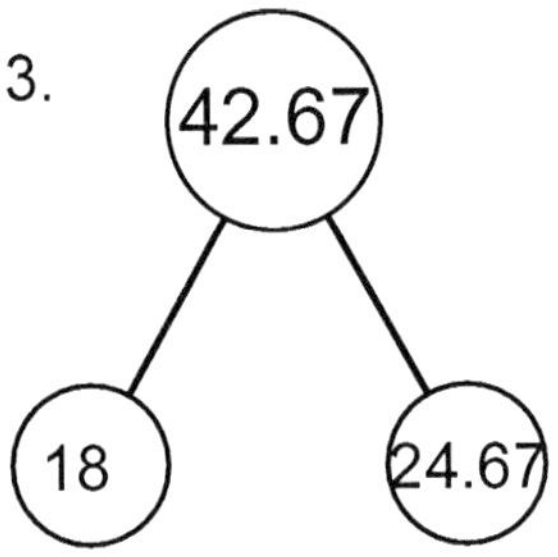

$18 + $24.67 = $42.67

4.

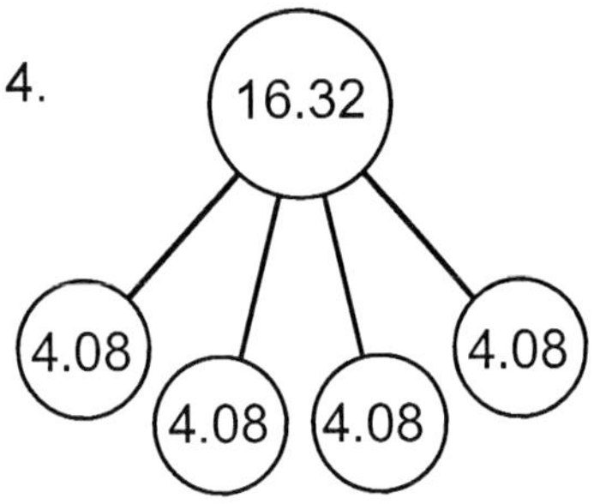

$16.32 ÷ 4 = $4.08

Problem 5. Larry spent 6 dollars and 43 cents for a gift. How much money does Larry have now? [$11.57]

Problem 6. Larry wants to buy 4 little gifts, which cost $2.85 each. Will he have enough money? [yes, total is $11.40]

5.

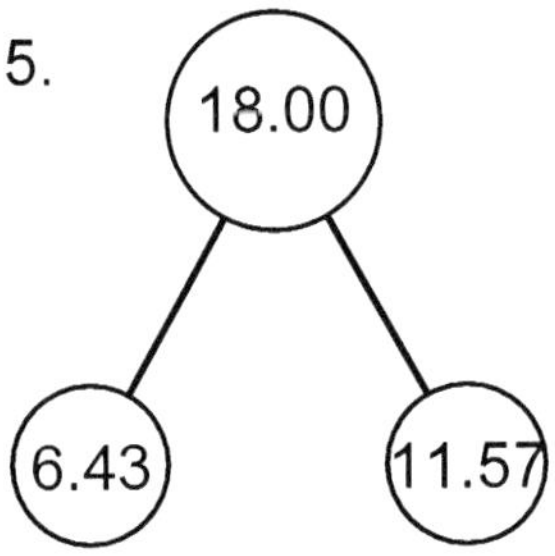

$18 – $6.43 = $11.57

6.

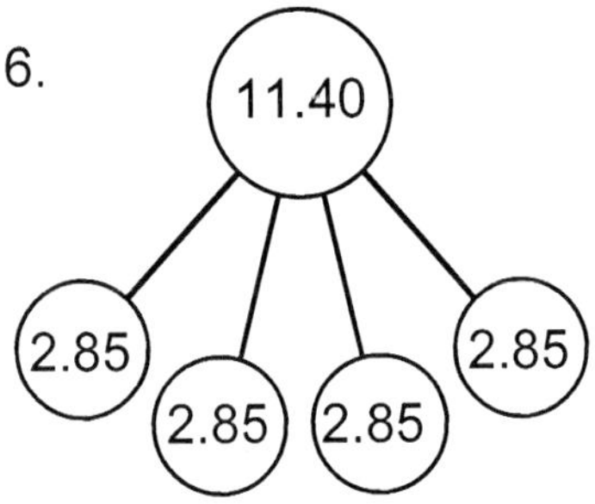

$2.85 × 4 = $11.40, yes

Lesson 107

Dividing With Multiples

OBJECTIVES
1. To connect multiplication and division through multiples
2. To practice division facts by playing Division Memory

MATERIALS
Worksheet 92, 96, or 104, Quick Practice
Worksheet 113, Dividing With Multiples-1
Cards for playing Division Memory

WARM-UP
Ask the child to solve the following problem in her journal: 5619 × 57. [320,283] The check numbers are (3) × (3) = (0). Ask a child to read the answer. [320 thousand 2 hundred 83]

ACTIVITIES
Facts practice. For multiplication, use Worksheet 92 (answers in Lessons 89 & 91) and Worksheet 96 (answers in Lessons 92-93). For subtraction, use Worksheet 104 (answers in Lessons 99-100).

Dividing by 2. Write the 2s multiples in 2 rows as shown.

2	4	6	8	10
12	14	16	18	20

Ask the child to look at the multiples to answer the following:

<u>How many 2s are in 8?</u> [4] <u>What is 8 divided by 2?</u> [4]

Ask her to point to the multiple that shows the answer. [The 4th multiple.]

<u>How many 2s are in 12?</u> [6] <u>What is 12 divided by 2?</u> [6]

Ask her to point to the multiple that shows the answer. [The 6th multiple.]

Continue with other facts; ask her to show the location.

Cover the 2s multiples. Ask her to see the multiples in her mind. Then ask the same or similar questions. Conclude with what is 2 divided by 2. [1]

Division Memory Game. Use the same cards that were needed for Skip Counting Memory or Multiplication Memory to play Division Memory.

Demonstrate a sample game. Emphasize turning over the multiplication card first and saying the equation before turning over the basic number card. If possible, have the child play two or more games, each time with different sets.

Worksheet. The worksheet asks for skip counting tables for 2s, 4s, 6s, and 8s. Then division problems are to be answered from those tables. The answers are on the next page.

3	10	1	8	2
6	7	9	4	5
3	1	5	7	8
6	9	4	2	10
2	4	9	3	7
10	1	8	6	5
10	1	7	9	8
6	3	5	2	4

DIVISION MEMORY

Purpose To help the players practice the division combinations for a particular divisor

Cards 1 set of multiplication cards, along with the multiples (on the envelope)
1 set of basic number cards 1 to 10
A small card with the divisor (multiples) number preceded by ÷ and followed by = as shown below

Number of players Two or two teams with the players sitting on the same side of the cards

Object of the game To collect the most cards by completing the division equation

Layout Lay the multiplication cards face down in 2 rows as shown below. To the right in separate rows lay the basic number cards. Place the card naming the divisor and the math symbols between the 2 groups of cards as shown below.

Play The first player turns over a multiplication card and states the fact. For example, if the 6s are being played and the card is 24, the player states, 24 divided by 6 equals 4. The player then decides where the 4 might be among the basic number cards. If the correct card is found, the player collects both cards and takes another turn. If it is the wrong card, both cards are returned face down in their original places and the other player takes a turn.

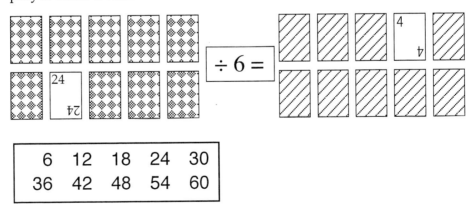

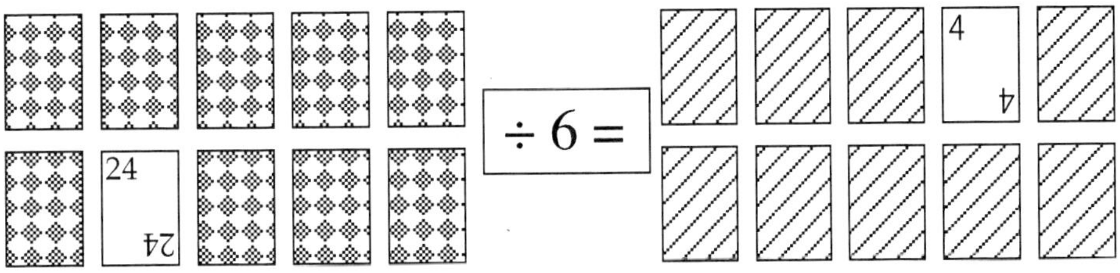

6	12	18	24	30
36	42	48	54	60

Lesson 108

Review

OBJECTIVE 1. To review and practice

MATERIALS Worksheet 114-A and 114-B, "Review" (2 versions)
Cards for playing games

ACTIVITIES ***Review worksheet.*** Give the child about 15 to 20 minutes to do the review sheet. The oral problems to be read as follows:

10 squared = 2 – 1½ = ½ of 18 =

Review Sheet 114-A.

1. Write only the answers to the oral questions. 100 $\frac{1}{2}$ 9

4. Write only the answers. 9 squared = 81 136 – 98 = 38 $3\frac{1}{2} + 3\frac{1}{2}$ = 7

Add parts and numbers to the part-whole circles sets. Then write the equations.

7. Four sisters are sharing a dozen cookies. How many cookies does each one get?

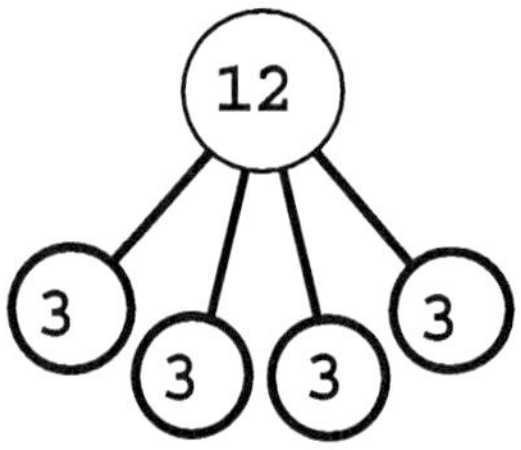

12 ÷ 4 = 3

9. Sammy pays 25¢ for a bus token. How many tokens can Sammy get for 75¢?

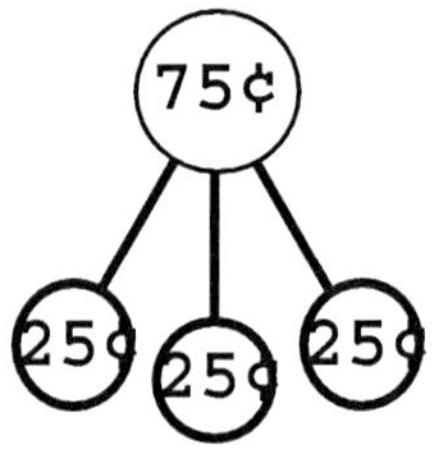

75¢ ÷ 25¢ = 3 tokens

11. Write the multiples of 3.

3 6 9
12 15 18
21 24 27
30

13. Use the multiples to find the quotients.

6 ÷ 3 = 2 21 ÷ 3 = 7
18 ÷ 3 = 6 27 ÷ 3 = 9
9 ÷ 3 = 3 24 ÷ 3 = 8
3 ÷ 3 = 1 15 ÷ 3 = 5

21.
```
   7608
  ×  74
  30432
 532560
 562,992
```

22.
```
  7561
+ 7649
 15,210
```

23.
```
  1232
 – 135
  1097
```

The oral problems to be read to the child are as follows:

9 squared = 1 – ½ = ½ of 16 =

Review Sheet 114-B.

1. Write only the answers to the oral questions. **81** $\frac{1}{2}$ **8**

4. Write only the answers. 1 squared = **1** 127 – 99 = **28** $2\frac{1}{2} + 3\frac{1}{2} =$ **6**

Add parts and numbers to the part-whole circles sets. Then write the equations.

7. Three brothers are splitting a dozen cookies. How many cookies does each one get?

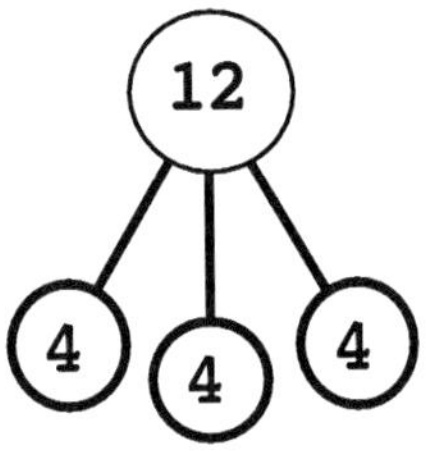

12 ÷ 3 = 4

9. Silvia pays 50¢ for a bus token. How many tokens can Silvia get for $2?

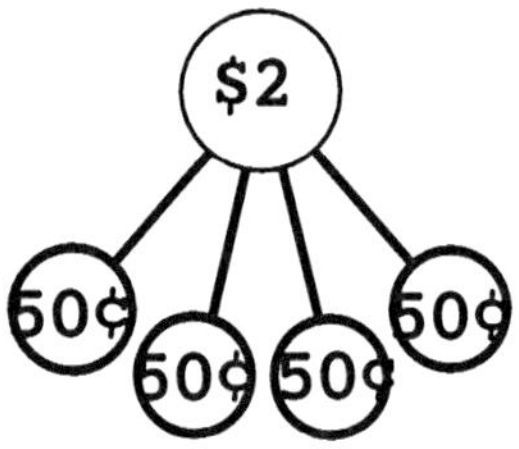

$2 ÷ 50¢= 4 tokens

11. Write the multiples of 4.

4 **8** **12** **16** **20**

24 **28** **32** **36** **40**

13. Use the multiples to find the quotients.

8 ÷ 4 = **2** 20 ÷ 4 = **5**

12 ÷ 4 = **3** 28 ÷ 4 = **7**

36 ÷ 4 = **9** 24 ÷ 4 = **6**

4 ÷ 4 = **1** 16 ÷ 4 = **4**

21. 8409 × 27 = **58863** + **168180** = **227,043**

22. 8349 + 5671 = **14,020**

23. 1483 – 493 = **990**

Games. Spend the remaining time playing games.

DIVISION. Division Memory (Lesson 107).

MULTIPLICATION. Multiplication Card Speed (*Math Card Games*, P31).

Lesson 109

Two-Step Problems Using Part-Whole

OBJECTIVES
1. To solve different types of problems
2. To relate the problems to part whole

MATERIALS Worksheet 92, 96, or 104, "Quick Practices"
Worksheet 115, "Two-Step Problems Using Part Whole"

WARM-UP Ask the child to solve the following problem in her journal: 7925 × 48 [380,400] The check numbers are (5) × (3) = (6). Ask the child to read the answer. [380 thousand 4 hundred]

Ask the child How many inches in a foot? [12] How many ounces in a quart? [32] How many hours in a day? [24] How many minutes in an hour? [60] How many degrees in a circle? [360] [100] How many eggs in a dozen? [12]

ACTIVITIES ***Facts practice.*** For multiplication, use Worksheet 92 (answers in Lessons 89 & 91) and Worksheet 96 (answers in Lessons 92-93). For subtraction, use Worksheet 104 (answers in Lessons 99-100).

Reviewing terms. Write the following:

$10 + 2 = 12$

$10 - 2 = 8$

$10 \times 2 = 20$

$10 \div 2 = 5$

What do we call the 12? [sum]

What do we call the 8? [remainder or difference]

What do we call the 20? [product]

What do we call the 5? [quotient]

Worksheet. Ask the child to work the first problem on the worksheet. She needs to complete the part whole circles and write the equations. Emphasize that in writing the equations she is to write the equation that tells how she found the solution. Then ask her to share her solution.

Explain there are two whole circles because she needs to do two operations to get the answer.

Problem 1. Four people order 2 pizzas, which are cut into 8 pieces. How many pieces does each person get?

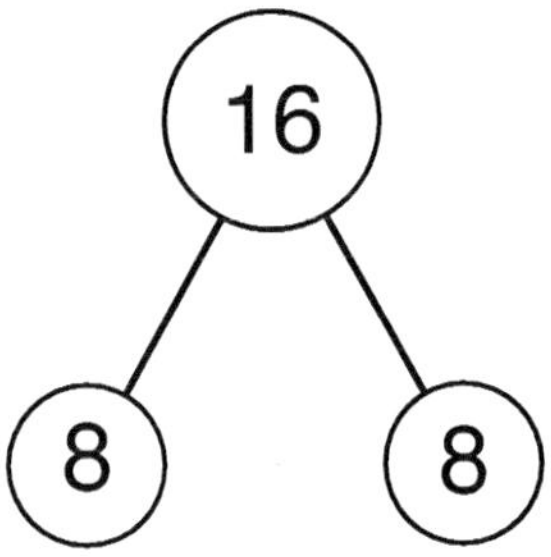

$8 \times 2 = 16$

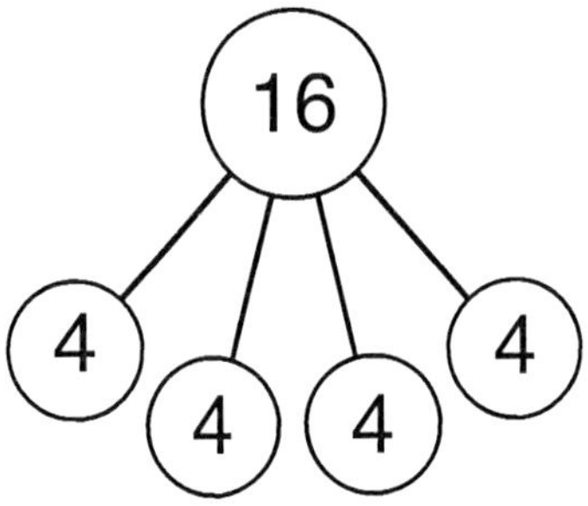

$16 \div 4 = 4$

Problem 2. Lynn practices the violin for one-half hour each day. Lynn also plays during the weekly lesson for 60 minutes. How many hours does Lynn play the violin in a week? [4½]

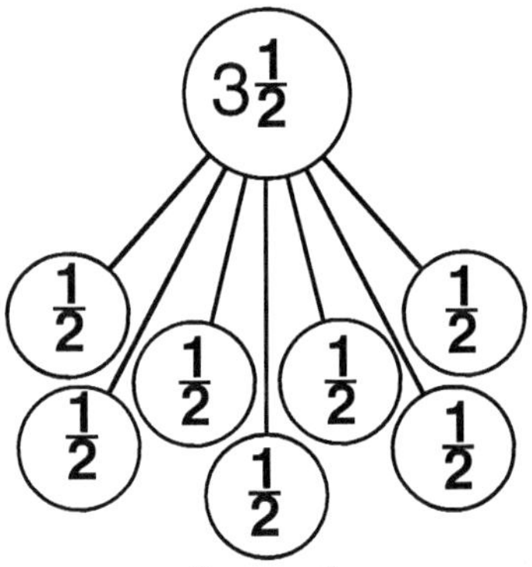

$7 \times \frac{1}{2} = 3\ \frac{1}{2}$

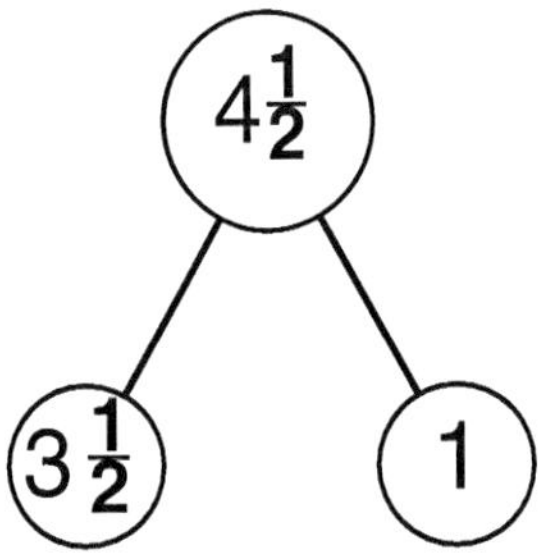

$3\frac{1}{2} + 1 = 4\ \frac{1}{2}$

Problem 3. Jo wants to read 100 pages. Jo read 22 pages on Monday, 29 pages on Tuesday, and only 17 pages on Wednesday. How many pages does Jo still need to read? [32]

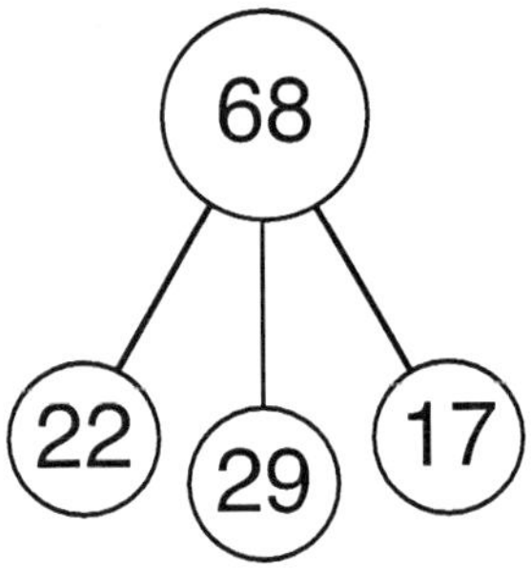

$22 + 29 + 17 = 68$

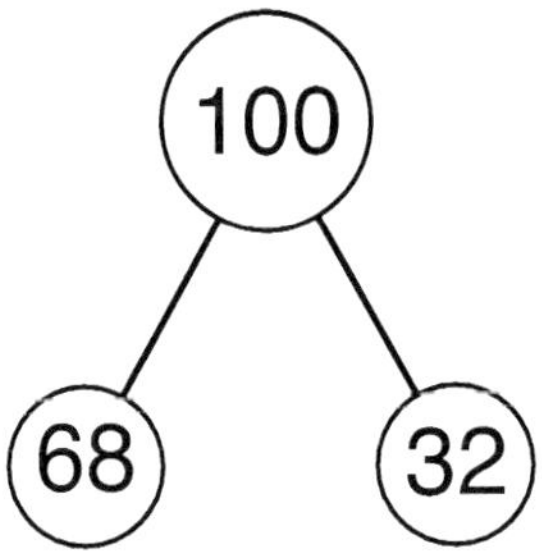

$100 - 68 = 32$

Lesson 110 (1 or 2 days)

Division Problems With Money

OBJECTIVES
1. To review the value of coins
2. To solve division money problems

MATERIALS
Worksheet 116, Dividing With Multiples
Plastic coins, optional
Worksheet 117, Division Problems With Money
Cards for Division Memory (Lesson 107)

WARM-UP
Ask the child to solve the following problem in his journal: 5614×85 [477,190] The check numbers are $(7) \times (4) = (1)$. Ask a child to read the answer. [477 thousand 1 hundred ninety]

ACTIVITIES
Dividing by 9. Ask the child to write the 9s multiples on the first worksheet. See below.

9	18	27	36	45
90	81	72	63	54

Ask him to look at the multiples to answer the following:

<u>How many 9s are in 36?</u> [4] <u>What is 36 divided by 9?</u> [4]

Ask him to describe where the multiple is that shows the answer. [the 4th multiple]

<u>How many 9s are in 63?</u> [7] <u>What is 63 divided by 9?</u> [7]

Ask him to point to the multiple that shows the answer. [the 7th multiple; that is, the second multiple in the second row]

Ask him to do the divisions by 9 on the first worksheet and look for patterns. [The quotient is 1 more than the number in tens place; for example, 36 ÷ 9 = **4.** The two quotients in each box sum to 11.] The answers are:

4	6	9	3	10
7	5	2	8	1

Dividing by 5. Ask the child to write the 5s multiples. Then ask him to find the multiple to answer the following:

<u>How many 5s are in 20?</u> [4] <u>What is 20 divided by 5?</u> [4, the 4th multiple.]

<u>How many 5s are in 30?</u> [6] <u>What is 30 divided by 5?</u> [6, the 6th multiple.]

Ask if those answers remind him of the clock. [the minute and hour numbers] Continue with other multiples. The answers are

5	6
1	9
4	7
10	2
8	3

When dividing a multiple of 10 by 5, the quotient will be twice the number of tens. For example, 30 divided by 5 is twice 3, which is 6.

Note: See Lesson 117 for a visual way of using the abacus to understand dividing by 5s.

Dividing by 3 and 7. Ask the child to complete the worksheet. The answers are:

7	5	8
10	1	2
9	4	6
3		

2	8	6
3	10	5
7	1	4
9		

Worksheet. The second worksheet has two money problems.

Problem 1. The following problem, which is on the worksheet, may be difficult for the child. Tell him he will have to think hard, but he can do it. Tell him he will have about 10 to 15 minutes to do it. Ask him to explain his work on the worksheet.

Abby, Brianna, and Carla bought a pizza for $4.68. How much money does each one need to pay? [$1.56]

The child's work might look something like the following

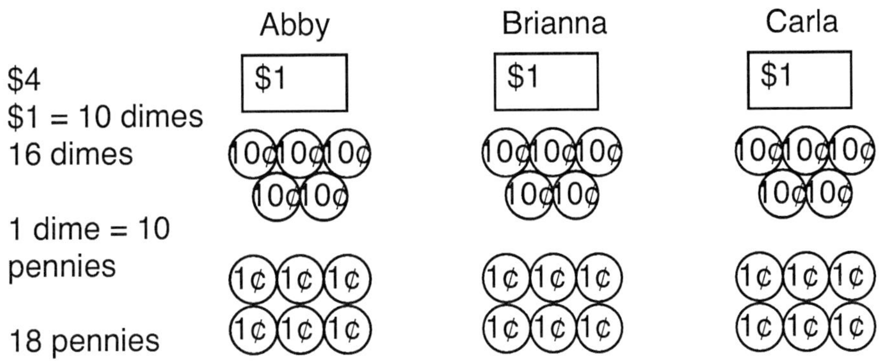

A solution to the problem of dividing $4.68 by 3.

Problem 2. Andy, Byron, Carlos, and Damian bought a pizza for $10.72. How much does each one owe? [$2.68]

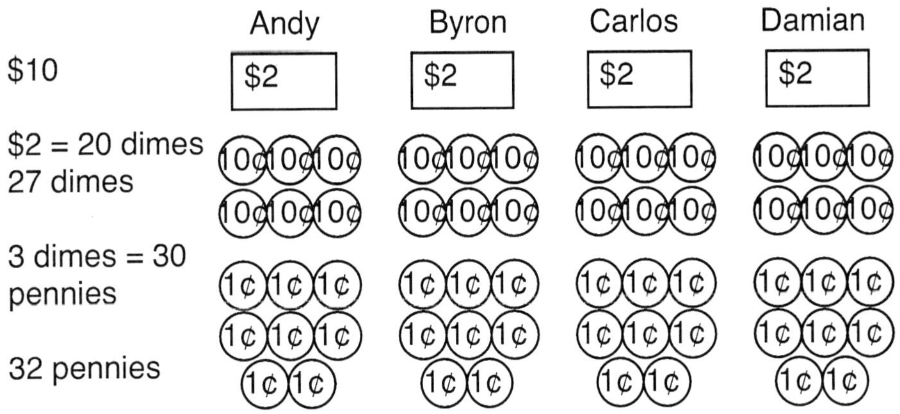

Division practice. Ask the child to play the Division Memory game (Lesson 107).

Lesson 111

The Dividing Line

OBJECTIVES
1. To use the calculator for dividing
2. To write division equations with the dividing line

MATERIALS Worksheet 118, "Quick Practice-16"
Worksheet 119, "The Dividing Line"

WARM-UP Ask the child to solve the following problem in her journal: 8125 × 96. [780,000] The check numbers are (7) × (6) = (6).

ACTIVITIES ***Facts practice.*** Take out Quick Practice-16 and ask the child to do the facts in the **first** rectangle. Ask her to correct them. The answers are:

6	2	9	4	3
7	8	1	10	5
6	8	5	1	3
2	7	4	9	10
7	3	2	4	8

Dividing on a calculator. Take out the calculator. Tell the child dividing on a calculator is easy. For a Casio calculator, the procedure to divide 18 by 3 is as follows:

Press 18 [÷] 3 [=]. (The answer 6 will show.)

Then give her the following problems to do on the calculator. There is no reason to write the equations.

<u>There are 56 crayons for 8 children. How many crayons is that for each child?</u> [7]

<u>Four children win a prize of $50. How much money does each one receive?</u> [$12.50] The calculator will show 12.5.

<u>There are 396 people living in an apartment building with 18 floors. If each floor has the same number of occupants, how many people are on each floor?</u> [22]

Practicing facts on a calculator. Write the following

$27 \div 9 =$ $\quad 81 \div 9 =$ $\quad 45 \div 9 =$

Challenge the child to see if she can figure out how to solve these problems without pressing [÷] 9 each time. For a Casio press 9 and [÷] [÷]. (A k will appear on the display.)

Then press 27 and [=], giving the answer 3.

Then press 81 and [=], giving the answer 9.

Then press 45 and [=], giving the answer 5.

Note: You might want to write the 9s on the board.

9s practice. Ask the child to find the quotient on the calculator as you say the multiples of 9 in random order. Challenge her to think of the answer before it shows on the calculator.

The dividing line. Ask the child to write the equation for 10 tiles divided among 5 people. [10 ÷ 5 = 2] Tell her there is another way to write division problems. Write

$$\frac{10}{5}$$

and say that it means 10 divided by 5. Ask what it equals [2]; and show her where to write the equals sign and the quotient.

$$\frac{10}{5} = 2$$

Ask her to write 6 divided by 2 and ask how much it is.

$$\frac{6}{2} = 3$$

Write

$$20 \div 5 = 4$$

and ask her to write it with the dividing line.

$$\frac{20}{5} = 4$$

Next tell her you have a tricky one for her; write

$$1 \div 2$$

and ask how to write it with a dividing line,

$$\frac{1}{2}$$

and ask her to read it. [1 divided by 2] Ask what it equals. [½] Ask if that makes sense. Tell her there is nothing else we can write that is simpler than one half, so we just leave it as one half.

Note: For complete understanding of fractions, the child must see them also as a type of division.

Ask how to write the equation for 5 cookies divided between 2 people.

$$\frac{5}{2} = 2\frac{1}{2}$$

Worksheet. The worksheet has similar problems, Answers follow:

7	5	4	2	9
3	1	8	10	6
6	7	3	2	10
8	9	1	4	5
5	$4\frac{1}{2}$	4	$3\frac{1}{2}$	3
$2\frac{1}{2}$	2	$1\frac{1}{2}$	1	$\frac{1}{2}$

$80 \div 10 = 8$ $30 \div 10 = 3$ $20 \div 2 = 10$

$\frac{80}{10} = 8$ $\frac{30}{10} = 3$ $\frac{20}{2} = 10$

$18 \div 3 = 6$ $21 \div 3 = 7$ $14 \div 2 = 7$

$\frac{18}{3} = 6$ $\frac{21}{3} = 7$ $\frac{14}{2} = 7$

Lesson 112

Non-Unit Fractions

OBJECTIVES
1. To review unit fractions
2. To understand the meaning of non-unit fractions, such as 2⁄3

MATERIALS
Worksheet 118, "Quick Practice-16"
Fraction strips cut apart with a scissors, 3 sets. (To keep the sets separate, code each set with a different letter on the backs.)
A fraction chart, kept whole
Worksheet 120, "Non-Unit Fractions"

WARM-UP
Ask the child to solve the following problem in his journal: 555 × 23. [12,765] The check numbers are (6) × (5) = (3).

ACTIVITIES
Facts practice. Ask the child to take out his Quick Practice sheet and ask him to do the facts in the **second** rectangle. Ask him to correct them. The answers are:

4	5	10	6	3
7	9	8	1	2
4	10	9	7	1
6	2	8	5	3
4	3	6	7	9

Unit fractions. Give the fraction set to the child. Write

$$\frac{1}{6}$$

and ask what it means. [1 divided by 6] Tell him when we use a dividing line to show division, the division is called a *fraction.*

Note: Ask the child to first review the facts in the first rectangle.

Then ask him the following:

1. <u>What fraction is 1 divided by 2?</u> Show the piece. [1⁄2 piece]
2. <u>What is 1 divided by 3?</u> Show the piece. [1⁄3 piece]
3. <u>What is 1 divided by 4?</u> Show the piece. [1⁄4 piece]
4. Show a tenth. [1⁄10 piece]

Fraction chart. Show the child the whole fraction chart and ask him to build the chart with the pieces. See the top of the next page.

Then ask the following questions.

1. <u>How many one-thirds are needed to make 1?</u> [3]
2. <u>How many fourths are needed to make 1?</u> [4]
3. <u>How many tenths make 1?</u> [10]
4. <u>Which is more, one-fifth or one-fourth?</u> [one-fourth]

Note: It is important to ask these questions in different ways.

Finding 2 divided by 3. Take out another set of fraction pieces, so the child has two sets.

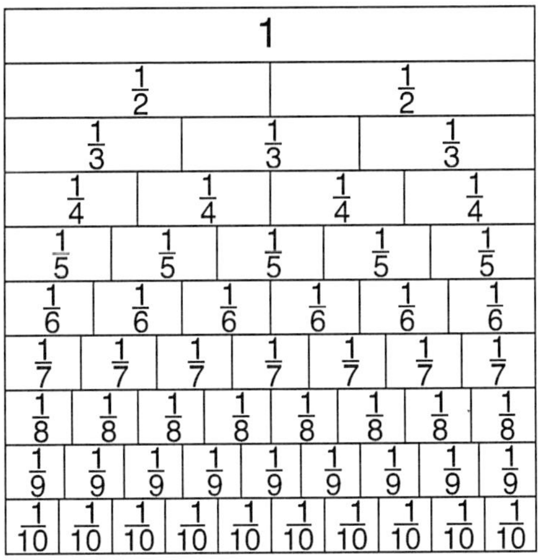

The fraction chart.

Write

$$\frac{2}{3}$$

and ask him to find 2 divided by 3, using one-thirds. [two ⅓s] Ask for explanations, including a figure to explain his reasoning. One approach is shown below. He could think that 1 divided by 3 is ⅓, so 2 divided by 3 must be twice as much, or ⅔.

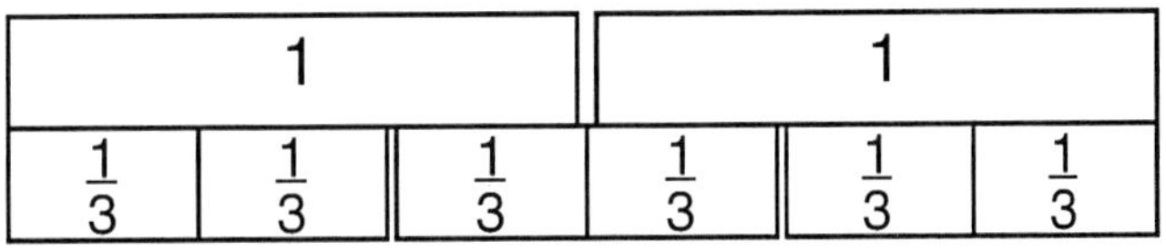

Showing 2 divided by 3 equals two ⅓s.

Summarize by asking, <u>How many one-thirds are in two-thirds?</u> [2]

Finding 3 divided by 4. Now ask the child to find 3 divided by 4, using one-fourths. See 1 solution below.

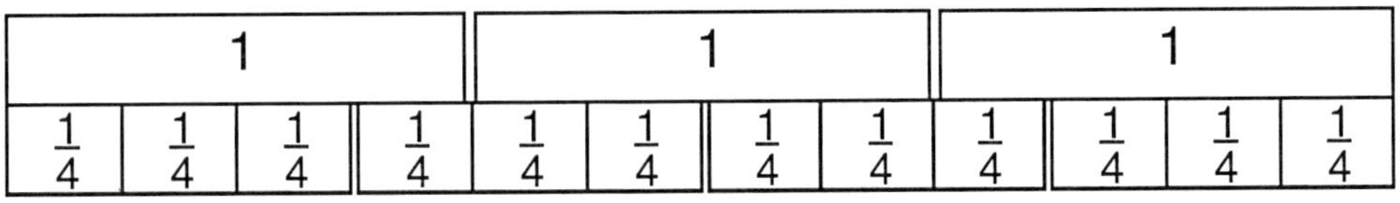

Showing 3 divided by 4 equal to three ¼s.

Worksheet. The first problem is the same as the first problem above. The second problem is dividing 3 circles by 4, similar to the second problem. The shaded areas below form the fourth portion.

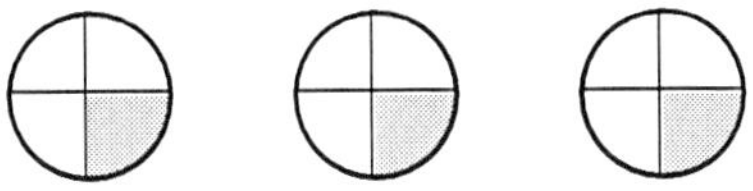

4 6 3 8 4

Lesson 113

Fractions Equaling One

OBJECTIVES
1. To review non-unit fractions
2. To focus on fractions equaling 1; for example, $\frac{1}{3}$ & $\frac{2}{3}$

MATERIALS
Worksheet 118, "Quick Practice-16"
Fraction strips
A fraction chart, kept whole
20 fraction cards for playing Concentrating on One: two $\frac{1}{2}$s and one of each of the following: $\frac{1}{3}$, $\frac{2}{3}$, $\frac{1}{4}$, $\frac{3}{4}$, $\frac{1}{5}$, $\frac{2}{5}$, $\frac{3}{5}$, $\frac{4}{5}$, $\frac{1}{6}$, $\frac{5}{6}$, $\frac{1}{8}$, $\frac{3}{8}$, $\frac{5}{8}$, $\frac{7}{8}$, $\frac{1}{10}$, $\frac{3}{10}$, $\frac{7}{10}$, and $\frac{9}{10}$
Concentrating on One Game (*Math Card Games*, F3)

WARM-UP
Ask the child to solve the following problem in her journal: 666 × 35 [23,310] The check numbers are (0) × (8) = (0).

Write the 5s in two columns as shown.

5	10
15	20
25	30
35	40
45	50

Ask, What is 20 ÷ 5? [4] What is 30 ÷ 5? [6] What is 40 ÷ 5? [8] What is 45 ÷ 5? [9] What is 25 ÷ 5? [5] What is 35 ÷ 5? [7]

ACTIVITIES
Facts practice. Ask the child to take out her Quick Practice sheet and ask her to do the facts in the **third** rectangle. Ask her to correct them. The answers are:

6	**3**	**1**	**2**	**8**
5	**3**	**4**	**4**	**5**
3	**7**	**6**	**9**	**7**
1	**2**	**7**	**9**	**6**
8	**9**	**10**	**4**	**10**

Fractions equaling 1. Take out the fraction set. Ask the child to find the 1 and to set it aside.

Draw a part whole circle set; write 1 in the whole and $\frac{3}{5}$ in a part circle. See the figure below on the left. Ask her to find pieces to represent it and to lay them under the 1. How much is needed to make 1? [$\frac{2}{5}$] See below.

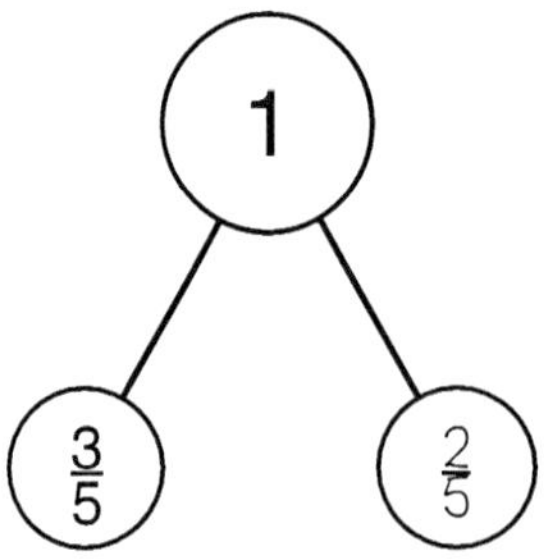

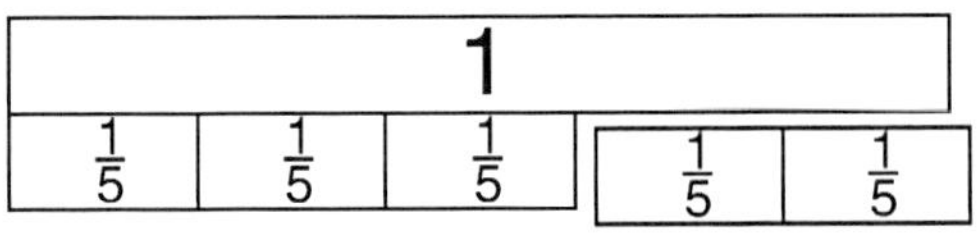

Showing that 2 fifths are needed with 3/5 to make 1.

What fraction goes in the other part circle? [$\frac{2}{5}$]

Change the left part circle to 3⁄8 and erase the right part circle. What is needed to make 1? [5⁄8] Ask her to show it with the fraction pieces. See the figures below.

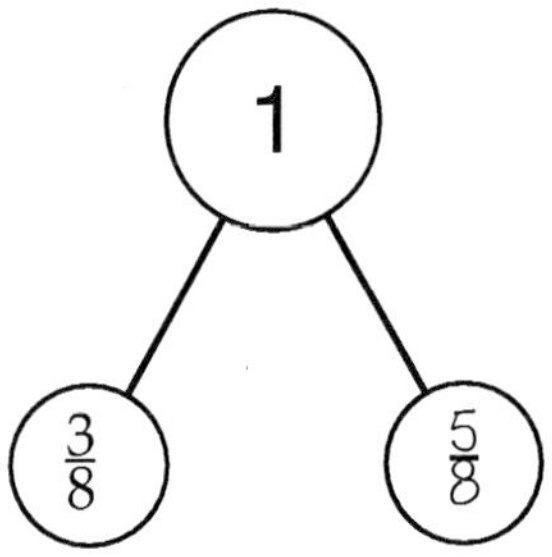

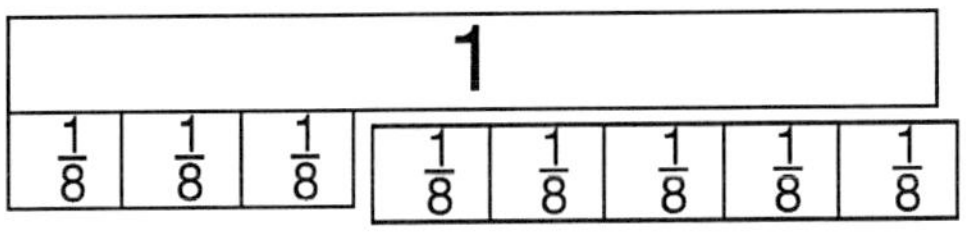

Showing that 5 eighths are needed with 3⁄8 to make 1.

Repeat for 1⁄10 [9⁄10] and 1⁄2. [1⁄2]

Matching pairs. Take out the sets of fraction cards needed for the Concentrating on One game. Ask the child to lay out the 20 fraction cards face up. Explain that she is to find the pairs that equal 1. Suggest she pick up a card, say what is needed to make 1, find it, and set the pair aside, face up on 2 piles.

Besides making the pairs visible, the 2 stacks make shuffling for the next activity or game unnecessary.

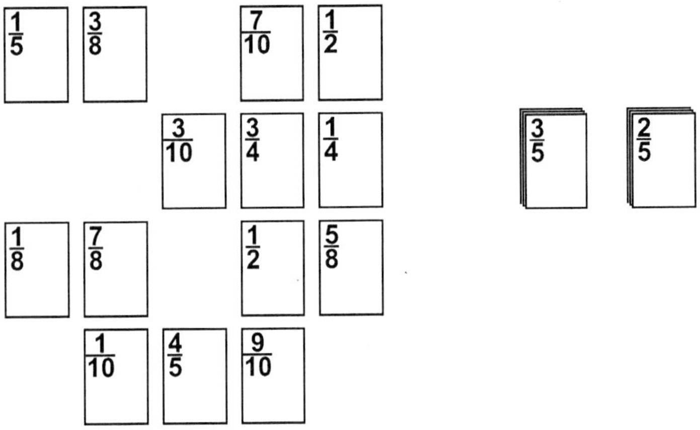

Collecting pairs that equal 1.

Concentrating on One game. Using the same cards, the child can play Concentrating on One, a memory game where the pairs are equal 1. See *Math Card Games* (F3) for the details.

Lesson 114

Review

OBJECTIVE 1. To review and practice

MATERIALS Worksheet 121-A and 121-B, "Review" (2 versions)
Cards for playing games

ACTIVITIES ***Review worksheet.*** Give the child about 15 to 20 minutes to do the review sheet. The oral problems to be read are as follows:

100 × 10 = 2 – 1½ = ⅕ of 10 =

Review Sheet 121-A.

1. Write only the answers to the oral questions. **1000** $\frac{1}{2}$ **2**

4. Write only the answers. $3\frac{1}{2} + 1\frac{1}{2}$ = **5** 141 – 13 = **128** 39 + 58 = **97**

7. Write the fractions in the rectangles.

1				
$\frac{1}{2}$	$\frac{1}{2}$			
$\frac{1}{3}$	$\frac{1}{3}$	$\frac{1}{3}$		
$\frac{1}{4}$	$\frac{1}{4}$	$\frac{1}{4}$	$\frac{1}{4}$	
$\frac{1}{5}$	$\frac{1}{5}$	$\frac{1}{5}$	$\frac{1}{5}$	$\frac{1}{5}$

22. From these fractions, write the pairs that equal 1.

$\frac{1}{2}$ $\frac{7}{8}$ $\frac{2}{3}$ $\frac{1}{3}$ $\frac{1}{2}$ $\frac{3}{8}$ $\frac{5}{8}$ $\frac{1}{8}$

$\frac{1}{2}$ $\frac{1}{2}$

$\frac{7}{8}$ $\frac{1}{8}$

$\frac{2}{3}$ $\frac{1}{3}$

$\frac{3}{8}$ $\frac{5}{8}$

23. Write the multiples of 3.

3 **6** **9**

12 **15** **18**

21 **24** **27**

30

33. Use the multiples to find the quotients.

24 ÷ 3 = **8** 27 ÷ 3 = **9**

30 ÷ 3 = **10** 9 ÷ 3 = **3**

6 ÷ 3 = **2** 18 ÷ 3 = **6**

15 ÷ 3 = **5** 21 ÷ 3 = **7**

34.
```
     3 5 6 8
       × 3 7
   2 4 9 7 6
 1 0 7 0 4 0
 1 3 2,0 1 6
```

35.
```
  2 8 5 6
 + 8 9 4
  3 7 5 0
```

36.
```
  2 8 0 2
 – 4 0 4
  2 3 9 8
```

The oral problems to be read to the child are as follows:

$200 \times 10 =$ $1\frac{1}{2} + 3\frac{1}{2} =$ $\frac{1}{3}$ of $30 =$

Review Sheet 121-B.

1. Write only the answers to the oral questions. **2000** **5** **10**

4. Write only the answers. $3\frac{1}{2} + 2\frac{1}{2} =$ **6** $180 - 34 =$ **146** $86 + 33 =$ **119**

7. Write the fractions in the rectangles.

1																			
$\frac{1}{2}$										$\frac{1}{2}$									
$\frac{1}{3}$							$\frac{1}{3}$							$\frac{1}{3}$					
$\frac{1}{4}$					$\frac{1}{4}$					$\frac{1}{4}$					$\frac{1}{4}$				
$\frac{1}{5}$				$\frac{1}{5}$				$\frac{1}{5}$				$\frac{1}{5}$				$\frac{1}{5}$			

22. From these fractions, write the pairs that equal 1.

$\frac{1}{2}$ $\frac{5}{6}$ $\frac{2}{5}$ $\frac{1}{2}$ $\frac{2}{4}$ $\frac{3}{5}$ $\frac{2}{4}$ $\frac{1}{6}$

$\frac{1}{2}$ $\frac{1}{2}$

$\frac{5}{6}$ $\frac{1}{6}$

$\frac{2}{5}$ $\frac{3}{5}$

$\frac{2}{4}$ $\frac{2}{4}$

23. Write the multiples of 3.

3 **6** **9**

12 **15** **18**

21 **24** **27**

30

33. Use the multiples to find the quotients.

$9 \div 3 =$ **3** $18 \div 3 =$ **6**

$27 \div 3 =$ **9** $21 \div 3 =$ **7**

$15 \div 3 =$ **5** $12 \div 3 =$ **4**

$30 \div 3 =$ **10** $6 \div 3 =$ **2**

34.
$$\begin{array}{r} 7246 \\ \times\ 92 \\ \hline 14492 \\ 652140 \\ \hline 666{,}632 \end{array}$$

35.
$$\begin{array}{r} 6942 \\ +\ 789 \\ \hline 7731 \end{array}$$

36.
$$\begin{array}{r} 3561 \\ -\ 864 \\ \hline 2697 \end{array}$$

Games. Spend the remaining time playing games.

DIVISION. Division Memory (Lesson 107) and Mixed-Up Quotients (D3) *(Math Card Games).*

FRACTIONS. Concentrating on One *(Math Card Games,* F3).

Lesson 115

Comparing Fractions

OBJECTIVES
1. To review the "<" and ">" symbols
2. To compare fractions

MATERIALS
Worksheet 122, "Quick Practice-17"
Fraction strips
A fraction chart, kept whole
Worksheet 123, "Comparing Fractions"

WARM-UP Ask the child to solve the following problem in her journal: 7777 × 68. [528,836] The check numbers are (1) × (5) = (5).

Ask the child to write the 3s with three in a column as shown.

3	6	9
12	15	18
21	24	27
30		

Ask, What is 9 ÷ 3. [3] What is 12 ÷ 3? [4] What is 27 ÷ 3? [9] What is 24 ÷ 3? [8] What is 30 ÷ 3? [10] What is 21 ÷ 3? [7]

ACTIVITIES ***Facts practice.*** Take out Quick Practice-17 and ask the child to do the facts in the **first** rectangle. Ask her to correct them. The answers are:

5	**9**	**10**	**8**	**3**
5	**2**	**4**	**7**	**6**
2	**4**	**1**	**8**	**6**
5	**7**	**10**	**7**	**9**
6	**4**	**1**	**9**	**3**

Greater than and less than symbols. Write

$$6 > 4$$

Ask the child how to read it. [6 is greater than 4] Use a piece of paper to cover the inequality. Slowly uncover it from left to right (in the direction of reading) until the ends of the symbol appear. See the figure below on the left.

Note: For a meaningful way to show how to write the < and > symbols, see Lesson 21.

Two points of the ">" symbol appear first when reading from left to right.

What do you see? [2 points] Remind her that that is how we can tell the greater than, or more than, symbol.

Now change the inequality to

$$6 < 8$$

Again cover the inequality until the single point shows as shown above on the right. How do we read this symbol? [is less than] Summarize by asking, What is the symbol if we see 2 points first as we are reading? [greater than] If we see 1 point? [less than]

Fractions equaling 1. Take out the fraction. Ask the child to find each unit fraction, 1/2, 1/3, and so forth and to place it under the 1. Explain that a unit fraction is a fraction with 1 in the numerator; that is, 1 is divided by another number. See the figure below.

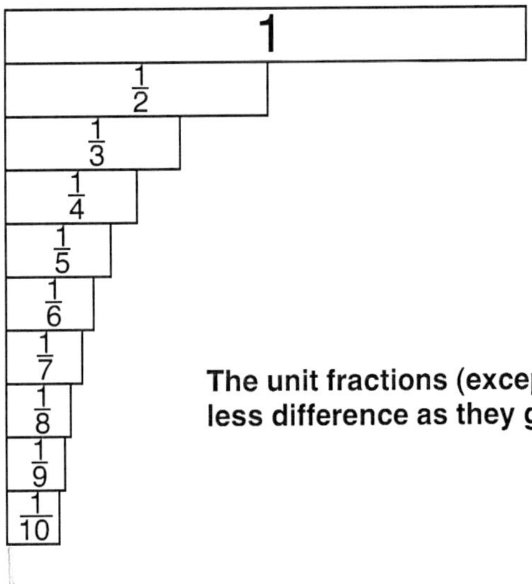

The unit fractions (except for 1), showing less difference as they get smaller.

If you are hungry, would you rather have 1/3 or 1/4 of a candy bar? Is there much difference? Is there much difference between 1/8 and 1/9? What pattern do you see? [As the fractions become smaller, there is less difference between them.] What happens when we divide 1 by a really large number? [The fraction become very small.]

Comparing fractions. Write

$$\frac{1}{6} > \_\_\_$$

Ask for answers from the unit fraction pieces. [1/7, 1/8, 1/9, 1/10]

Repeat for 1/4 < _ . [1/3, 1/2]

Worksheet. The worksheet has questions comparing fractions in various formats. The answers are given below.

<	>	<
>	<	>
>	<	>

10. 2/6
11. **same**
12. **same**
13. 9/10
14. 1/6
15. 2/3
16. 1/6
17. 7/8
18. **same**

Lesson 116

The Ruler Chart

OBJECTIVES
1. To gain facility in reading a ruler to eighths
2. To practice comparing fractions

MATERIALS
Worksheet 122, "Quick Practice-17"
Fraction strips
Worksheet 124, "The Ruler Chart"
Fraction War (*Math Card Games*, F7)

WARM-UP
Ask the child to solve the following problem in his journal: 3584 × 44 [157,696] The check numbers are (2) × (8) = (7).

Ask the child to write the 3s with three in a row as shown.

3	6	9
12	15	18
21	24	27
30		

Ask, What is 12 ÷ 3. [4] What is 21 ÷ 3? [7] What is 15 ÷ 3? [5] What is 24 ÷ 3? [8] What is 27 ÷ 3? [9] What is 18 ÷ 3? [6]

ACTIVITIES
Facts practice. Ask the child to take out his Quick Practice sheet and ask him to do the facts in the **second** rectangle. Ask him to correct them. The answers are:

2	**5**	**4**	**2**	**3**
1	**10**	**7**	**6**	**4**
9	**10**	**9**	**5**	**6**
7	**8**	**4**	**9**	**3**
1	**6**	**7**	**8**	**5**

Fractions on a ruler. Ask the child to build the fraction chart, but with only the one, halves, fourths, and eighths. See below.

<table>
<tr><td colspan="8">1</td></tr>
<tr><td colspan="4">$\frac{1}{2}$</td><td colspan="4">$\frac{1}{2}$</td></tr>
<tr><td colspan="2">$\frac{1}{4}$</td><td colspan="2">$\frac{1}{4}$</td><td colspan="2">$\frac{1}{4}$</td><td colspan="2">$\frac{1}{4}$</td></tr>
<tr><td>$\frac{1}{8}$</td><td>$\frac{1}{8}$</td><td>$\frac{1}{8}$</td><td>$\frac{1}{8}$</td><td>$\frac{1}{8}$</td><td>$\frac{1}{8}$</td><td>$\frac{1}{8}$</td><td>$\frac{1}{8}$</td></tr>
</table>

The fraction chart with 1s, halves, fourths, and eighths.

Discuss the following questions:

1. How many fourths in 1? [4]
2. How many eighths in a whole? [8]
3. How many fourths in one-half? [2]
4. How many eighths in one-half? [4]
5. Which is more, $2/8$ or $1/4$? [same]
6. Which is more, $5/8$ or $1/2$? [$5/8$]
7. What fraction is the same as $6/8$? [$3/4$]
8. What fraction is the same as $4/8$? [$1/2$, or $2/4$]
9. Which is more, 1 or $7/8$? [1]

Ruler chart. Take out the worksheet. Ask the child to write in the fractions on the chart.

While he is writing in the fractions, draw the following chart; write 0 above the left line and 1 above the right line.

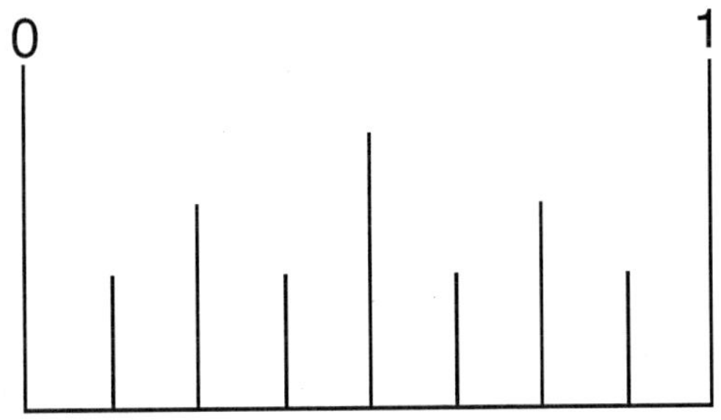

The ruler chart without values.

Then ask him to write the fractions on the chart and to give his explanations. What pattern do you see? [The highest line is half, next highest is fourths, and lowest is eighths.] See the figure below.

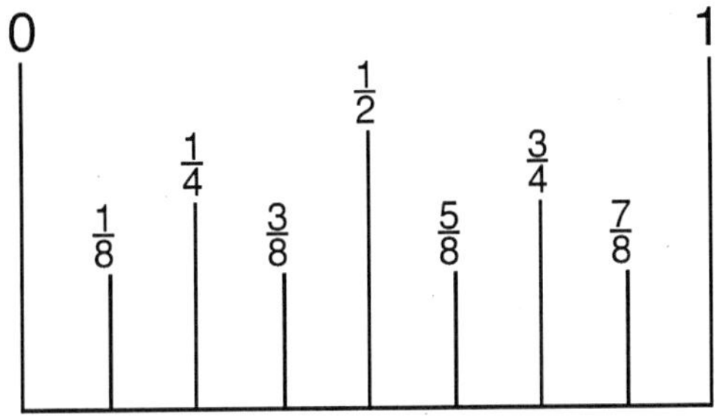

The ruler chart with values.

Does this chart remind you of anything? [a ruler] If desired, show him a ruler marked with eighths.

Fraction War. The Fraction War is a "war" game, where the players compare the fractions on the ruler. Tell the child to use his chart on the worksheet. This game is a good way for the child to become proficient with these fractions. He will need to play it several times.

Worksheet. The second half of the worksheet asks the child to compare these fractions. The answers are below.

<	>	=	3/4
=	>	>	2/2
=	<	=	

1/8

3/4, 7/8, 1

3/4

3/8

5/8, 3/4

Lesson 117

Fraction Problems

OBJECTIVES
1. To work with a fraction line
2. To solve some fraction problems informally

MATERIALS
Worksheet 122, Quick Practice-17
Worksheet 125, Fraction Problems

WARM-UP
Ask the child to solve the following in her journal: 1929 × 64. [123,456] The check numbers are (3) × (1) = (3).

Ask the child to write the 9s multiples on the first worksheet. See below.

9	**18**	**27**	**36**	**45**
90	**81**	**72**	**63**	**54**

Then ask for various division facts.

ACTIVITIES
Facts practice. Ask the child to take out her Quick Practice sheet and ask her to do the facts in the **third** rectangle. Ask her to correct them. The answers are:

2	**9**	**1**	**9**	**6**
5	**6**	**5**	**5**	**8**
4	**10**	**7**	**6**	**10**
2	**1**	**7**	**9**	**3**
7	**3**	**4**	**8**	**4**

5s patterns review. Enter 20 on the abacus and ask how she could see 20 ÷ 5. [There are 2 groups of 5 in a row, so 4.] See the figure below on the left.

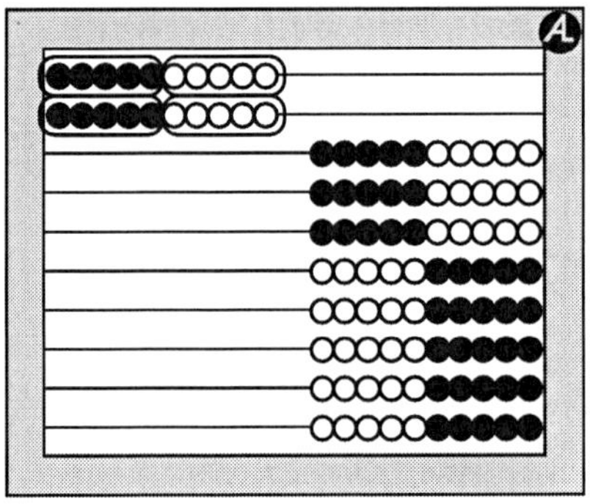

Seeing 20 ÷ 5 as 4 groups of 5.

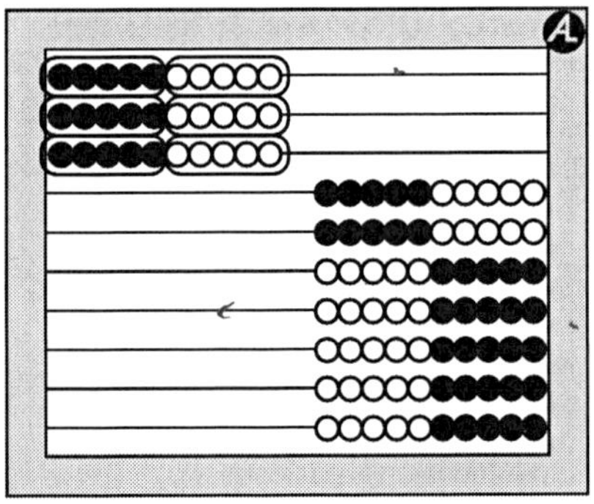

Seeing 30 ÷ 5 as 6 groups of 5.

Repeat for 30 ÷ 5. See above on the right.

Write

10 ÷ 5 = 2   20 ÷ 5 = 4   30 ÷ 5 = 6   40 ÷ 5 = 8   50 ÷ 5 = 10

Ask if she sees a pattern that could help her remember the 5s. [Double the number of tens.]

Next write

15 ÷ 5 = 3   25 ÷ 5 = 5   35 ÷ 5 = 7   45 ÷ 5 = 9

Ask what pattern will help for remembering these. [One more than the ten divided by 5. Also, they are always odd.]

Fraction line. Draw the following fraction line, (which also looks a ruler). Ask the child to write an l at 3½.

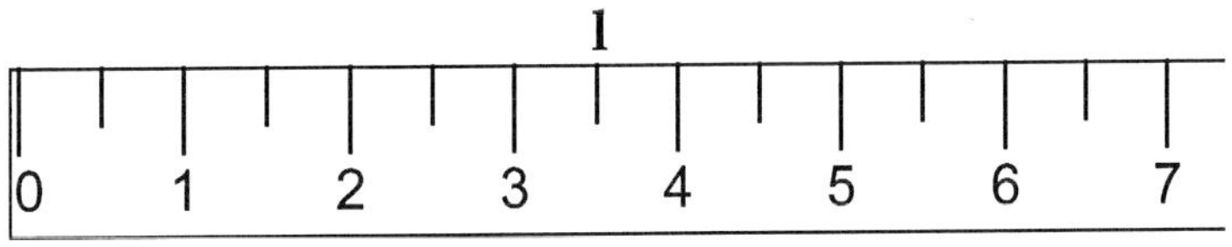

Ask her to travel two spaces to the right and write an *r*.

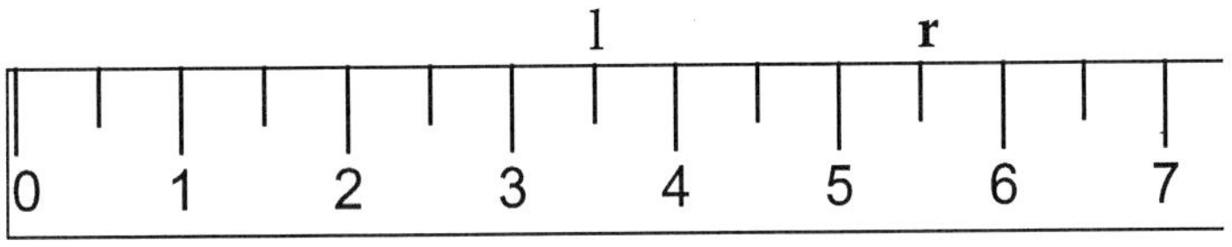

Ask her to travel 2½ spaces left and write a *u*.

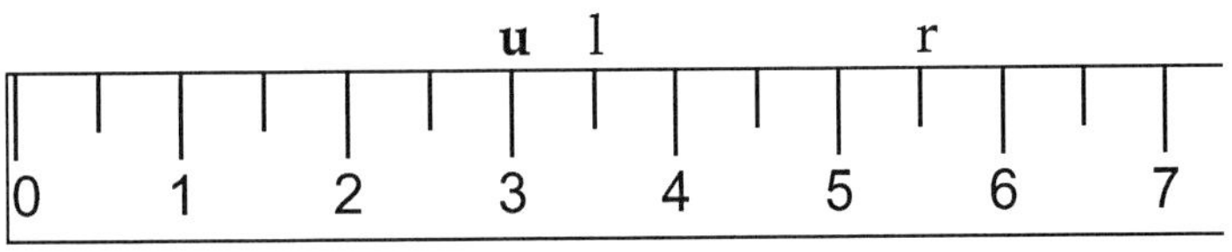

Ask her to travel 1½ spaces right and write an *e*.

Ask her to travel 2½ spaces left and write an *r*.

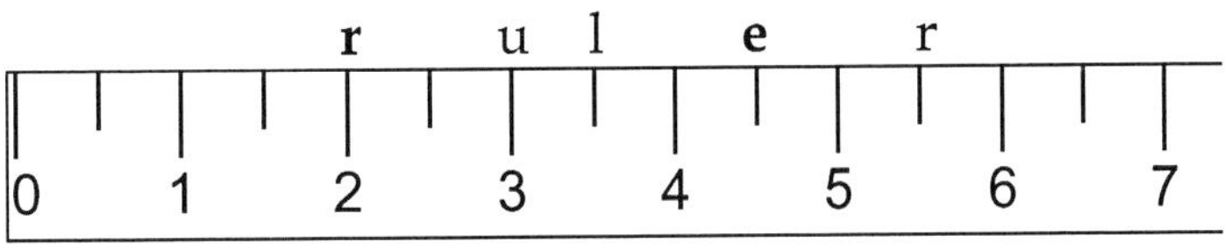

Problem 1. Take out the worksheet. Ask the child to solve the problems. Ask if she can solve them in more than one way. When she is finished, discuss various solutions.

> **Note:** It is very important that the child be able to solve these types of problems long before tackling algorithms.

Kelsey spent ¾ of her allowance. What fraction did she have left? [¼] Her allowance is $4. How much did she have left? [$1] How much did she spend? [$3]

Problem 2. Kyle had ⅞ of his math problems correct. What fraction did he have wrong? [⅛] There are 24 problems. How many did he have right? [21] How many did he have wrong? [3]

Problem 3. Keith wants to know his pulse–how many times his heart beats in a minute. He counted 20 beats in a fourth of a minute. How many beats is that per minute? [80]

The final worksheet problem is similar to the one above:

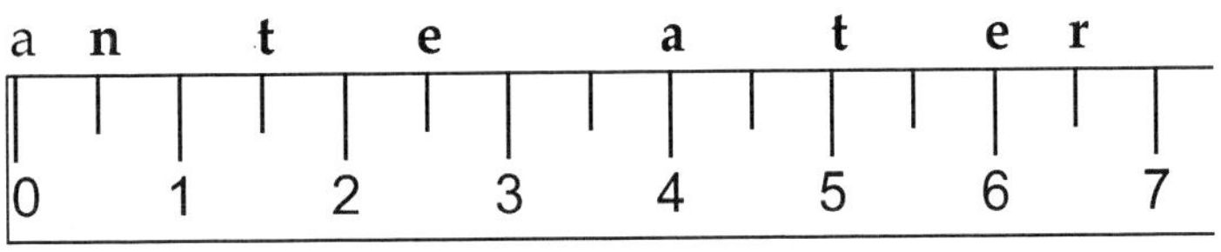

Lesson 118

The Division "House"

OBJECTIVES
1. To practice the 3s division facts
2. To learn to write division problems using a division "house"

MATERIALS Worksheet 126, "Quick Practice-18"
Worksheet 127, "The Division 'House'"

WARM-UP Ask the child to solve the following problem in his journal: $10.36 × 15. [$155.40] The check numbers are (1) × (6) = (6).

Write the 3s in three columns as shown.

3	6	9
12	15	18
21	24	27
30		

Ask, What is 9 ÷ 3? [3] What is 18 ÷ 3? [6] What is 27 ÷ 3? [9] What is 30 ÷ 3? [10] What is 12 ÷ 3? [4] What is 24 ÷ 3? [8]

ACTIVITIES ***Facts practice.*** Take out Quick Practice-18 (Worksheet 126) and ask the child to do the facts in the **first** rectangle when you say start. Ask him to correct them. The answers are:

6	**4**	**8**	**4**	**7**
5	**5**	**9**	**2**	**1**
10	**3**	**2**	**10**	**3**
8	**6**	**9**	**1**	**7**

Reviewing names for answers. Review the terms for the answer found in calculations. Write the following.

9 + 3 = [12]
9 – 3 = [6]
9 × 3 = [27]
9 ÷ 3 = [3]

Which number is a sum? [12] Which number is a product? [27] Which number is a difference or remainder? [6] Which number is a quotient? [3] Review definitions as necessary.

Note: Note that this division format cannot be used in equations. It is useful only in computations.

The $\overline{)}$ division sign. Ask the child for 2 ways to write 27 divided by 3 = 9. [27 ÷ 3 = 9, $^{27}/_{3}$ = 9]

If he does not mention it, tell him there is a third way to write it. Demonstrate it as follows. Write and say 27; make the division sign (make the symbol by starting at the upper left hand corner, first making a curved line and then a horizontal line) and say, divided by; write and say 3; and write 9 above and say, equals 9.

27	$3\overline{)27}$	$\begin{array}{r}9\\3\overline{)27}\end{array}$
twenty-seven	divided by 3	equals 9

Tell him there is no name for this symbol, but he can call it a division house if he likes.

Ask him to write the 3s division facts with the new symbol in his journal, from 3 divided by 3 to 30 divided by 3.

Write several division problems, such as the following and ask the child to read them.

$$3\overline{)30} \qquad\qquad 3\overline{)9}$$

[30 divided by 3 and 9 divided by 3]

Practice. Ask the child to write 15 divided by 3 three different ways as shown below.

$$15 \div 3 = 5 \qquad \frac{15}{3} = 5 \qquad \begin{array}{r}5\\3\overline{)15}\end{array}$$

Repeat for 21 divided by 3 and 24 divided by 3.

Remainder problem. Lauren has 10 pencils to divide among 3 children. How many does each one get? [3 r1] (The answer cannot be 3⅓ because we do not split pencils.) Show the child how to write it as follows.

$$\begin{array}{r}3\text{ r1}\\3\overline{)10}\end{array}$$

Change the problem so Lauren has 17 pencils. Ask him to write it.

$$\begin{array}{r}5\text{ r2}\\3\overline{)17}\end{array}$$

Worksheet. The worksheet has 3s division practice and asks the child to write division problems in three formats. The answers are boldfaced.

2	**4**	**9**	**10**
7	**6**	**5**	**3**
8	**1**	**7**	**6**

40 ÷ 5 = 8	$\mathbf{\frac{40}{5}} = 8$	$\begin{array}{r}8\\5\overline{)40}\end{array}$
18 ÷ 2 = **9**	$\mathbf{\frac{18}{2}} = 9$	$\begin{array}{r}9\\\mathbf{2\overline{)18}}\end{array}$
21 ÷ 3 = 7	$\frac{21}{3} = 7$	$\begin{array}{r}7\\\mathbf{3\overline{)21}}\end{array}$
18 ÷ 3 = 6	$\mathbf{\frac{18}{3}} = 6$	$\begin{array}{r}6\\3\overline{)18}\end{array}$
50 ÷ 5 = **10**	$\mathbf{\frac{50}{5}} = \mathbf{10}$	$\begin{array}{r}10\\\mathbf{5\overline{)50}}\end{array}$
14 ÷ 2 = 7	$\frac{14}{2} = 7$	$\begin{array}{r}7\\\mathbf{2\overline{)14}}\end{array}$
25 ÷ 5 = 5	$\mathbf{\frac{25}{5}} = 5$	$\begin{array}{r}5\\5\overline{)25}\end{array}$
12 ÷ 2 = **6**	$\mathbf{\frac{12}{2}} = 6$	$\begin{array}{r}6\\\mathbf{2\overline{)12}}\end{array}$
12 ÷ 3 = 4	$\frac{12}{3} = 4$	$\begin{array}{r}4\\\mathbf{3\overline{)12}}\end{array}$
24 ÷ 3 = 8	$\mathbf{\frac{24}{3}} = 8$	$\begin{array}{r}8\\3\overline{)24}\end{array}$

Lesson 119

Division Remainders in Context

OBJECTIVE 1. To solve division problems with remainders in context

MATERIALS Worksheet 126, "Quick Practice-18"
Worksheet 128, "Division Remainders in Context"

WARM-UP Ask the child to solve the following in her journal: $36.86 × 98. [$3612.28] The check numbers are (5) × (8) = (4).

Ask the child to write the 4s multiples on the first worksheet. See below.

4	**8**	**12**	**16**	**20**
24	**28**	**32**	**36**	**40**

Ask, What is 4 ÷ 4. [1] What is 12 ÷ 4? [3] What is 24 ÷ 4? [6] What is 36 ÷ 4? [9] What is 16 ÷ 4? [4] What is 20 ÷ 4? [5]

ACTIVITIES ***Facts practice.*** Ask the child to take out her Quick Practice sheet and ask her to do the facts in the **second** rectangle. Ask her to correct them. The answers are:

1	2	8	2	5
5	6	10	1	7
10	8	7	3	3
6	9	4	4	9

Division problem #1. Take out the worksheet and ask the child to read the problem. Thirteen people are going to a concert. Four will fit in a car. How many cars will be needed? [4 cars]

Ask her to explain what it means. Tell her to explain her work with pictures and words.

Then ask her to show the solution.

These types of problems force the child to think about the situation. They cannot blindly apply a rule or an operation. They are difficult for a child who does not want to think, but only apply a rule.

Division problem #2. Patsy is planting petunias in rows, with only 4 in a row. She has 13 plants. How many rows can she plant? [3 rows] One child, after solving this problem, remarked that Patsy would have to throw away the extra plants.

Use the same procedure as above. This problem required rounding down; whereas, the first problem required rounding up.

Division problem #3. Four girls earn $13. They divide it evenly. How much does each one get? [$3.25] Here the remainder must be divided also.

Division problem #4. Four boys are splitting 13 miniature candy bars. They divide them evenly. How much does each one get? [3¼ pieces of candy] For this problem the remainder is split as a fraction.

Division problem #5. Jack is packing cookies 4 to a bag. He has 13 cookies to pack. He gets to eat any leftover cookies. How many cookies does he eat? [1 cookie] This problem is only concerned with the remainder.

Question #6. This summary is important. What is the same about these 5 problems? [The numbers and the operation (division) are the same.] What is different about them? [Each answer is different depending upon the context.]

Lesson 120

Review

OBJECTIVE 1. To review and practice

MATERIALS Worksheet 129-A and 129-B, "Review" (2 versions)
Cards for playing games

ACTIVITIES ***Review worksheet.*** Give the child about 15 to 20 minutes to do the review sheet. The oral problems to be read are as follows:

1000 × 10 = 1 – ⅓ = ¼ of 40 =

Review Sheet 129-A.

1. Write only the answers to the oral questions. **10,000** $\frac{2}{3}$ **10**

4. Write only the answers. $\frac{1}{2}$ of 3 = $1\frac{1}{2}$ 142 – 71 = **71** 89 + 91 = **180**

7. Fourteen children are eating at a banquet. Four will fit at a table. How many tables will be needed?

14 ÷ 4 = 3 r2, so
4 tables are needed

9. Paul is building stools, with three legs each. He has 19 legs. How many stools can he build?

19 ÷ 3 = 6 r1, so
Paul can build 6.

11. Paula wants to know her pulse–how many times her heart beats in a minute–after running. She counted 30 beats in a fourth of a minute. How many beats is that per minute?

There are 4 fourths in a min, so
30 × 4 = 120 beats per min

13. Write the missing fractions.

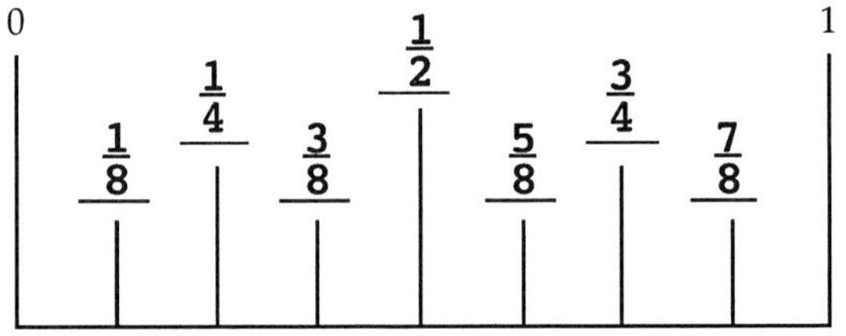

21. Write <, >, or = in the circles.

$\frac{1}{2}$ (>) $\frac{3}{8}$ $\frac{3}{8}$ (<) $\frac{3}{4}$

$\frac{1}{4}$ (<) $\frac{3}{4}$ $\frac{1}{2}$ (=) $\frac{2}{4}$

25. Write each division problem and answer three ways.

14 ÷ 2 = **7**	$\frac{14}{2}$ = 7	$2\overline{)14}$ = 7
45 ÷ 5 = 9	$\frac{45}{5}$ = 9	$5\overline{)45}$ = 9
24 ÷ 3 = 8	$\frac{24}{3}$ = 8	$3\overline{)24}$ = 8

The oral problems to be read to the child are as follows:

$2000 \times 10 =$ $1 - 2/3 =$ $1/3$ of $30 =$

Review Sheet 129-B.

1. Write only the answers to the oral questions. 20,000 $\frac{1}{3}$ 10

4. Write only the answers. $\frac{1}{2}$ of $5 =$ $2\frac{1}{2}$ $185 - 34 =$ 151 $78 + 82 =$ 160

7. Seventeen children are riding on a chairlift. Three will fit on a chair. How many chairs will they needed?

17 ÷ 3 = 5 r2, so
6 chairs are needed

9. Roberta is building tables, with four legs each. She has 21 legs. How many tables can she build?

21 ÷ 4 = 5 r1, so
Roberta can build 5.

11. Roberto wants to know his pulse–how many times his heart beats in a minute–after running. He counted 41 beats in a third of a minute. How many beats is that per minute?

There are 3 thirds in a min, so
41 × 3 = 123 beats per min

13. Write the missing fractions.

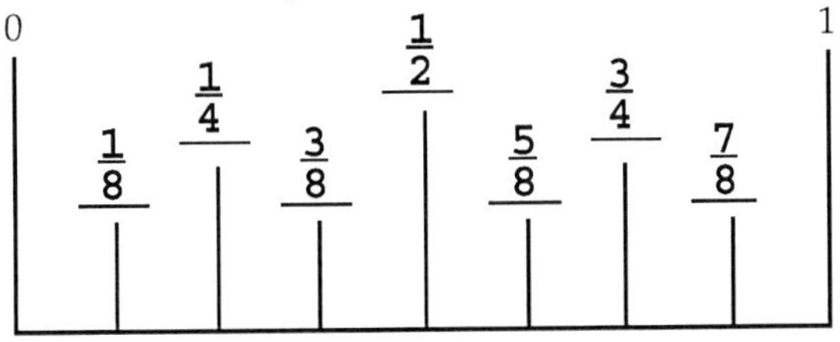

21. Write <, >, or = in the circles.

$\frac{1}{2} = \frac{3}{6}$ $\frac{3}{5} < \frac{3}{4}$

$\frac{3}{8} > \frac{1}{8}$ $\frac{7}{8} > \frac{1}{10}$

25. Write each division problem and answer three ways.

15 ÷ 3 = 5	$\frac{15}{3} = 5$	$3\overline{)15}$ = 5
18 ÷ 2 = 9	$\frac{18}{2} = 9$	$2\overline{)18}$ = 9
25 ÷ 5 = 5	$\frac{25}{5} = 5$	$5\overline{)25}$ = 5

Games. Spend the remaining time playing games.

DIVISION. Division Memory (Lesson 107) and Mixed-Up Quotients (D3) *(Math Card Games).*

FRACTIONS. Fraction War (F7) *Math Card Games).*

Lesson 121

Graphing Growth

OBJECTIVES
1. To work with fractions
2. To graph the results

MATERIALS
Worksheet 126, "Quick Practice-18"
Worksheet 131, "Graphing Growth"
Crayons or colored pencils

WARM-UP
Ask the child to solve the following problem in her journal: 9999 × 46. [459,954] The check numbers are (0) × (1) = (0). (If one check number is (0), it is not necessary to calculate the other check number since (0) times any check number = (0).) What is special about the sum? [It can be read the same forward and backward.]

Ask the child to write the 4s multiples as shown below.

4	8	12	16	20
24	28	32	36	40

Ask, What is 4 ÷ 4. [1] What is 12 ÷ 4? [3] What is 24 ÷ 4? [6] What is 36 ÷ 4? [9] What is 16 ÷ 4? [4] What is 20 ÷ 4? [5]

ACTIVITIES
Facts practice. Take out Quick Practice-18 and ask the child to do the facts in the **third** rectangle when you say start. Ask her to correct them. The answers are:

7	**10**	**8**	**10**	**3**
7	**4**	**6**	**4**	**9**
8	**1**	**5**	**1**	**3**
2	**2**	**5**	**6**	**9**

Worksheet. Take out the worksheet. Ask the child to read the problem.

Aster and Pansy are pets, which grow every day. Aster is ¼ inch long and Pansy is 2 inches long on Day 1. Aster doubles its length every day. Pansy grows a ½ inch a each day.

Fill in the table to show how the pets grow. Then graph the first 6 days.

Ask her to explain the problem.

Ask her to fill in the table, which is given below.

	Day 1	Day 2	Day 3	Day 4	Day 5	Day 6	Day 7	Day 8
Aster	**¼**	**½**	**1**	**2**	**4**	**8**	**16**	**32**
Pansy	**2**	**2½**	**3**	**3½**	**4**	**4½**	**5**	**5½**

Ask how she calculated the values for the table. What do you notice about the values? [Aster starts out smaller, but overtakes Pansy after the fourth day.]

Graphing. What do the numbers along the left side mean? [inches] Are they actual inches? [no, smaller] Ask her to write "inches" along the left edge. Also ask her to write a title at the bottom.

Suggest she choose a color and draw bar graphs for Aster. Remind her to leave space for drawing the bar graphs for Pansy. See the graph below with only Aster's bars.

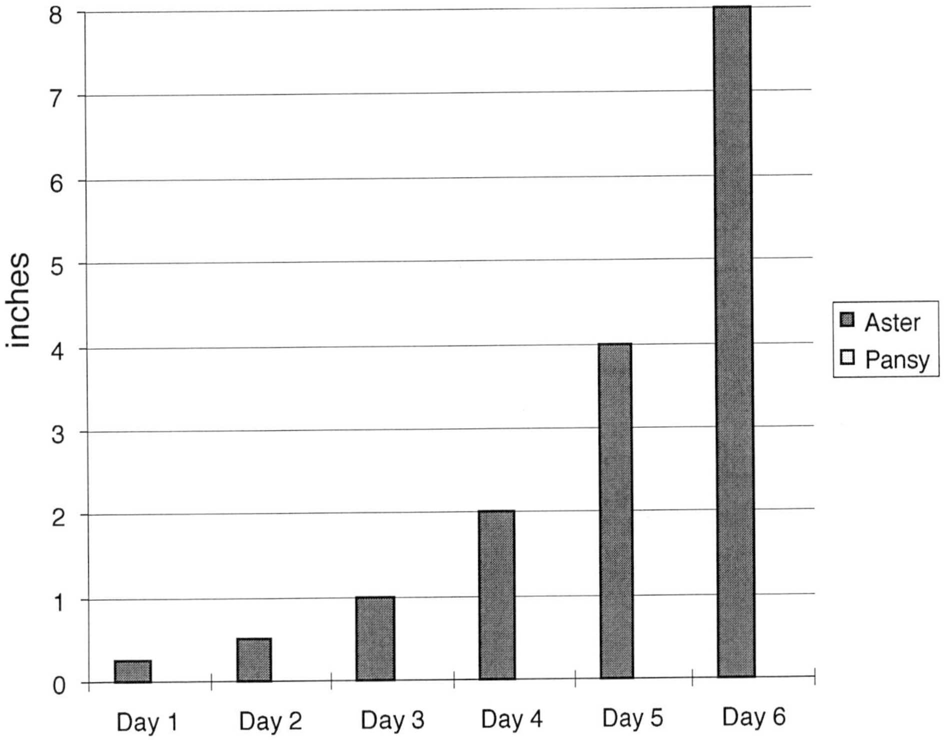

Then ask her to add Pansy's bars. Discuss the graph. Which is easier to use to find information, the table or the graph?

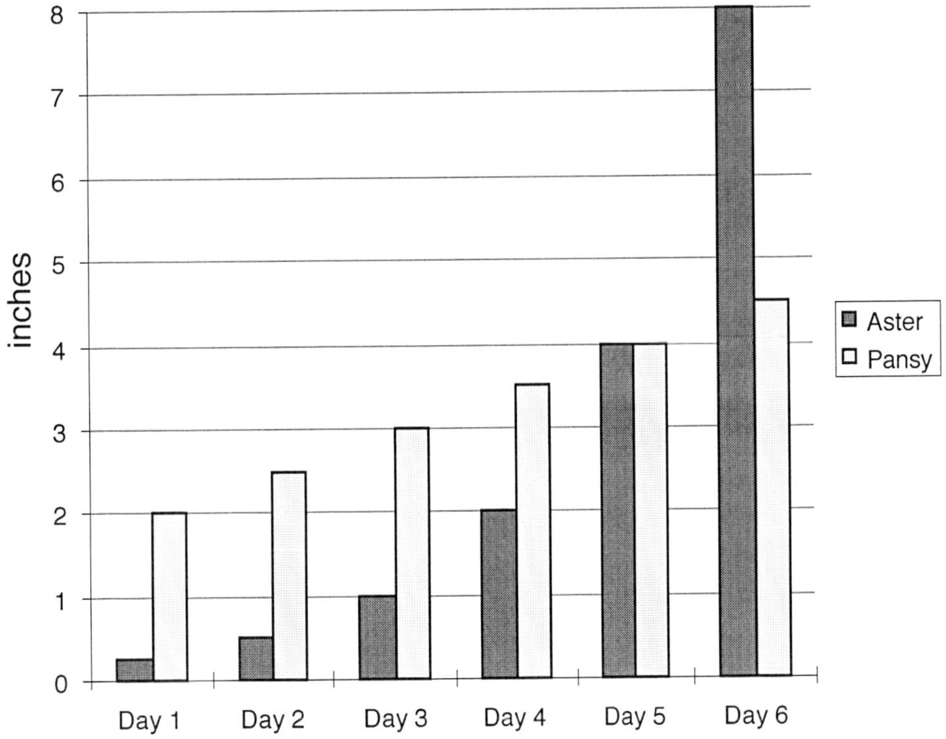

Home Educators

RIGHTSTART™ MATHEMATICS

developed by Joan A. Cotter, Ph.D.

There are 20 lessons before the end of this text. The following are books and manipulatives needed to move into RightStart™ Mathematics Level E Lessons.

Shaded areas are new REQUIRED items. Unshaded REQUIRED areas are items being used in the current curriculum.

STATUS	ITEM	CODE
REQUIRED	***Level E Lessons***	T-E
REQUIRED	***Level E Worksheets***	W-E
REQUIRED	***Math Card Games*** book	M4
RECOMMENDED; choice of abacus	Classic AL Abacus - 8-1/2" x 9-1/2" hardwood frame & beads	A-CL
	Standard AL Abacus - 7-1/2" x 9-1/2" plastic frame & beads	A-ST
	Junior AL Abacus - 5-1/4" x 6" plastic frame & beads	A-JR
RECOMMENDED	Place Value Cards	P
RECOMMENDED	Abacus Tiles	AT
REQUIRED	Cards, Six Special Decks needed for Games	C
RECOMMENDED	Fraction Charts	F
REQUIRED	Basic Drawing Board Geometry Set	DS
RECOMMENDED	Wooden Cubes, 20-1" cubes in set	RH13
REQUIRED	Colored Tiles, apx 200 in set	RH2
REQUIRED	Casio Calculator SL-450	R4
REQUIRED	Centimeter Cubes, 100 in set	R8
RECOMMENDED	Wood Geometry Solids, 12 in set	R14
RECOMMENDED	Math Balance (Invicta)	R7
REQUIRED	4-in-1 Ruler	R10
REQUIRED	Folding Meter Stick	R15
REQUIRED	Goniometer (Angle Measure)	R11

Lesson 122

Reading a Graph on Population

OBJECTIVES 1. To review reading numbers in the millions
2. To read a graph

MATERIALS Worksheet 130, "Quick Practice-19"
Worksheet 132, "Reading a Graph on Population"

WARM-UP Ask the child to solve the following problem in his journal: 6006 × 37. [222,222] The check numbers are (3) × (1) = (3).

Ask the child to write the 4s multiples as shown below.

4	8	12	16	20
24	28	32	36	40

Ask, What is 16 ÷ 4. [4] What is 24 ÷ 4? [6] What is 32÷ 4? [8] What is 36 ÷ 4? [9] What is 12 ÷ 4? [3] What is 20 ÷ 4? [5]

ACTIVITIES ***Facts practice.*** Take out Quick Practice-19 and ask the child to do the facts in the **first** rectangle when you say start. Ask him to correct them. The answers are:

2	**5**	**6**	**4**	**10**	**3**
9	**9**	**2**	**4**	**6**	**10**
5	**1**	**9**	**6**	**8**	**3**
4	**8**	**8**	**1**	**10**	**2**
7	**3**	**1**	**5**	**7**	**7**

Reading the graph. Take out the worksheet. Ask the child what he thinks the graph tells. Why do you think abbreviations were used on the graph? [not enough space for the whole name]

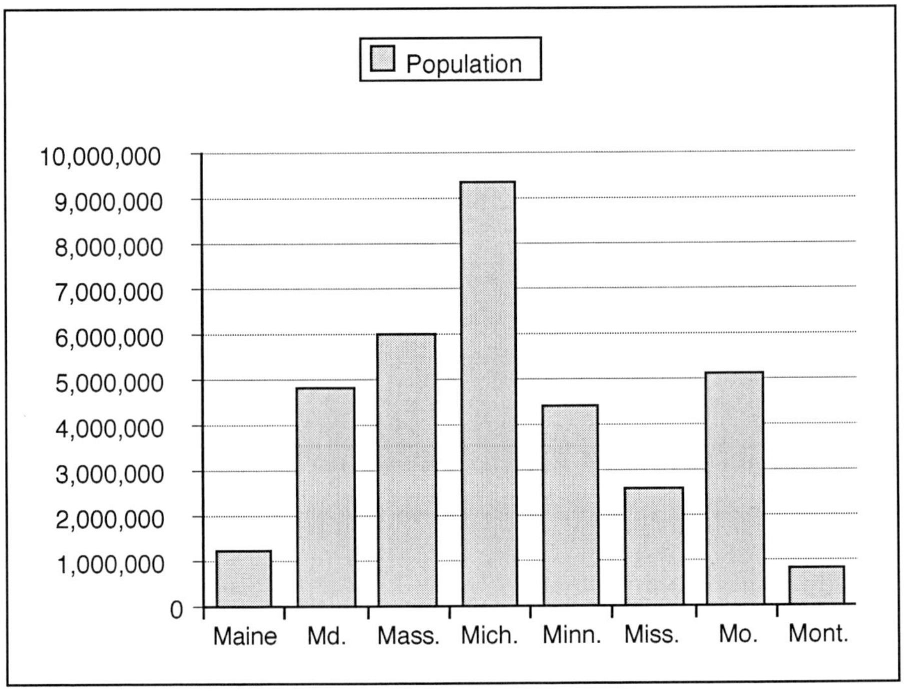

The population graph on the worksheet.

Discuss that it is shows the number of people living in the eight states listed. Ask the following questions and discuss the answers:

1. Does the graph show the exact number of people living in those states? [no] Tell the child that the numbers are taken from the 1990 census.

2. Do we know the exact population of a state? [no] It is constantly changing because of babies being born, people dying, and people moving in and out of a state.

3. What is special about the states listed on the chart? [All start with the letter *M*.]

4. Are they in any particular order? [alphabetical] Refer to the table that gives the actual spelling.

5. Which state has 4 million 400 thousand people? [Minnesota]

6. Which state has 4 million 800 thousand people? [Maryland]

7. Which states have a population greater than Missouri? [Massachusetts and Michigan] What are their populations? [6,000,000 and 9,300,000]

8. What is the population of Maine? [1 million 200 thousand]

9. What state has about twice the population of Maine? [Mississippi]

10. What is the population of Michigan? [9 million 300 thousand]

11. Which state has the lowest population? [Montana] About how many? [800,000]

12. Which states have populations close to 5 million and what are their populations? [Maryland, 4,800,000 and Missouri, 5,100,000]

Worksheet. The worksheet asks the child similar questions to those above. The questions and answers are as follows.

Note: The child's answers may vary by 100,000.

1. Which state has a population of 1,200,000? **Maine**

2. Which state has a population of 4,300,000? **Minnesota**

3. Which state has a population less than 1 million and what is the population? **Montana, 800,000**

Note: The child will need this worksheet for the next lesson.

4. What is the population of the state with the greatest population? **Michigan, 9,300,000**

5. What is the population of Mississippi? **2,600,000**

6. What state has twice the population of Mississippi and what is the population? **Missouri, 5,100,000**

7. Which state has a population closest to Missouri and what is the population? Maryland, **4,800,000**

8. What is the difference in population between Massachusetts and Michigan? **3,200,000**

Lesson 123

Reading a Graph on Area

OBJECTIVES
1. To work with fractions
2. To graph the results

MATERIALS
Worksheet 130, Quick Practice-19
Worksheet 133, Reading a Graph on Area
Worksheet 132, Reading a Graph on Population, from the previous lesson
A map of the U.S.

WARM-UP
Ask the child to solve the following problem in her journal: 2749 × 38. [104,462] The check numbers are (4) × (2) = (8).

Ask the child to write the 4s multiples as shown below.

4	8	12	16	20
24	28	32	36	40

Ask, What is 40 ÷ 4. [10] What is 28 ÷ 4? [7] What is 32 ÷ 4? [8] What is 36 ÷ 4? [9] What is 20÷ 4? [5] What is 8 ÷ 4? [2]

ACTIVITIES
Facts practice. Ask the child to take out her Quick Practice sheet and ask her to do the facts in the **second** rectangle when you say start. Ask her to correct them. The answers are:

6	**1**	**3**	**2**	**8**	**7**
4	**9**	**9**	**1**	**5**	**2**
10	**9**	**10**	**3**	**7**	**4**
6	**1**	**6**	**7**	**3**	**8**
2	**8**	**5**	**5**	**10**	**4**

Comparing areas. Refer to a U.S. map. If appropriate, ask the child to identify several or all of the states starting with the letter *M*, Maine, Maryland, Massachusetts, Michigan, Minnesota, Mississippi, Missouri, and Montana. Which is larger, Maine or Montana? [Montana] Does perimeter or area tell us which is larger? [area]

Reading the graph. Take out the worksheet. The graph is shown on the next page.

Ask the child what she thinks the graph tells. What do you think a square mile is? Discuss that it is a large square with sides of 1 mile. What is a mile? Ideally, it would be good to be able to tell her that a certain landmark is 1 mile from home. Then ask her to imagine how large a square mile is. For a rural student familiar with acres, say that 640 acres fit in a square mile.

Discuss the following questions:

1. Which state has 48,000 square miles? [Mississippi]
2. Which state has 84,000 square miles? [Minnesota]
3. Which state has 8,000 square miles? [Massachusetts]
4. Which state has the largest area? [Montana] Does it have the largest population? [no, smallest]

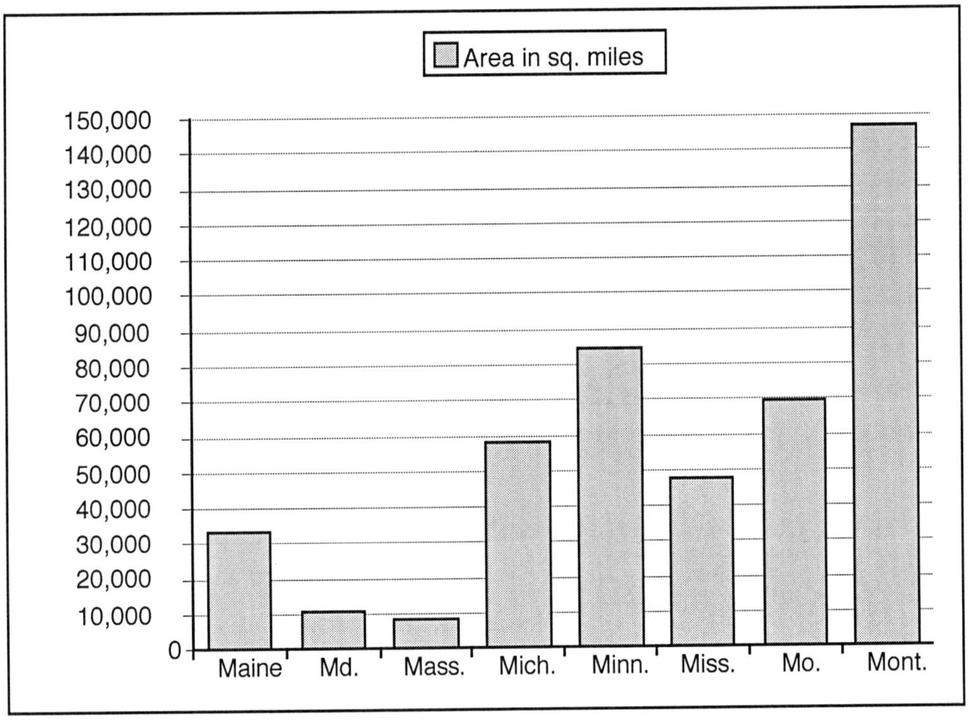

The area graph on the worksheet.

5. Which state has the largest population? [Michigan] Does it have a large area? [no, in the middle]

6. Which state is 7 times larger than Maryland? [Missouri]

7. Which state is closest in size to Maine? [Mississippi]

8. The population graphs keep changing. Do the area graphs also keep changing? [no]

9. How many of the smaller states could fit inside Montana at the same time? [Massachusetts, Maine, Maryland, Mississippi] (8,000 + 33,000 + 10,000 + 48,000 = 99,000; adding Mich. with 59,000 is 158,000, more than Mont.)

Note: The child's answers may vary by 1,000.

Worksheet. The worksheet has similar questions, which are given below along with the answers.

1. Which are the 2 smallest states and what are their sizes? **Maryland, 10,000, Massachusetts, 8,000**

2. Which state is closest to having 50,000 square miles and what is its size? **Mississippi, 48,000**

3. What is the second largest state and what is its size? **Minnesota, 84,000**

4. Would Missouri and Michigan fit in Montana at the same time? Explain. **70,000 + 59,000 > 129,000, yes**

5. What state is nearest in size to Maine and Maryland together? **Mississippi**

Lesson 124

Drawing Rectangles on a Drawing Board

OBJECTIVES

1. To review (or learn) how use a drawing board and tools
2. To draw a rectangle, find its area, and length of the diagonals

MATERIALS

Worksheet 134, "Drawing Lines and Rectangles"
Tiles, optional
A drawing board with Worksheet 134 taped onto it.
A T-square and a 45° triangle 
Ruler
Sharp pencil and eraser

> **Note:** The best tape to use is 3M's Removable tape.

WARM-UP

Make the following arrangement on either paper or with tiles.

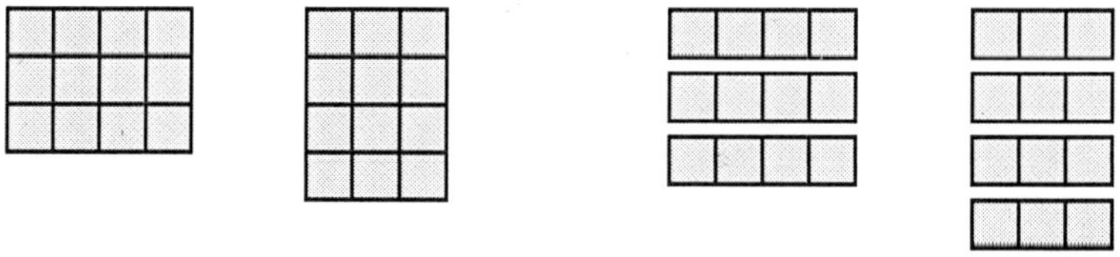

Showing 4 × 3 and 3 × 4. **Showing 12 ÷ 3 and 12 ÷ 4.**

> **Note:** Quick Practice sheets have been discontinued for the remainder of the year.

Then ask the child to write 2 multiplication equations and 2 division equations. [4 × 3 = 12, 3 × 4 = 12, 12 ÷ 3 = 4, and 12 ÷ 4 = 3] It may help to show the groups for the division equations, as shown above on the right.

ACTIVITIES

Preparation. To be certain the worksheet is correctly aligned on the drawing board, tape it on before lesson time. Check a horizontal line with the T-square to be sure the paper is straight." See the figure below. At the end of the lesson, you might ask the child to leave the tape on the board; it frequently can be reused.

> **Note:** Teaching geometry with a drawing board, T-square, triangles, and compass helps a child learn the informal, practical side of geometry. It can also help develop coordination, which improves with practice.

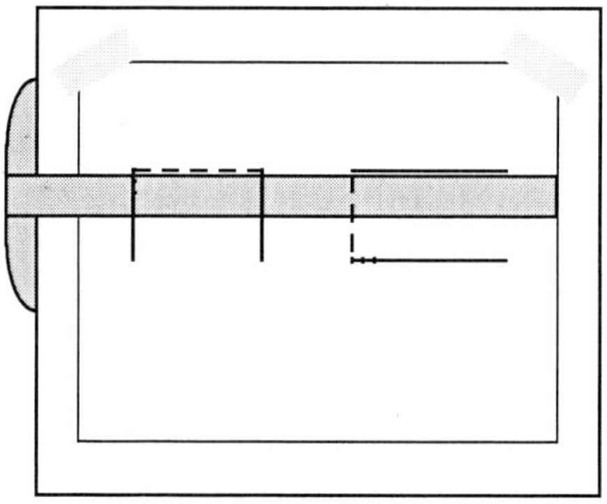

Aligning the paper and taping it to the drawing board.

Tape only the top corners.

Drawing horizontal lines on the drawing board. Draw a horizontal line and ask the child if he remembers what we call it. [horizontal line] Repeat for a vertical line.

Take out the drawing board and T-square. Show the T-square and tell the child that it is called a T-square.

To make the 1/4-inch marks, ask him to use his ruler. Then ask him to make marks 1/4-inch apart as shown on the next page.

Note: Lines are drawn from a starting point, which are the little marks in this example,

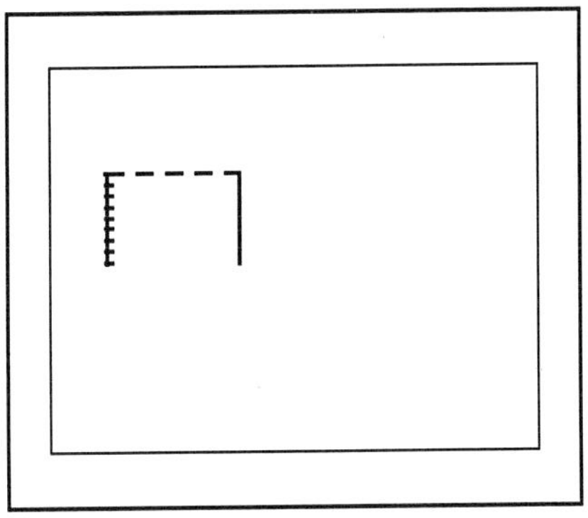

Make the tick marks ¼" apart.

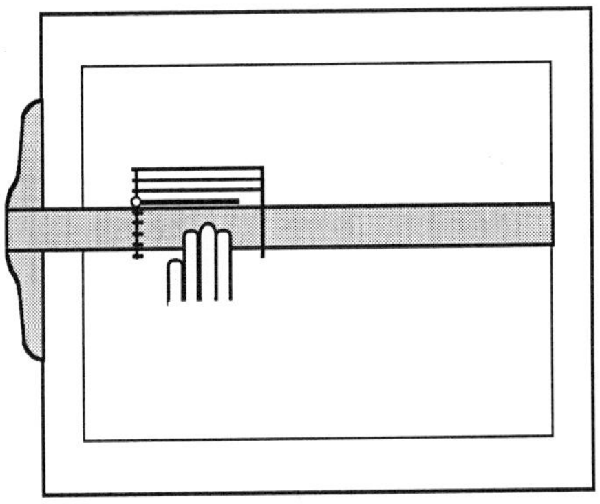

Draw horizontal lines ¼" apart.

Next he is to draw parallel lines ¼ inch apart as shown above on the right. For a right-handed user, the T-square is placed along the left side as shown. A left-handed user places the T-square along the right side. The T-square is held in position with the fingertips of the non-writing hand. Tell the child to check before drawing each line to be sure the T-square is against the edge of the board.

Drawing vertical lines. To draw vertical lines the triangle needs to touch the T-square while the T-square still touches the board. The T-square needs to be at least a half inch below the lower line.

To hold the tools in place, demonstrate as follows: a) hold the T-square with the left hand, b) with the right hand move the triangle to the correct place against the T-square, c) the right hand holds both tools (see the lower left figure), until d) the left hand takes over holding both tools (see the lower right figure).

Draw the lines from bottom to top.

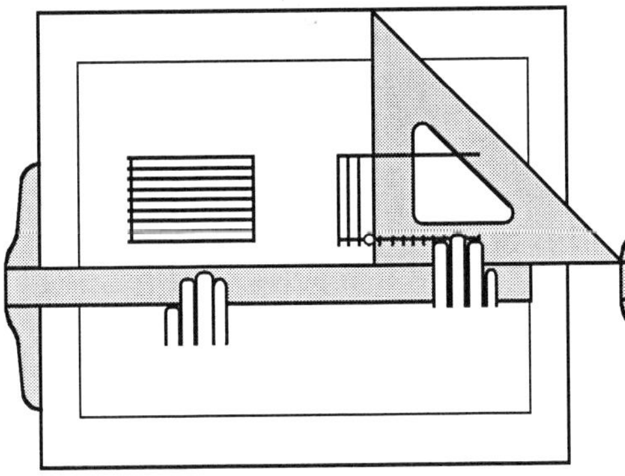

Hold both the triangle and T-square with the writing fingers.

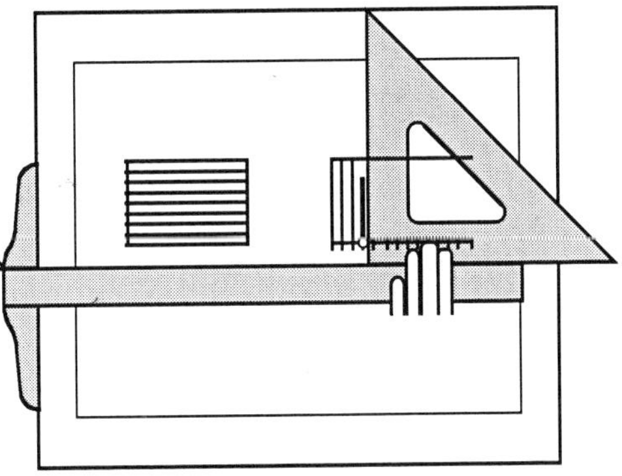

Hold both tools with the nonwriting fingers, ready to draw.

Drawing a rectangle, opt. Next the child draws a 4 × 2 rectangle with square inches. If necessary, review the term *diagonal.* To draw the diagonal, he will need to use the triangle as a straightedge. See the figure on the right.

A = 8 sq. in.; diag = 4½ in.

Note: Use this problem for a child who has worked with the drawing board in Level C.

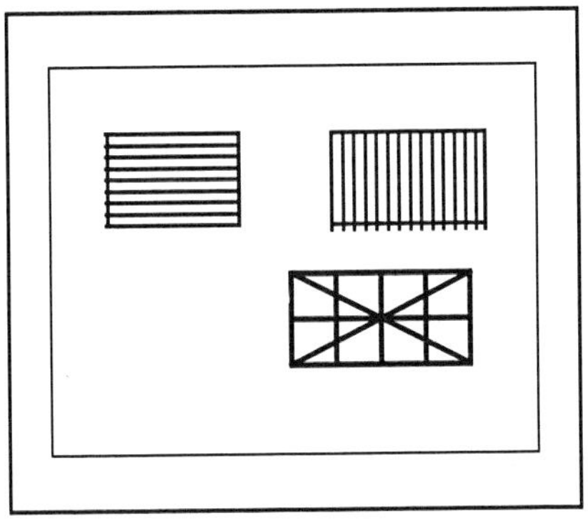

Lesson 125

Drawing Diagonals

OBJECTIVES
1. To measure with centimeters
2. To learn the term *right-angled triangle*
3. To draw diagonals

MATERIALS
Worksheet 135, "Drawing Diagonals"
A drawing board with Worksheet 135 taped onto it
A T-square and a 45° triangle
4-in-1 ruler, which has both inches and centimeters
Sharp pencil and eraser

WARM-UP
Ask the child to solve the following problem in her journal: \$25.39 × 28. [\$710.92] The check numbers are (1) × (1) = (1).

Make a 5 by 3 arrangement similar to the previous lesson and ask the child to write two multiplication equations and two division equations. [5 × 3 = 15, 3 × 5 = 15, 15 ÷ 3 = 5, and 15 ÷ 5 = 3]

ACTIVITIES

> **Note:** Using a separate centimeter ruler is preferable to using the ruler on the T-square.

> **Note:** Some children learn best by discussing the concepts first and then working independently. Others need step-by-step direction .

> **Note:** Drawing a square using the diagonal will be taught in a later lesson.

Drawing a square. Take out the drawing board with the attached worksheet and the 45 triangle. What are you asked to draw on the dotted line? [a square that is 5 centimeters on a side] Which is longer, an inch or a centimeter? [inch] Where are the centimeters on your ruler?

Explain that she is to make a small mark, called a *tick* mark, on the line to show where 5 centimeters is. Tell her that we **never** draw on the bottom of the T-square. Then ask her to draw the line.

Where is the starting point to draw a side of the square? [at the end of the line or the tick mark.] How do we know how long the sides are? She can measure. An advanced child might use the 45 triangle to find the height of the triangle as shown below on the right.

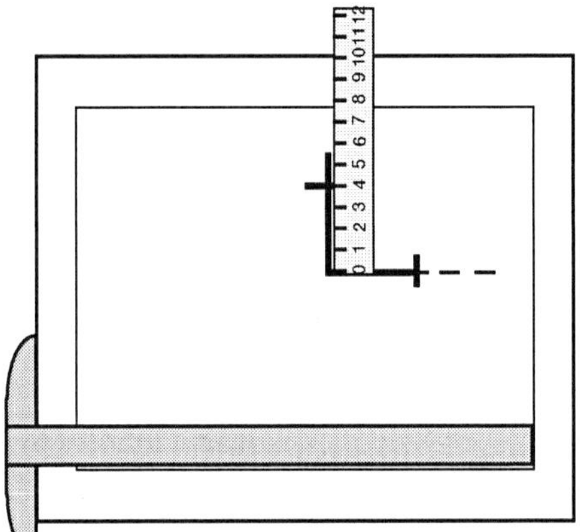

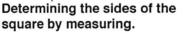

Determining the sides of the square by measuring.

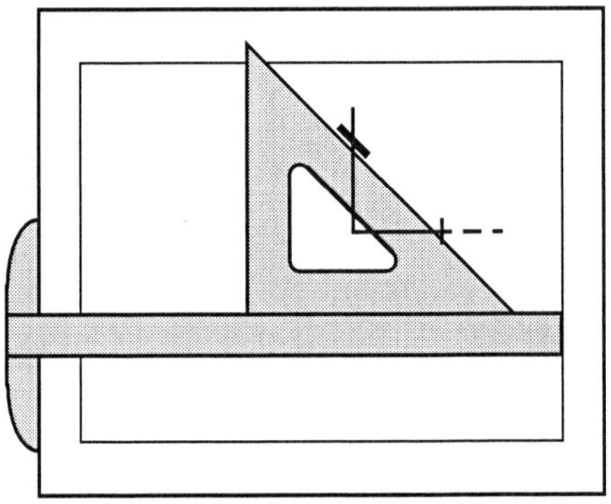

Using the 45 triangle to find the height of the side.

Ask her to complete the square using the tools. Do we need to do any more measuring? [no] Remind her erasing is part of this work.

Next she is to calculate the area in square inches [**4 sq. in.**] and in square centimeters. [**25 sq. cm**]

Reviewing right angle. <u>What is a right angle?</u> Ask the child to draw one. Draw a triangle with a right angle. <u>What do you think we call this kind of triangle?</u> [a right-angled triangle] See below.

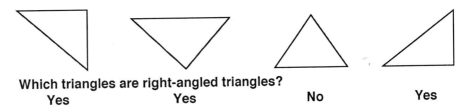

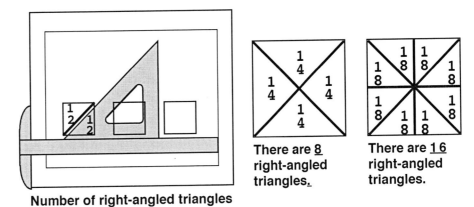

A right angle.

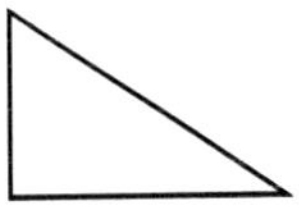

A right-angled triangle.

Draw other triangles, such as those drawn below. <u>Which triangles are right-angled triangles?</u>

Which triangles are right-angled triangles?
Yes Yes No Yes

Drawing diagonals in squares. Ask the child to draw the diagonals in the bottom row according to the directions. Stress that she is to use her drawing tools. If necessary, explain that the square is 1 whole and she is to write the fractions made by the diagonal(s). The solutions are below.

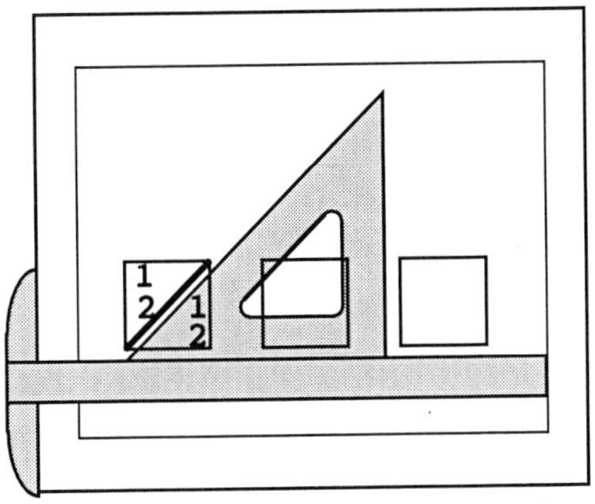

Number of right-angled triangles (with 1 diagonal) is <u>2</u>.

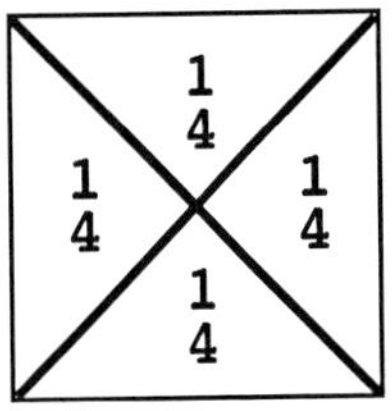

There are <u>8</u> right-angled triangles.

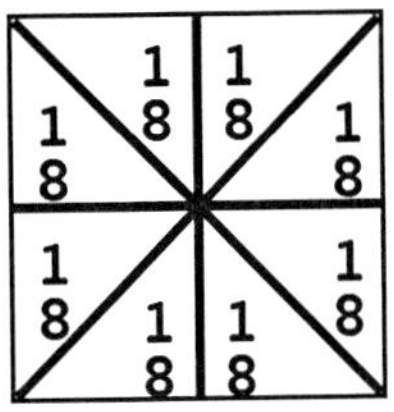

There are <u>16</u> right-angled triangles.

The number of right-angled triangles is a bit tricky in the second square. [8] There are two more large triangles formed by the second diagonal and four medium-sized triangles.

The diagonals must be drawn first and then vertical and horizontal lines drawn through the center. In addition to the 8 right-angled triangles previously found, there are 8 small ones, for a total of 16.

Problem (optional). If time remains, give her this problem. <u>Lou is coloring half of the second square by coloring two fourths. How many different ways could Lou do it?</u> [6, top & bottom, left & right, top & right, left & bottom, top & left, and right & bottom]

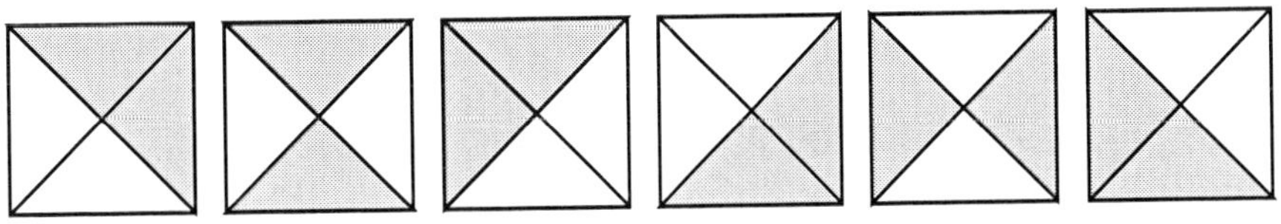

Lesson 126

Review

OBJECTIVE 1. To review and practice

MATERIALS Worksheet 136-A and 136-B, "Review" (2 versions)
Cards for playing games

ACTIVITIES ***Review worksheet.*** Give the child about 15 to 20 minutes to do the review sheet. The oral problems to be read are as follows:

10,000 × 10 = 2 – 1⁄4 = 3⁄4 + 1⁄4 =

Review Sheet 136-A.

1. Write only the answers to the oral questions. 100,000 $1\frac{3}{4}$ 1

4. Write only the answers. 34 × 10 = 340 152 – 26 = 126 199 + 3 = 202

3. Write the missing fractions.

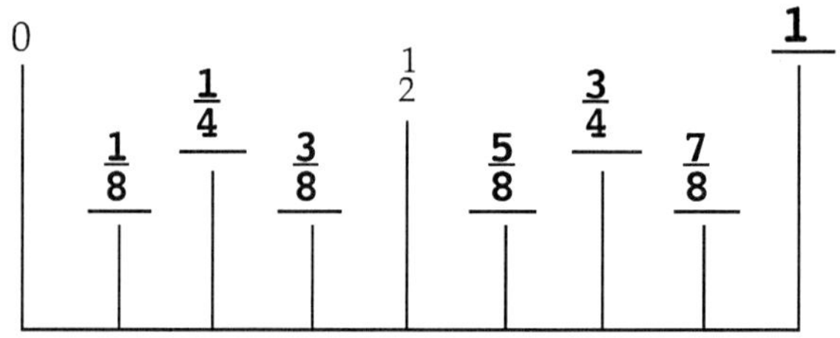

11. Write <, >, or = in the circles.

$\frac{1}{2}$ (>) $\frac{3}{8}$ $\frac{3}{8}$ (<) $\frac{3}{4}$

$\frac{1}{4}$ (<) $\frac{3}{4}$ $\frac{1}{2}$ (=) $\frac{2}{4}$

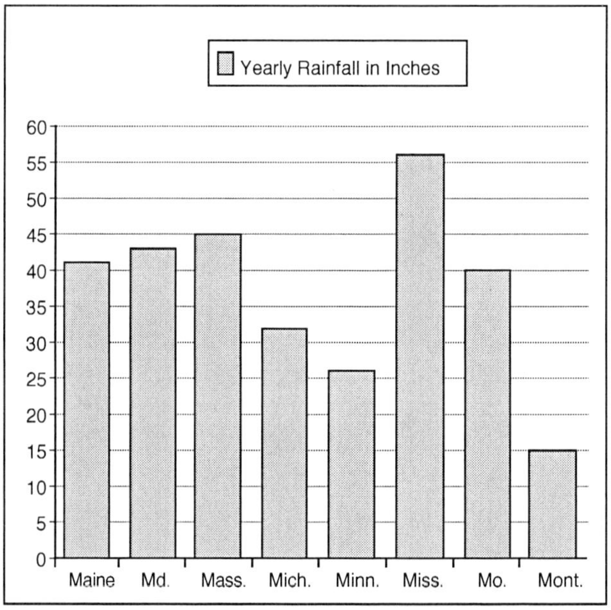

15.

18 ÷ 2 = 9 8 ÷ 2 = 4

2 ÷ 2 = 1 15 ÷ 3 = 5

45 ÷ 5 = 9 20 ÷ 5 = 4

15 ÷ 5 = 3 40 ÷ 5 = 8

23. Draw two diagonals in the rectangle.

Use the chart to find the answers.

24. Which state receives 43 inches of rain? Md.

25. How much rain does Minn. get? 26 in.

26. Which state gets about twice as much rain as Mont.? Mich.

The oral problems to be read to the child are as follows:

20,000 × 10 = 3 – ¼ = ⅔ + ⅓ =

Review Sheet 136-B.

1. Write only the answers to the oral questions. **200,000** **$2\frac{3}{4}$** **1**

4. Write only the answers. 65 × 10 = **650** 175 – 38 = **137** 198 + 4 = **202**

3. Write the missing fractions.

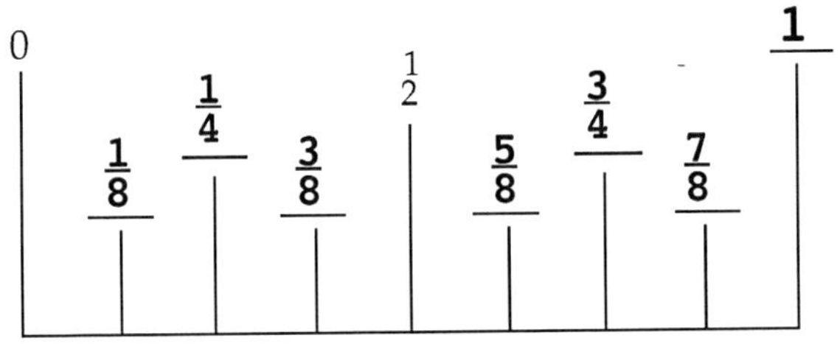

11. Write <, >, or = in the circles.

$\frac{1}{2}$ (<) $\frac{7}{8}$ $\frac{2}{8}$ (=) $\frac{1}{4}$

$\frac{1}{4}$ (>) $\frac{1}{8}$ $\frac{1}{2}$ (<) 1

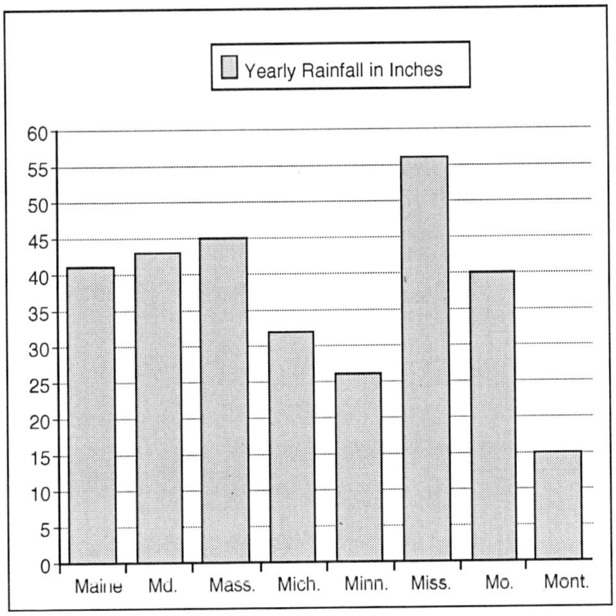

15.
18 ÷ 3 = **6** 30 ÷ 5 = **6**
12 ÷ 3 = **4** 21 ÷ 3 = **7**
45 ÷ 5 = **9** 35 ÷ 5 = **7**
12 ÷ 2 = **6** 18 ÷ 2 = **9**

23. Draw two diagonals in the quadrilateral.

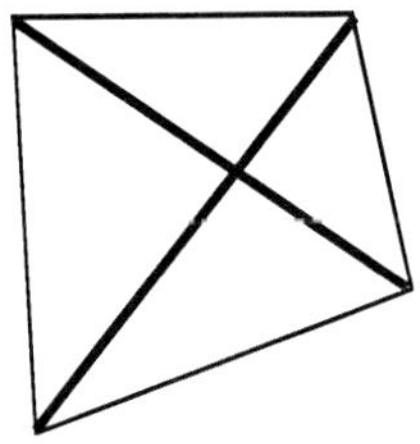

Use the chart to find the answers.

24. Which state receives 41 inches of rain? **Maine**

25. How much rain does Miss. get? **56 in.**

26. Which state gets 3 times as much rain as Mont.? **Mass.**

Games. Spend the remaining time playing games.

DIVISION. Mixed-Up Quotients (D3) *(Math Card Games).*

FRACTIONS. Fraction War (F7) *Math Card Games).*

Lesson 127

Drawing Octagons

OBJECTIVES
1. To review some geometry terms
2. To copy designs using the drawing tools
3. To draw octagons
4. To connect geometry and art through drawing original designs

MATERIALS Worksheet 137, Drawing Octagons
A drawing board with Worksheet 137 taped onto it
A T-square and a 45 triangle 
Crayons or colored pencils

WARM-UP Ask the child to solve the following problem in her journal: $56.78 × 32. [$1816.96] The check numbers are (8) × (5) = (4).

Write the 4s in 2 rows. What is 16 ÷ 4? [4] and so forth.

ACTIVITIES ***Reviewing terms.*** Draw the following figures.

Note: The term quadrilateral was introduced in kindergarten.

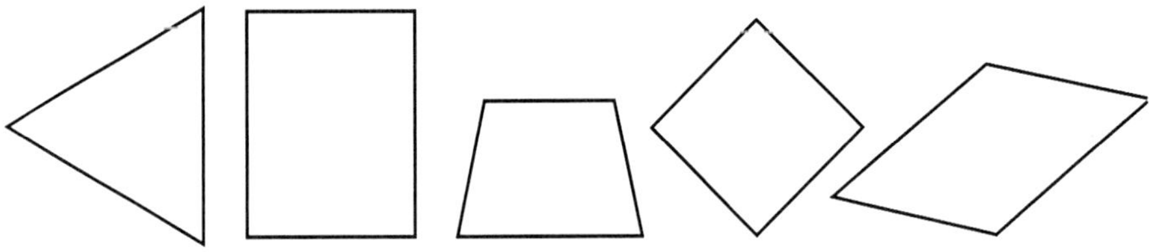

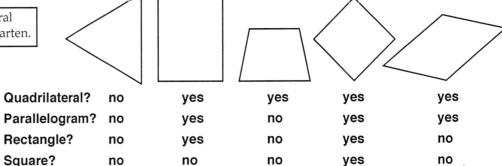

Quadrilateral?	no	yes	yes	yes	yes
Parallelogram?	no	yes	no	yes	yes
Rectangle?	no	yes	no	yes	no
Square?	no	no	no	yes	no

What is a quadrilateral? [A closed figure with four lines.] Point to each figure in turn and ask, Is this a quadrilateral? The answers are below the figure.

What is a parallelogram? [a quadrilateral with two sets of parallel lines] Point to each figure in turn and ask, Is this a parallelogram? The answers are below the figure. What is a parallelogram? [a quadrilateral with 2 sets of parallel lines]

What is a rectangle? [a quadrilateral with 4 right angles]Point to each figure in turn and ask, Is this a rectangle? The answers are below the figure. What is a rectangle? [a quadrilateral with 4 right angles]

What is a square? [a rectangle with equal sides] Point to each figure in turn and ask, Is this a square? The answers are below the figure above.

Note: Most children will not understand the concept of equal angles. It will not be mastered until Level E.

Names of regular polygons. Draw the following figures. Tell the child that they are called *polygons.* Refer to the previously drawn figures. Are those polygons? [yes] Also tell her that if all the sides and angles are equal, they are *regular polygons.* Are the figures below regular polygons? [yes]

The word *octagon* comes from Latin, meaning *eight.* Octopus is another example. So is October, the eighth month when the year started in March.

Note: These same figures will be needed in the next lesson.

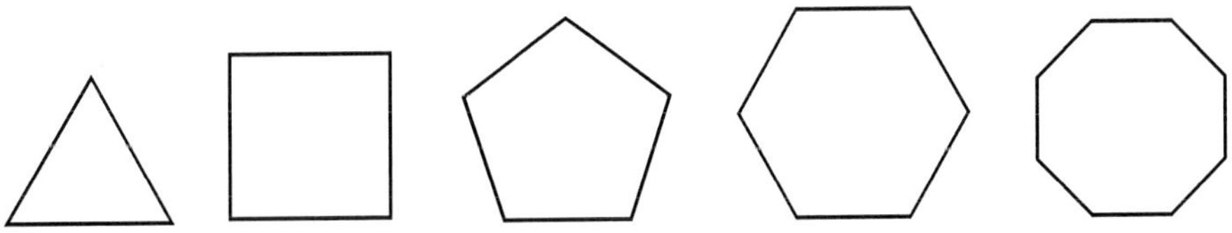

equilat tri | square | pentagon | hexagon | octagon

Worksheet. Take out the drawing board with the worksheet and the 45 triangle. Ask the child to complete it. Stress he must use the drawing tools. Then discuss his construction and the answers to the questions.

For the second circle ask, How can you find the starting points to draw the lines? [Repeat the work for the first circle, using tick marks and not lines.] Then the lines are drawn with the tools. The results are shown below; two squares and an octagon emerge.

Note: Sometimes children forget that to draw a horizontal line they need only the T-square and not a triangle.

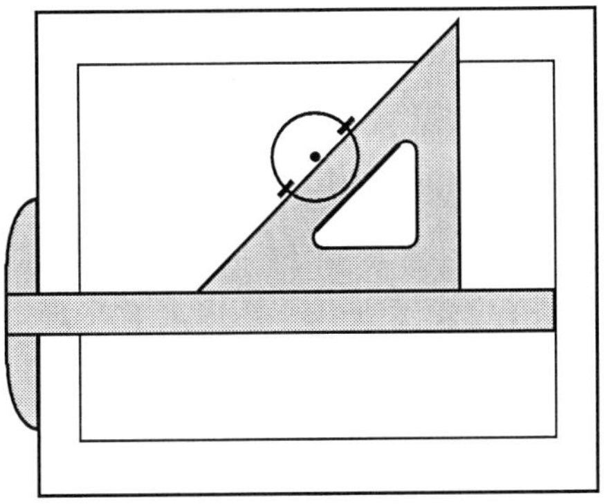

Drawing some tick marks for the second circle.

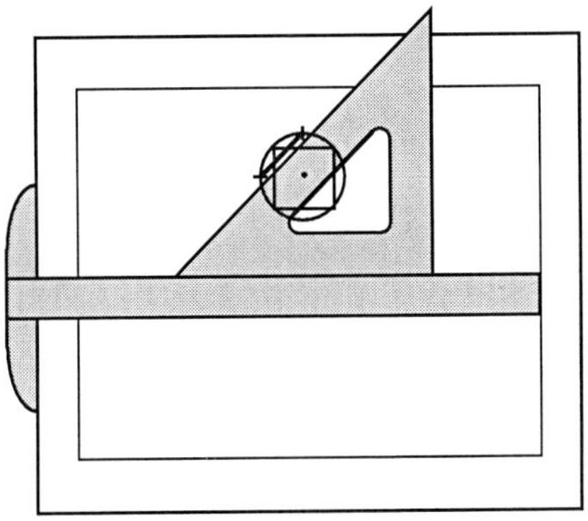

Drawing a line for the second circle.

The third circle has tick marks, which only need to be connected with the T-square and triangle to form the octagon.

In the fourth circle, every third tick mark is connected, again with drawing tools. There are many ways it can be colored. Ask if he sees the two squares and the octagon.

For the last two circles, he is to make and color his own designs similar to the other designs, using the tools.

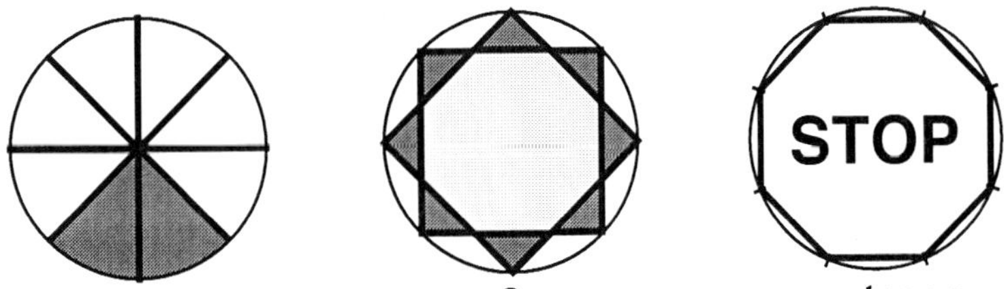

$\frac{3}{8}$ 2 octagon

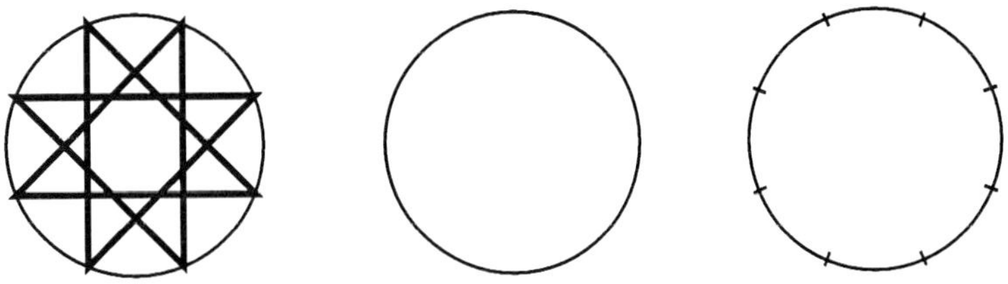

4

Lesson 128

Drawing Hexagons

OBJECTIVES
1. To copy designs using the drawing tools
2. To draw hexagons
3. To connect geometry and art through drawing original designs

MATERIALS
Worksheet 130, "Quick Practice-19"
Worksheet 138, "Drawing Hexagons"
A drawing board with Worksheet 138 taped onto it
A T-square and a 30-60 triangle 
Crayons or colored pencils

WARM-UP
Ask the child to solve the following problem in his journal: \$34.98 × 56. [\$1958.88] The check numbers are (6) × (2) = (3).

Write the 4s in two rows. What is 24 ÷ 4? [6] and so forth.

ACTIVITIES
Facts practice. Ask the child to take out her Quick Practice sheet and ask him to do the facts in the **third** rectangle when you say start. Ask him to correct them. The answers are:

10	9	5	4	6	2
3	3	10	3	1	4
5	2	7	8	10	5
7	1	8	2	9	4
9	6	1	8	7	6

Reviewing names of polygons. Draw the following figures, which are the same as those from the previous lesson. Ask the child to identify the figures. The names are below each figure.

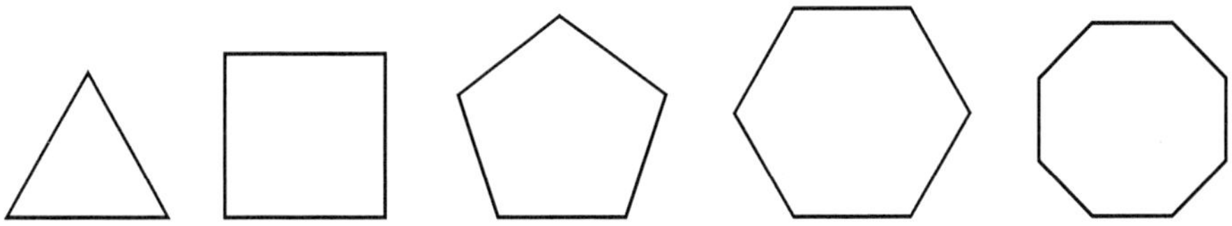

equilateral triangle **square** **pentagon** **hexagon** **octagon**

Ask the following questions.

1. What name applies to all of the figures? [polygon]
2. Which figure is a quadrilateral? [the second figure]
3. Which figure is an octagon? [the last one] How many sides does it have? [8]
4. Which figure is an hexagon? [the fourth one] How many sides does it have? [6]
5. What kind of triangle is the first figure ? [equilateral]
6. Which figures have parallel lines? [square, hexagon, octagon]
7. How many sets of parallel lines do they have? [square, 2; hexagon, 3; and octagon, 4]
8. Which figures have right angles? [square]

Worksheet. Take out the drawing board with the worksheet and the 30-60 triangle. Tell the child there are 2 angles on this triangle; sometimes he will need one angle and sometimes the other.

Do the first construction together. What lines are easiest to draw first? [probably the horizontal and vertical lines] Ask him to draw those and then the other lines. Ask him to explain how to do it. The figures on the next page show drawing lines with the two angles.

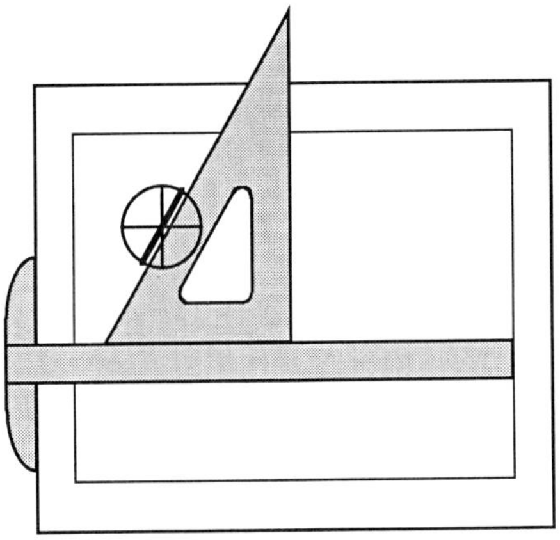

Drawing with 60-degree angle.

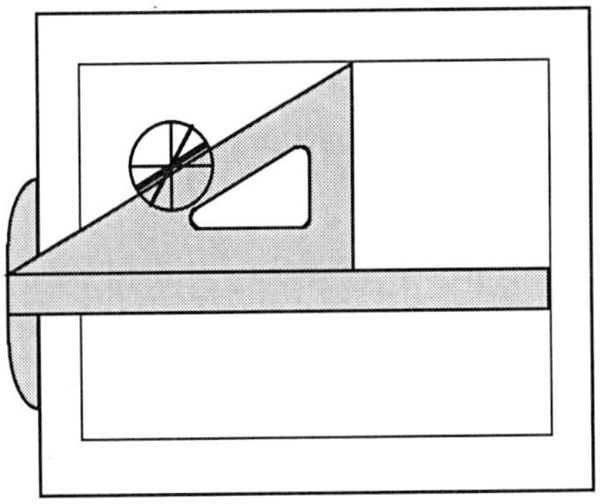

Drawing with 30-degree angle.

Flip the triangle horizontally to draw the remaining lines. Then ask him to read the directions for coloring and to answer the questions. Remind him to check that the T-square is touching the edge of the drawing board and the triangle is touching the T-square before drawing each line.

See the figures below for the solutions.

For the fourth circle, ask the child to think carefully and to count the number of triangles. Ask for explanations.

In the fifth circle there three sets of parallel lines for the larger triangles and three for the smaller, making a total of six.

Note: When fractions are truly understood, the extra lines, sometimes called *perceptual distractors,* will not be a problem.

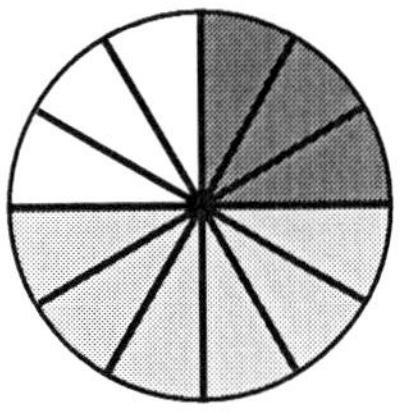

$\mathbf{\frac{1}{4}}$

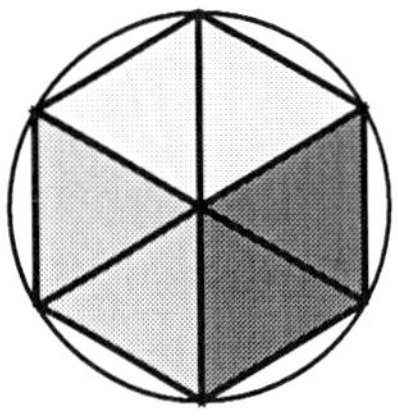

triangles, hexagon

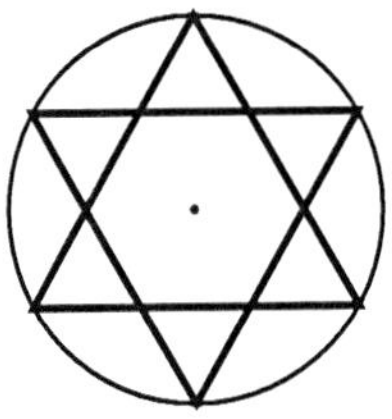

small: **6**
large: **2**

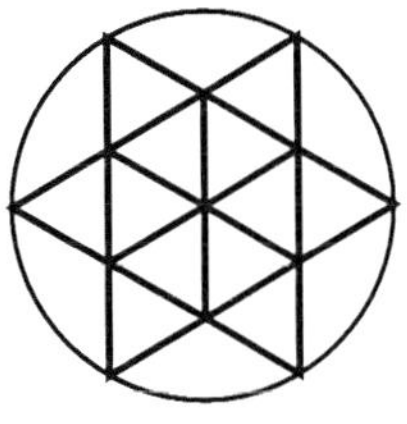

small: **12**
medium: **6**
large: **2**

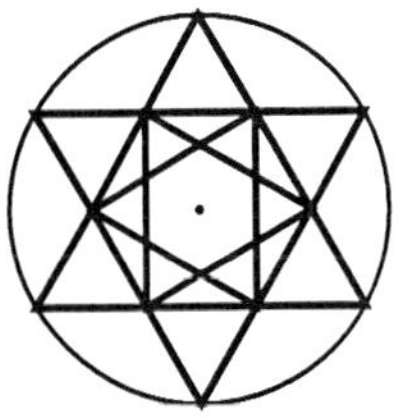

sets of parallel lines: **6**

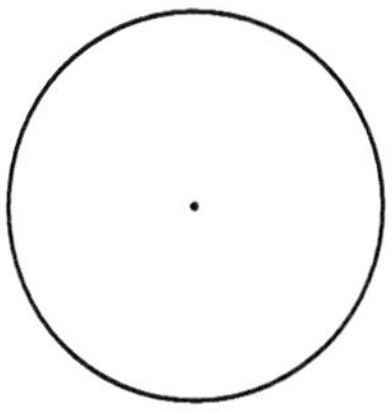

Lesson 129

Drawing Congruent Copies

OBJECTIVES
1. To learn the term *congruent*
2. To construct congruent figures

MATERIALS
Worksheet 139, "Drawing Congruent Copies"
A piece of paper for cutting out a triangle (see preparation below)
A T-square, 45 triangle, and 30-60 triangle 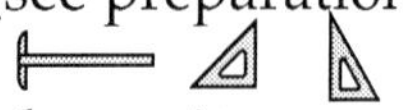
A drawing board with Worksheet 139 taped onto it

WARM-UP
How many sides in a parallelogram? [4] How many sides in a quadrilateral? [4] How many sides in a hexagon? [6] How many sides in an octagon? [8]

ACTIVITIES

Preparation. Cut out a triangle similar to the one drawn on the left in the figure below. Then draw the figures below; use the cutout triangle for tracing. The first triangle is similar, but smaller (trace the upper part and then draw a horizontal line). To draw the second triangle , flip the cutout vertically and trace. To draw the third one, flip it back and then flip it horizontally and trace. For the fourth go back to the original position and rotate it 90° to the left.

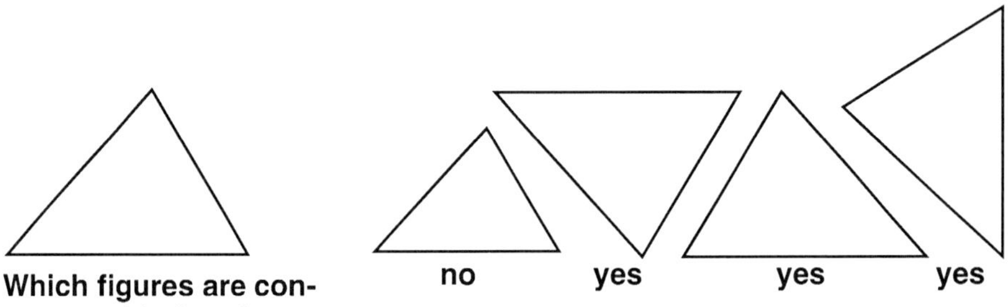

Which figures are congruent with the triangle above?

The term congruent. Tell the child that two figures are *congruent* if they fit exactly on top of the other. Point to the above figures and ask in turn if each figure is congruent with the first. Ask for explanations.

The first figure is too small. Show with the cutout that it doesn't fit. Flip or rotate the figure to show the others are congruent with it.

Worksheet. Take out the drawing board and tools. These 3 problems require thinking. Ask the child to read the directions at the top and the first problem. What does it ask you to do? [Draw the figure on the dotted line exactly as the one above it.]

Ask her to think about how she could do it, using only the tools. Ask her to explain her procedure. She will need to draw 2 tick marks; the simplest way is to draw the edges of the triangle as shown below.

She will also need to find the center of the baseline, which can be done after the largest triangle is drawn. See the figure on the next page on the right. When she has completed the figure, ask how she did it.

Note: Encourage the child to work carefully without guessing. Set high expectations. Ask her to do any inadequate work over.

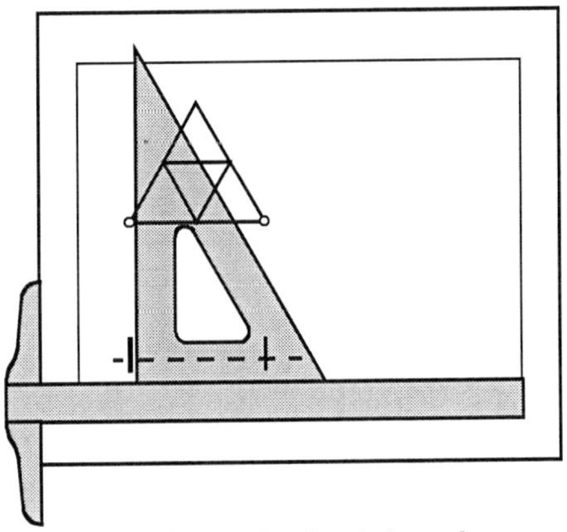

Draw tick marks to determine the width of the triangle.

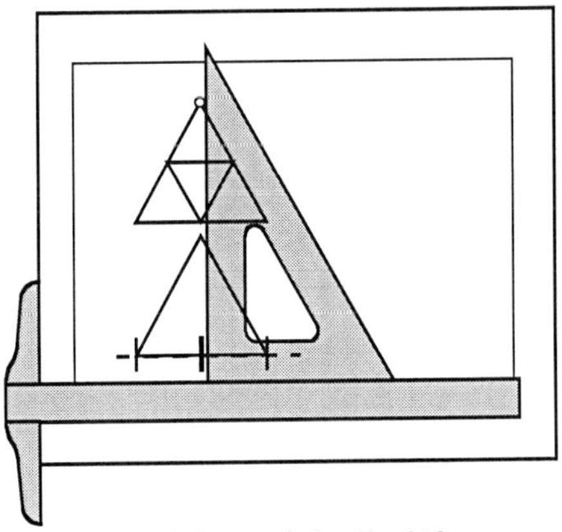

Draw a tick mark to find the center for drawing the other lines.

Encourage the child to plan her work before starting. Explain that these problems are like puzzles. After it is completed, ask for an explanation of how she did it.

For problem 2, the diagonal of the square determines the length of the base line.

For problem 3, make tick marks to show where to begin drawing the sides of the triangles. The tick mark for the largest triangle is shown in the figure below.

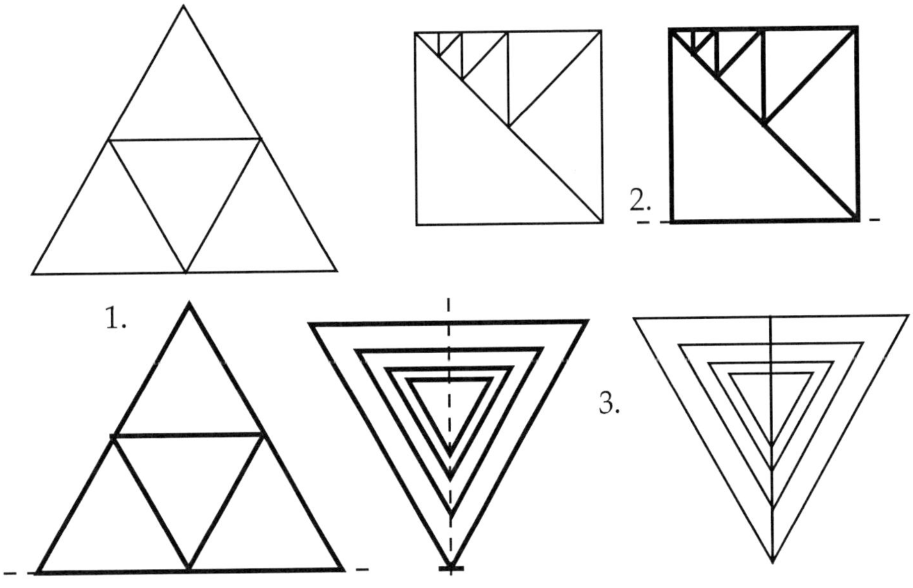

The solutions for Worksheet 139.

Lesson 130

Drawing New Fractions

OBJECTIVES
1. To construct the whole, or other fraction, when a geometric part is given
2. To learn or review the terms *clockwise* and *counterclockwise*

MATERIALS A T-square, 45 triangle, and 30-60 triangle 
Worksheet 140, "Drawing New Fractions"
A drawing board with Worksheet 140 taped onto it

WARM-UP Ask the child to stand and face you. Turn to the right 90°. Turn to the right 180°. Turn to the right 90°. [He will now be facing you.]

Turn 90° clockwise. Remind him that clockwise means in the same direction that the hands on a clock move. Turn another 90° clockwise. Turn counterclockwise 180°. If necessary, remind him that counterclockwise means in the opposite direction that the hands on a clock move. [He will now be facing you.]

Turn clockwise 270°. Turn counterclockwise 90°. Turn clockwise 360°. Turn counterclockwise 180°. [He will be facing you.]

ACTIVITIES ***Finding the whole.*** Draw a rectangle as shown below on the left. This represents 1⁄4 of a candy bar. Copy the 1⁄4 of the candy bar in your math journal and draw the whole candy bar. Ask the child for his solution and discuss it. Help him connect to his visual image of 1⁄4 on the fraction chart.

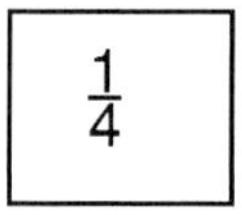

Showing 1⁄4 of the candy bar.

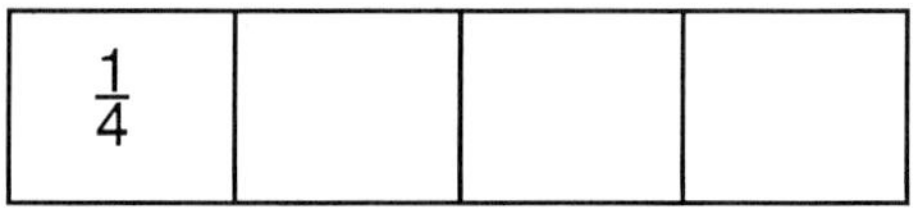

Showing the whole candy bar.

Draw a rectangle and write 2⁄3 in it. See below on the left. Copy the rectangle and draw the whole. See below. Ask an advanced child to find several solutions. First he needs to find 1⁄3 and either add it to the 2⁄3 or draw three 1⁄3s. See 3 solutions given below.

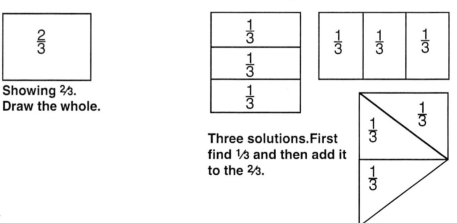

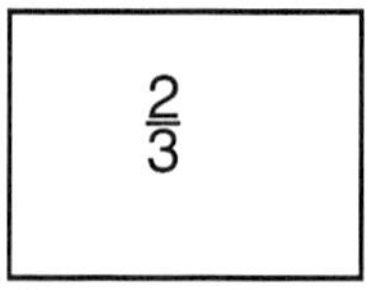

Showing 2⁄3.
Draw the whole.

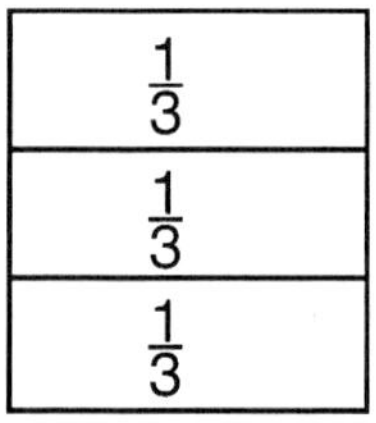

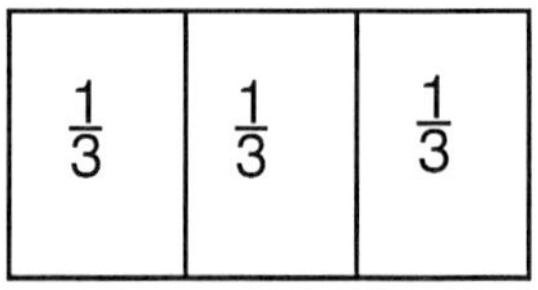

Three solutions.First find 1⁄3 and then add it to the 2⁄3.

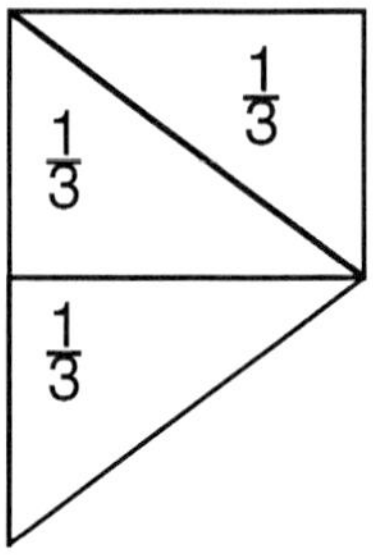

Finding the new fraction. Draw a triangle and write 1⁄4 on it as shown on the next page. Copy and draw the whole, or 1. There are several other solutions besides the one shown.

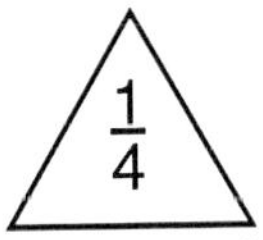

Showing ¼.

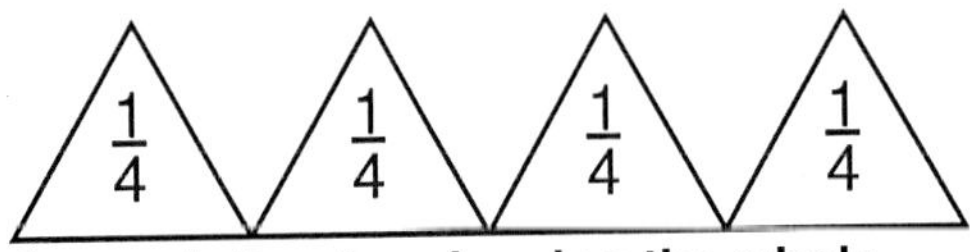

One solution for showing the whole.

Draw a rectangle different in size from those above and write ⅕ in it as shown below. Copy and draw ⅖. See below.

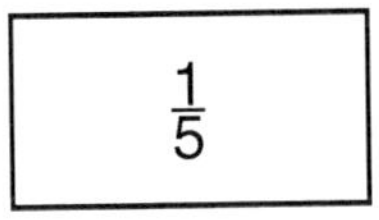

A rectangle representing ⅕.

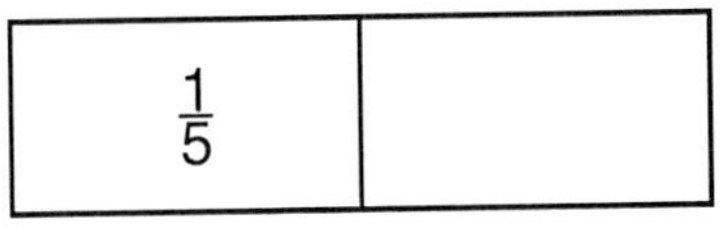

Showing ⅖ when ⅕ is given.

Next draw another rectangle and label it ½. Copy and draw ¼. See below.

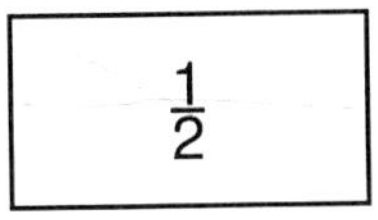

A rectangle representing ½.

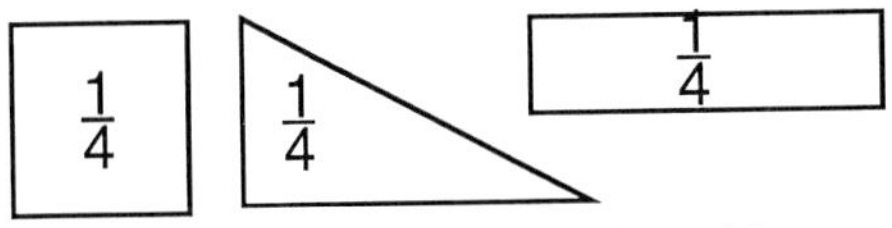

Three solutions for showing ¼ when ½ is given.

Lastly, draw a half square with a diagonal, as shown; label it 2⁄4. Copy and draw ¾. See below.

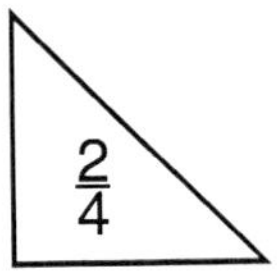

Showing 2⁄4.

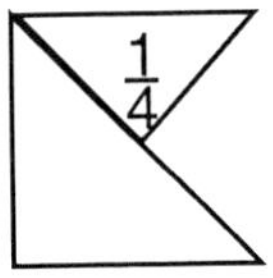

Showing ¾.

Worksheet. For this worksheet the child is given a figure. He is to construct a new fraction. The answers are below. The lines in boldface show the constructions.

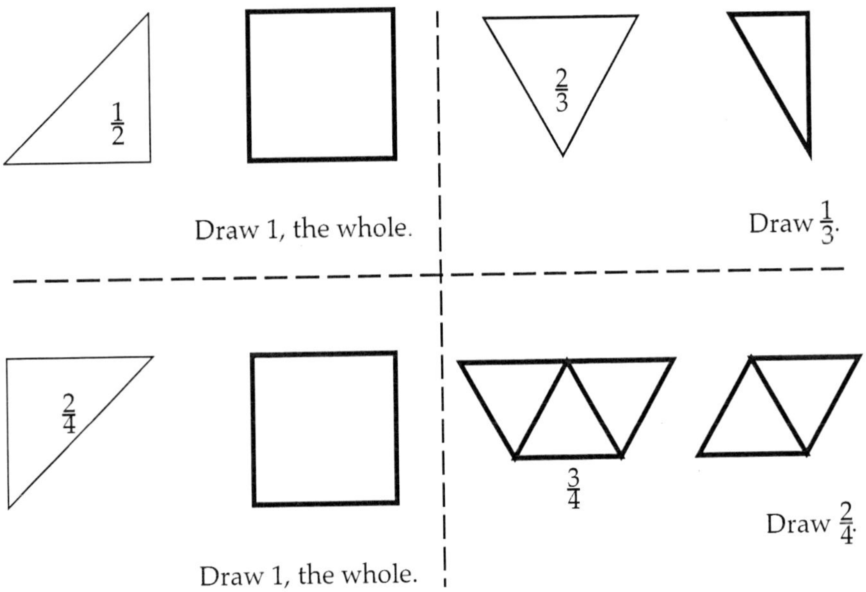

Lesson 131

Drawing Symmetrical Figures

OBJECTIVES 1. To review the terms *symmetry* and *line of symmetry*
2. To learn the term *reflected*

MATERIALS 2 pieces of paper (1 piece can be half size) and scissors
Mirror, if available
A T-square, 45 triangle, and 30-60 triangle 
Worksheet 141, "Drawing Symmetrical Figures"
A drawing board with Worksheet 141 taped onto it

WARM-UP Ask the child to write the multiples of 9.

9	18	27	36	45
90	81	72	63	54

While she is looking at the multiples, ask her the following.

What is 36 divided by 9? [4] What is 18 divided by 9? [2]

What is 45 divided by 9? [5] What is 9 divided by 9? [1]

What is 27 divided by 9? [3] What is 90 divided by 9? [10]

What is 54 divided by 9? [6] What is 63 divided by 9? [7]

What is 81 divided by 9? [9] What is 72 divided by 9? [8]

ACTIVITIES ***Lines of symmetry.*** Take out the paper and scissors. Ask the child to fold the piece of paper in half and to cut out a random design. What will you see when your paper is opened? See the figures below.

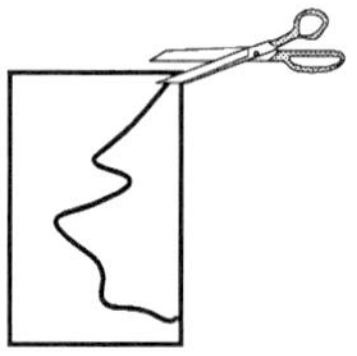

Fold paper in half.

Cut out a design.

Unfold the paper.

Do you see any congruent parts? [yes, the 2 halves] What word do we use to describe these congruent parts? [symmetry] What is the mathematical name for the fold line? [line of symmetry] Tell her it is also called *axis of symmetry*. Tell her the plural of *axis* is *axes*, like the plural of parenthesis is parentheses and oasis, oases.

Give the child the mirror to explore with her design. What happens when the mirror is placed at the line of symmetry? [The whole design is seen.] Explain that her design is *reflected* across the axis of symmetry. Explain that when she looks into a mirror or a still puddle of water, she sees herself reflected.

Now ask her to take the other piece of paper and to fold it into fourths and cut out a random design. See below.

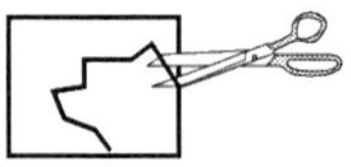

Fold paper in fourths.

Cut out a design.

Unfold the paper twice.

Do you have an axis of symmetry? [yes, 2] Where are they? [the fold lines] How can you be sure you have 2 axes of symmetry? [by folding the design on the axes of symmetry]

Where can you place the mirror to see the whole design reflected? [on either axis of symmetry]

Worksheet. Take out the board with the worksheet. Explain she is to reflect the design across the line of symmetry. There are four designs, each to be reflected in a different way: left, right, top, and bottom. She can use the mirror to check her work.

A good technique for drawing some of these lines is to first match the correct angle and then flip the triangle to draw the new line. See below.

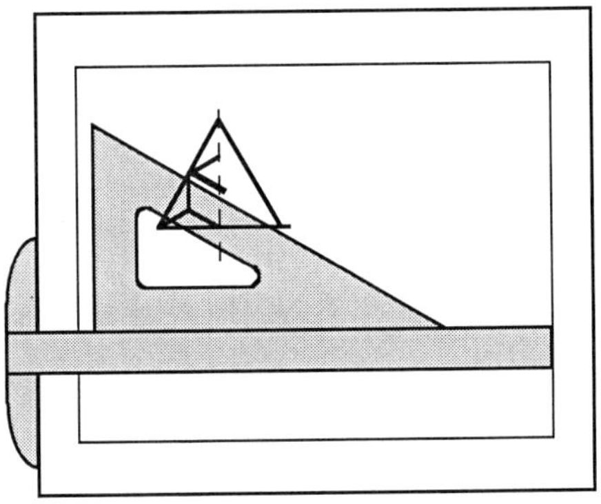

Finding the correct angle.

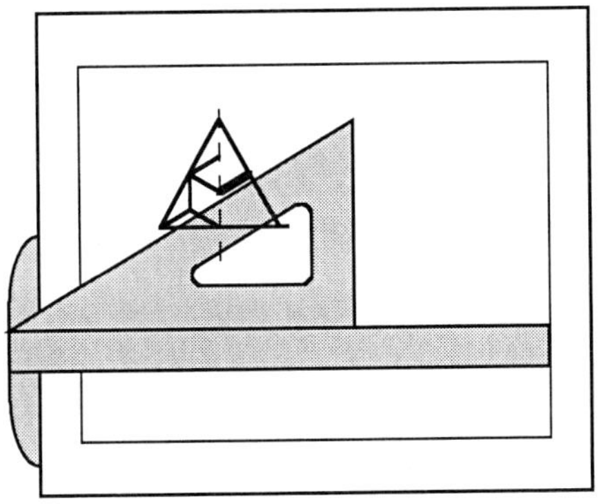

Flipping the triangle and drawing the reflected line.

Remind her that she is to use only the top edge of the T-square. She may be tempted to use the bottom edge for the first design in the second row, the star.

For the last two figures, the child is to make her own symmetrical designs. The designs and their reflections are shown below. The lines in boldface show the constructions.

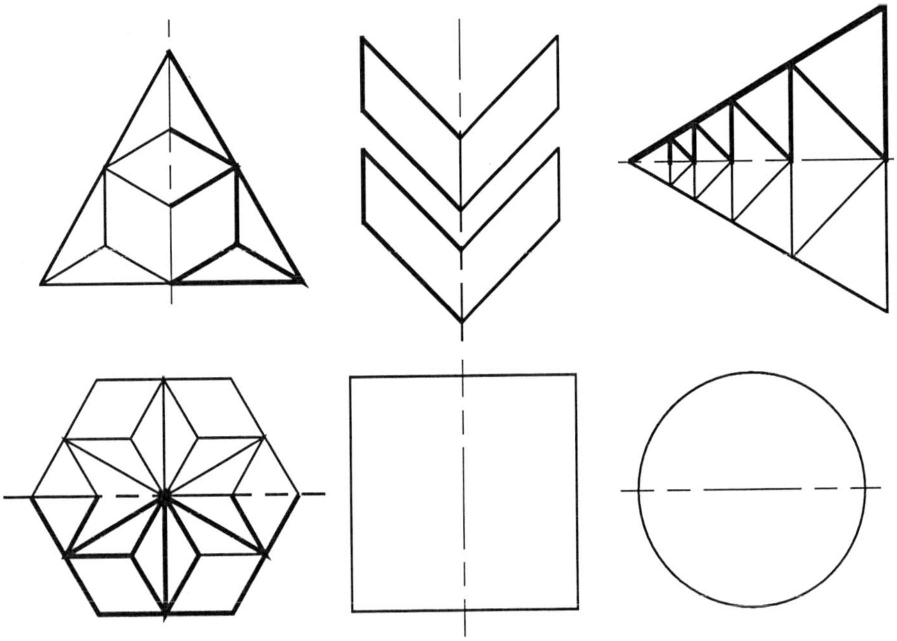

Lesson 132

Review

OBJECTIVE 1. To review and practice

MATERIALS Worksheet 142-A and 142-B, "Review" (2 versions)
Cards for playing games

ACTIVITIES ***Review worksheet.*** Give the child about 15 to 20 minutes to do the review sheet. The oral problems to be read are as follows:

$3{,}000 \times 10 =$ $\quad 199 + 201 =$ $\quad \frac{1}{2}$ of $48 =$

Review Sheet 142-A.

1. Write only the answers to the oral questions. **30,000** **400** **24**

4. Write only the answers. $2\frac{1}{4} + 1\frac{3}{4} =$ **4** $\quad 135 - 56 =$ **79** $\quad 23 \times 10 =$ **230**

Use the letter of the figure to answer the questions below.

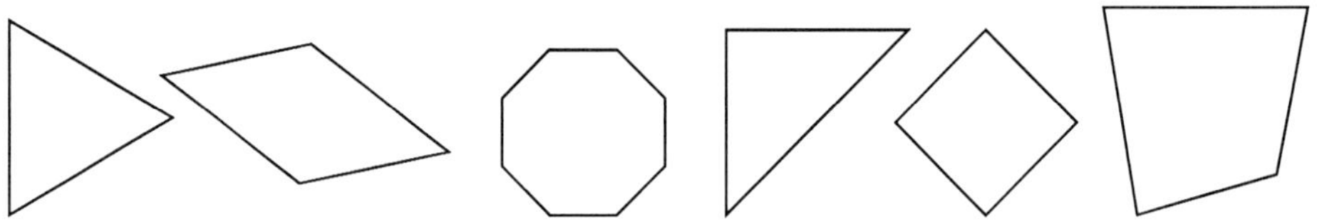

a b c d e f

7. Which figure is an octagon? **c**

8. Which figures are quadrilaterals? **b,e,f**

9. Which figure are parallelograms? **b,e**

10. Which figure is a rectangle? **e**

11. Which figures have parallel lines? **b,c,e**

12.
$3\overline{)24}$ = **8** $\quad 2\overline{)14}$ = **7**
$5\overline{)35}$ = **7** $\quad 3\overline{)18}$ = **6**
$2\overline{)18}$ = **9** $\quad 5\overline{)45}$ = **9**
$9\overline{)18}$ = **2** $\quad 5\overline{)30}$ = **6**

20. Apples cost 89¢ a pound and grapes cost \$1.67 a pound. What is the total cost of 3 pounds of apples and 4 pounds of grapes?

\$.89	**\$1.67**	**\$2.67**
× 3	**× 4**	**6.68**
\$2.67	**\$6.68**	**\$9.35**

23. Draw freehand the figure representing 1.

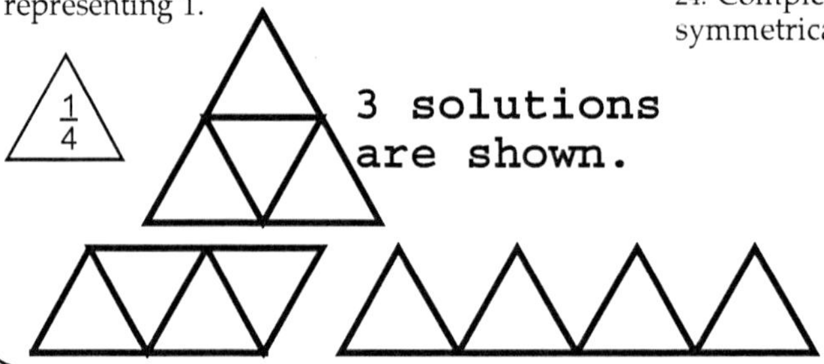

3 solutions are shown.

24. Complete the figure freehand so it is symmetrical about the dotted line.

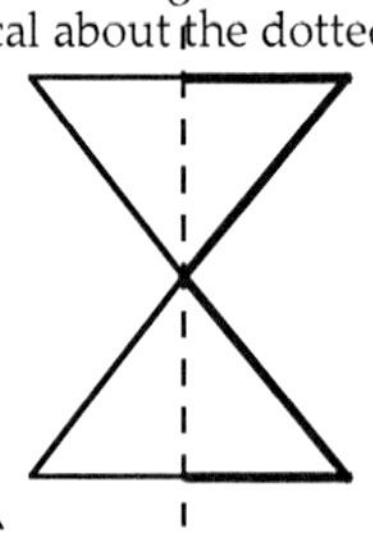

The oral problems to be read to the child are as follows:

$8{,}000 \times 10 =$ $298 + 302 =$ $\frac{1}{2}$ of $26 =$

Review Sheet 142-B.

1. Write only the answers to the oral questions. **80,000** **600** **13**

4. Write only the answers. $2\frac{1}{3} + 2\frac{2}{3} =$ **5** $141 - 85 =$ **56** $39 \times 10 =$ **390**

Use the letter of the figure to answer the questions below.

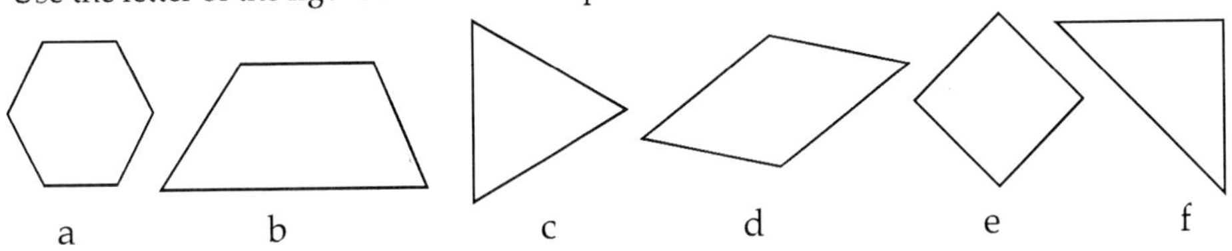

7. Which figure is an rectangle? **e**

8. Which figures are quadrilaterals? **b,d,e**

9. Which figures are parallelograms? **d,e**

10. Which figure is a hexagon? **a**

11. Which figures have parallel lines? **a,b,d,e**

12. $5\overline{)25}$ **5** $3\overline{)27}$ **9**
$9\overline{)90}$ **10** $2\overline{)14}$ **7**
$4\overline{)4}$ **1** $3\overline{)15}$ **5**
$2\overline{)16}$ **8** $5\overline{)40}$ **8**

20. Bananas cost 39¢ a pound and peaches cost \$1.85 a pound. What is the total cost of 4 pounds of bananas and 5 pounds of peaches?

\$.39 × 4 = \$1.56   **\$1.85 × 5 = \$9.25**   **\$1.56 + 9.25 = \$10.81**

23. Draw freehand the figure representing 1.

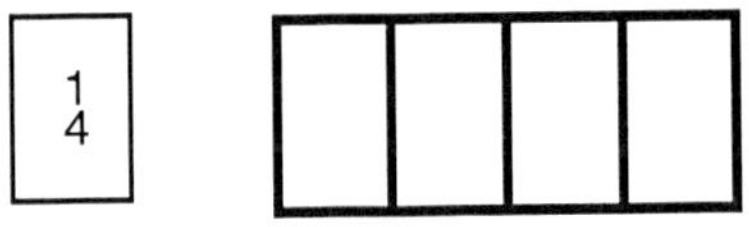

24. Complete the figure freehand so it is symmetrical about the dotted line.

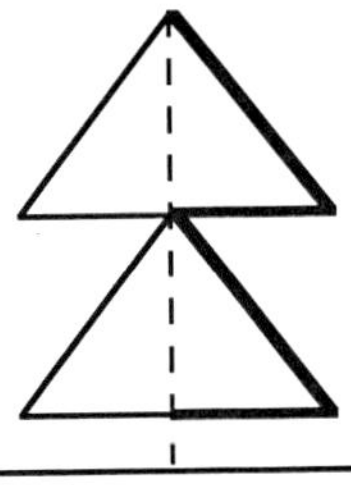

Games. Spend the remaining time playing games.

DIVISION. Mixed-Up Quotients (D3) *(Math Card Games).*

FRACTIONS. Fraction War (F7) and One (F6) *(Math Card Games).*

Lesson 133

Regular Polygons From Paper

OBJECTIVES

1. To construct some regular polygons through paper folding
2. To discover that the diagonal in a square makes a 45° with the sides of the square

MATERIALS

5 sheets of 8½ × 11 inch paper (standard size)
A 45 triangle 
Scissors

WARM-UP

What is a polygon? [a closed figure made with straight lines]

Ask the child to write the multiples of 9, with the second row reversed, as shown.

9	18	27	36	45
90	81	72	63	54

While she is looking at them, ask her division combinations, such 27 ÷ 9 [3] and so on.

ACTIVITIES

Constructing a square. Take out the sheets of paper. Tell the child we want to construct a square from the piece of paper. What is special about a square? [The sides are congruent.]

How can we make sure the left side and the bottom sides are congruent? [Fold them together and cut off the extra strip.] See the 2 left figures below. Finally, open the paper.

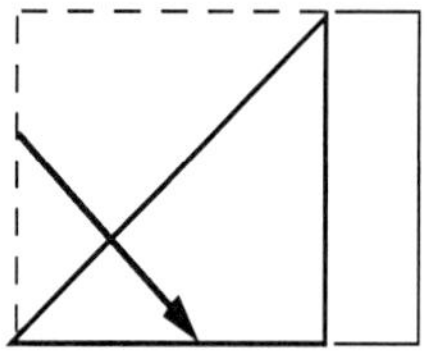

Making the two sides equal.

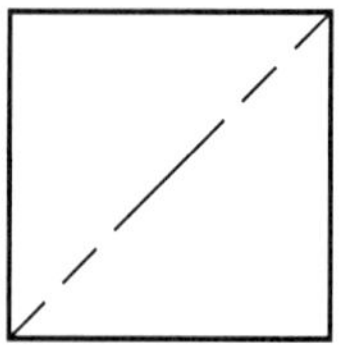

The extra piece cut off and the paper opened.

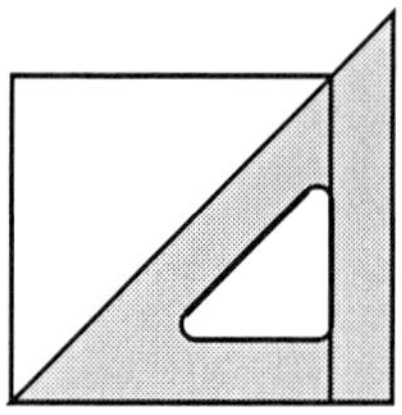

The diagonal makes a 45 angle with the side.

Now ask her to compare the diagonal with the 45 triangle. See the right figure above. Could this help us in drawing a square? [yes]

Ask her to set that square aside since it will be needed in the next lesson, as will the other figures she will construct.

Constructing an octagon. Tell the child that next she will make an octagon, which takes 2 squares. Ask her to make 2 more squares. Tell her to fold 1 square with both diagonals as shown in the first figure below. Tell her to make both folds on the same side of the paper.

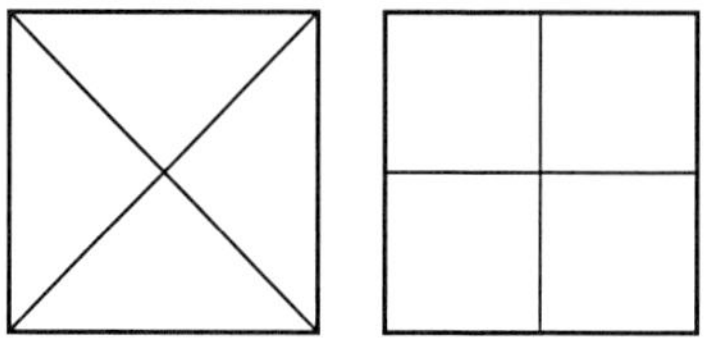

Construct 2 squares and fold them into fourths as shown.

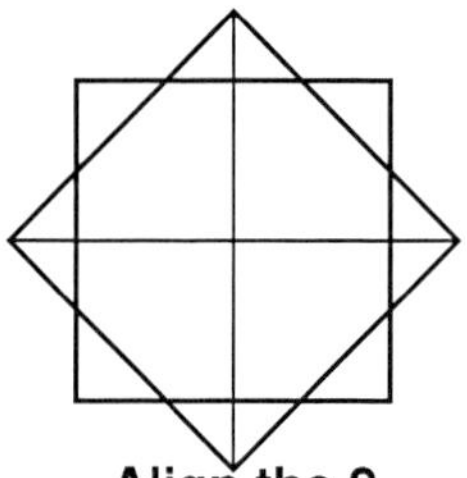

Align the 2 squares.

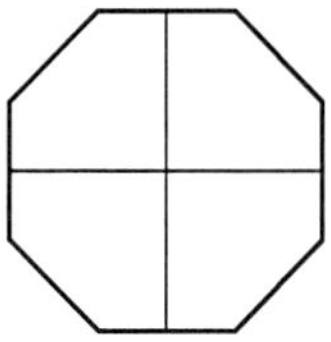

Cut off the points.

What fraction is the square divided into? [fourths] How could you fold the second square into fourths without using diagonals? [See the second figure on the previous page.] Ask her to fold the second square that way, again with both folds on the same side on the paper.

Then tell her to place the squares on top of each other so the folds line up. Do you see the octagon? The last step is to remove the eight points. How could you remove the points? [fold them down and cut them off] She will have two octagons. Save one for the next lesson.

Constructing an equilateral triangle. To construct an equilateral triangle, first fold the paper in half the long way. Next carefully pull the lower left corner to the fold line until the upper left corner is folded exactly at the corner. See the first figure.

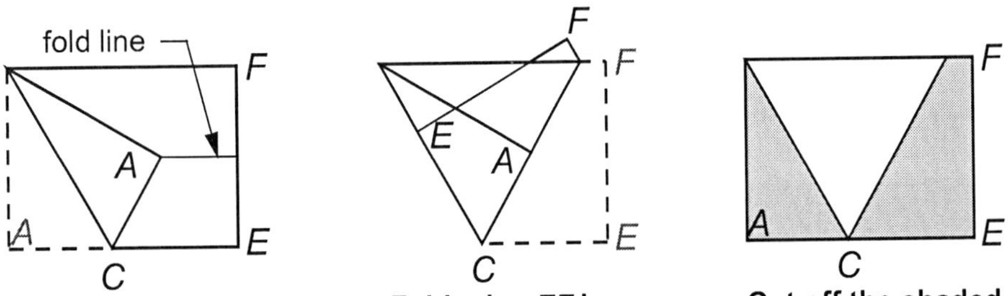

Overlap so corner *A* is on the fold line. **Fold edge *EF* behind edge *AC*.** **Cut off the shaded areas.**

What does the folded part looks like? [half an equilateral triangle] Where is the other half? [along the top] Fold the paper along the right side behind the triangle as shown in the second figure. Cut off the extra pieces, the shaded area, as shown in the third figure.

Constructing a regular hexagon. To construct a regular hexagon, first fold the paper in half twice the long way. Then unfold it once, making sure the folded edge is on top. See the first figure. Pull the lower left corners to the midline until the upper left corner *B* is folded exactly at the corner.

Note: These figures will be needed in the next lesson.

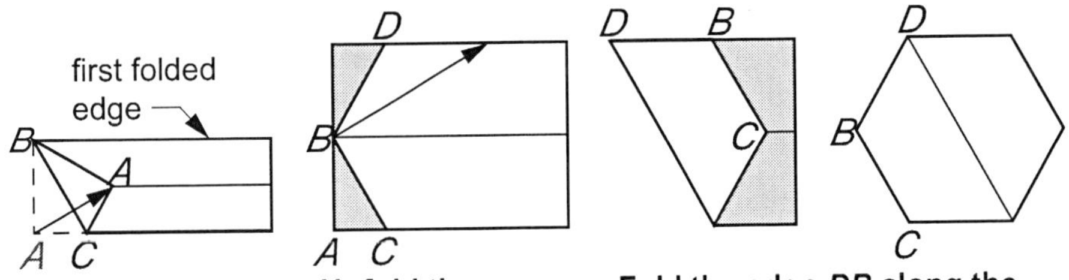

Overlap so corner *A* is on the quarter fold. **Unfold the paper and cut off the shaded area.** **Fold the edge *DB* along the top edge. Cut off the shaded area and unfold.**

Unfold the paper completely and cut off the shaded area as shown in the second figure.

Ask if she sees two equal sides. To get another equal side, fold as shown in the third figure above. Then cut off the extra paper and open it up.

Lesson 134

Tenths of a Centimeter

OBJECTIVES
1. To measure and calculate using tenths of a centimeter
2. To determine and compare the angles of the regular polygons

MATERIALS
Worksheet 143, "Tenths of a Centimeter"
4-in-1 ruler
The paper polygons made during the previous lesson
A 45 triangle and a 30-60 triangle 

WARM-UP
How many centimeters in a meter? [100]

How many degrees are in a right angle? [90°]

Robin is a cook and is buying 24 jars of spaghetti sauce and 24 boxes of pasta. The sauce costs $3 per jar and the pasta costs $2 per box. What is the total cost? [either 24 × 5 = $120 or 24 × 3 + 24 × 2 = $120]

What is 20 ÷ 2 ÷ 2? [5] What is 20 ÷ 4? [5] What is 40 ÷ 2 ÷ 2? [10] What is 40 ÷ 4? [10] What is 24 ÷ 2 ÷ 2? [6] What is 24 ÷ 4? [6] What is 48 ÷ 2 ÷ 2? [12] What is 48÷ 4? [12] What is 400 ÷ 2 ÷ 2? [100] What is 400 ÷ 4? [100] Which is easier?

ACTIVITIES
Divisions in an inch. Take out the worksheet and ask the child to read and answer the questions down to the table. The answers are as follows.

fourths **eighths**

by the number of spaces between the lines

tenths

$6\frac{8}{10}$ **$9\frac{2}{10}$**

Ask him for his response to how he can tell. Take out the ruler. What are the inches divided into?

Measuring in tenths. Ask the child to draw a line that is 7 and $\frac{3}{10}$ cm long. Then ask him to draw another line touching the first line that is 4 and $\frac{6}{10}$ cm long. See the figure below.

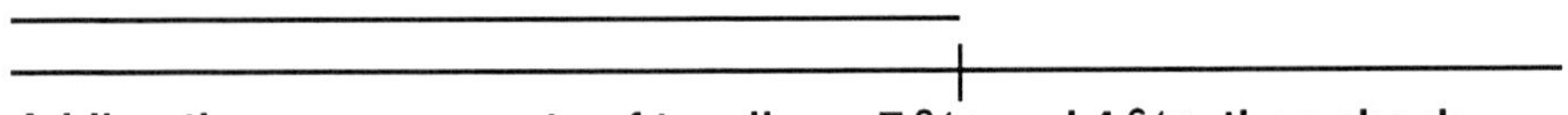

Adding the measurements of two lines, 7 $\frac{3}{10}$ and 4 $\frac{6}{10}$, then checking by measuring. [11 $\frac{9}{10}$]

Note: Tenths of a centimeter will be written in fraction form for now, rather than the usual decimal notation.

How long will the new line be? [11 and $\frac{9}{10}$ cm] Ask him how he added the two numbers. [added the ones and the tenths separately]

Repeat for lines that are 8 and $\frac{6}{10}$ cm long and 6 and $\frac{8}{10}$ cm long. Ask how he added the 2 numbers. He probably noticed that $\frac{18}{10}$ equals 1 and $\frac{8}{10}$. Then ask him to measure the new line and check its length. See the figures on the next page.

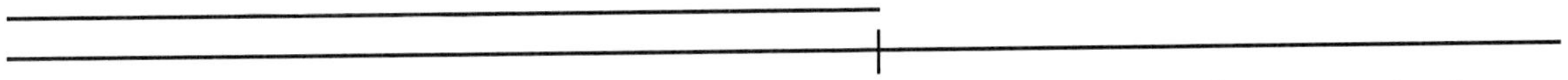

Adding the measurements of two lines, 8 6⁄10 and 6 8⁄10, then checking by measuring. [15 4⁄10]

Problem. What is the perimeter of an equilateral triangle if a side is 7 and 7⁄10 cm? [21 and 2 1⁄10 cm, which is 23 1⁄10 cm]

Measuring the paper polygons. Ask the child to guess which polygon has the greatest perimeter. Ask him to measure the sides of the triangle. Ask for various measurements. Why aren't all the sides the same? [hard to make the polygons exactly right] Let him decide how to handle the problem. He could choose a measurement in the middle. Given below are the calculated lengths, assuming the paper was 8½ inches wide to start.

Ask him to fill in the remaining boxes.

Polygon	Sketch	Length	No. of Sides	Perimeter
Equilateral triangle		**25 cm**	3	**75 cm**
Square		**21 6⁄10 cm**	4	**86 4⁄10 cm**
Hexagon	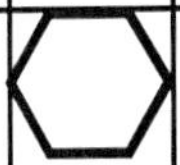	**12 5⁄10 cm**	6	**75 cm**
Octagon	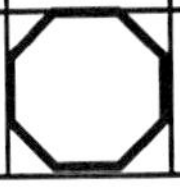	**8 9⁄10 cm**	8	**71 2⁄10 cm**

Lesson 135

Building a Box

OBJECTIVES
1. To construct a box pattern
2. To review the term *cube*

MATERIAL
A box, made from the Box Pattern (Appendix pg. 6)
Worksheet 144, "Building a Box"
A drawing board with the worksheet taped on it
A T-square and 45 triangle  
Scissors and removable tape
A 1-inch cube 

WARM-UP
What is a diagonal? [a line drawn between 2 vertices that are not next to each other in a polygon]
How many sides in a hexagon? [6]
How many sides in a octagon? [8]
What is a square inch? [a square with the sides measuring 1 inch]
Ask the child to add 468,657 + 326,978 in her journal. [795,635]

Note: The child should be familiar with both *vertices* and *vertexes* as the plural of vertex.

ACTIVITIES
Drawing a square. Explain that you want to draw a square. What do we draw first? [a horizontal line] What do we draw next? [the vertical lines] See the figure below on the right.

How do we know where to draw the last line without measuring? Remind the child of the square she made by paper folding. Draw a square with a diagonal and shade as shown below on the right. What fits here? [the 45 triangle] How could that help us? Draw the diagonal as shown below in the third figure. How do we draw the last line? See the fourth figure below.

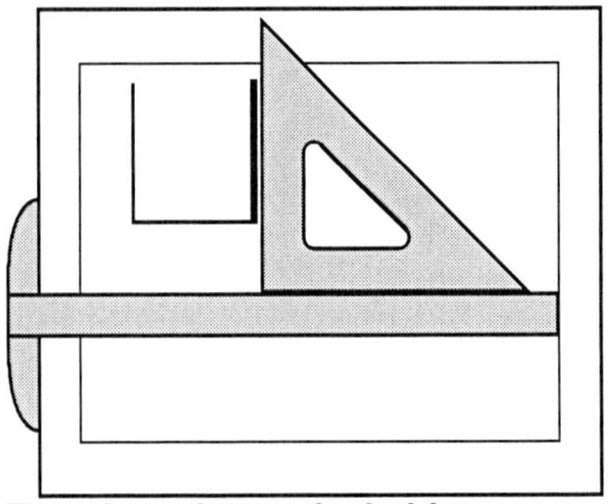
First draw the vertical sides.

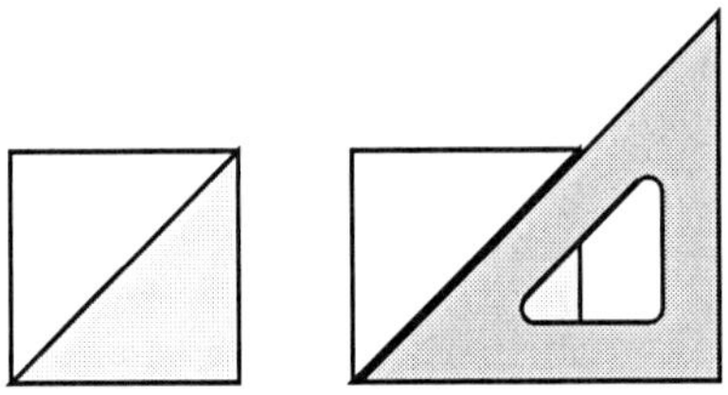
Noticing how the 45 triangle matches the diagonal.

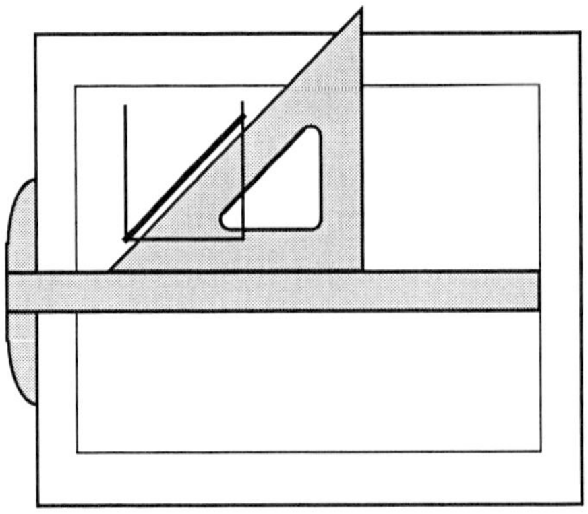
Drawing the diagonal.

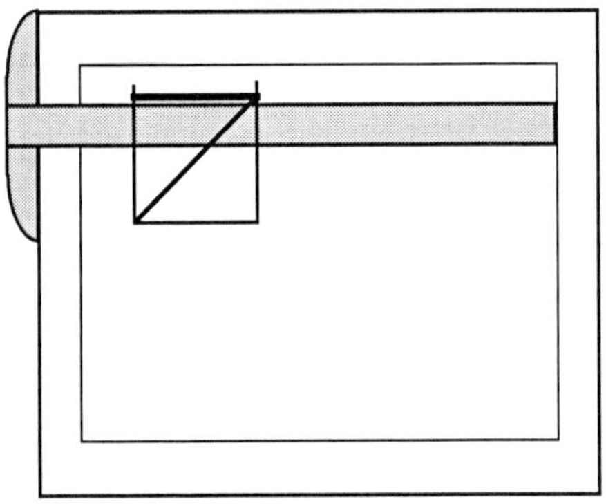
Drawing the last line of the square.

Note: The child may prefer to draw only tick marks, rather than the complete diagonal.

Drawing a box. Take out the worksheet and drawing tools. Ask the child to read the directions. Challenge her to make the box without seeing how it looks, but have the constructed box available. The steps for constructing the box pattern are shown below. Encourage her to work on her own; do not lead her step-by-step.

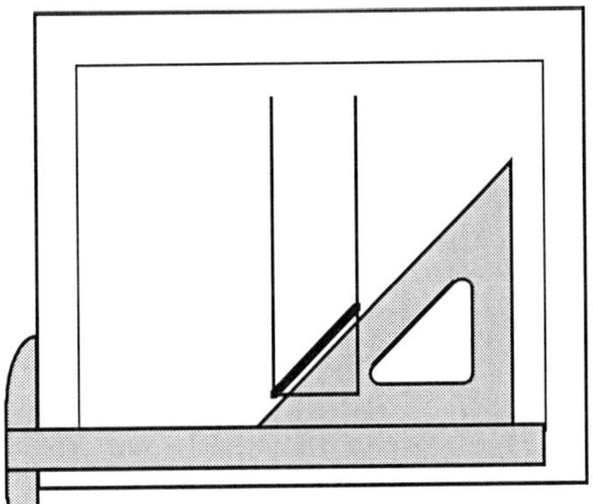

Drawing the diagonal to find the height of the lower square.

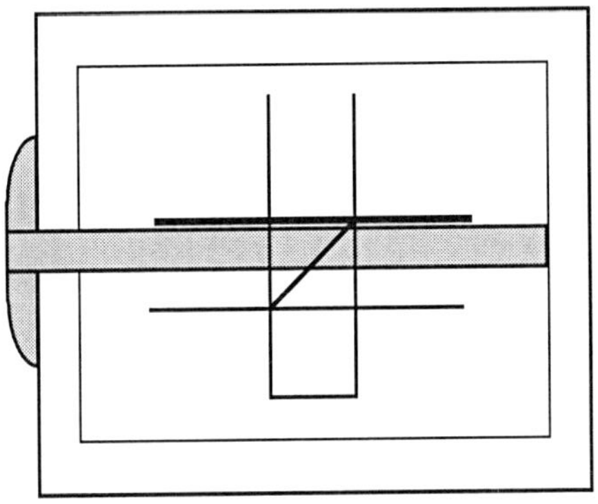

Drawing the diagonal and line above the center square.

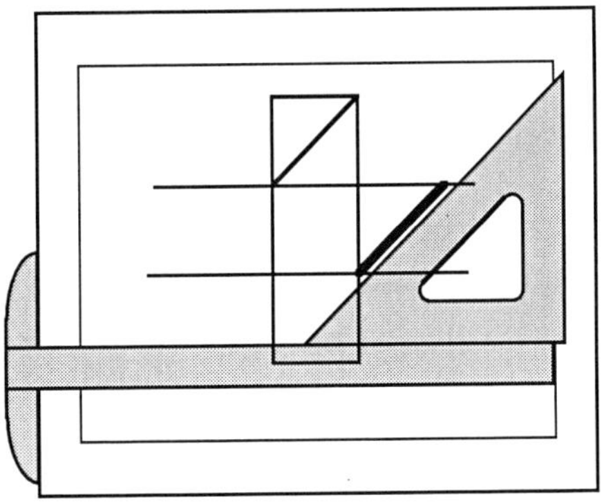

Drawing the diagonal to find the right side of the right square.

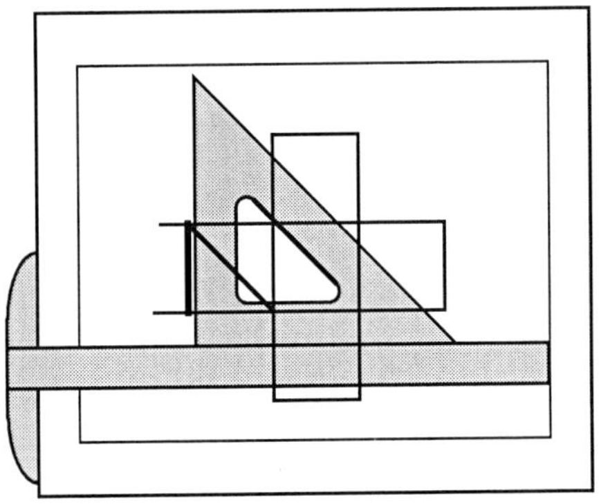

Drawing the diagonal and vertical line for the left square.

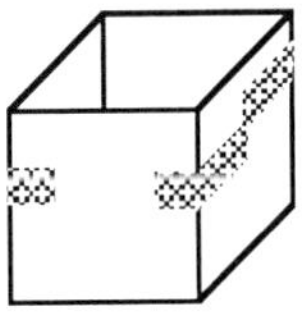

The completed box.

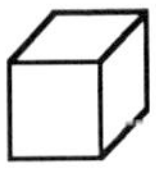

The 1-inch cube.

Cubes. When the child's box is complete, take out the 1-inch cube. Ask her to turn her box upside down. Ask the following questions.

What is the name of the shape? [cube]
How many square sides do they have? [5, (1 side is missing)]
How long is an edge of the large cube? [2 in.] small cube? [1 in.]
What is the area of a side of the large cube? [4 sq in] small cube? [1 sq in.]
How many small cubes will fit in the large cube? [8]

Tell her that a cube that measures 1 inch on a side is called a *cubic inch*. How many cubic inches does the large cube hold? [8 cu in.]

If the child is advanced ask, How many cubic inches would a cube hold if a side of the cube was 3 inches? [27 cu in.]

Lesson 136

Congruent Shapes

OBJECTIVE 1. To practice recognizing congruent shapes

MATERIALS Worksheet 145, "Congruent Shapes"
Crayons or colored pencils
Two 30-60 triangles for demonstration 

WARM-UP What coins do you need to make 31¢? [1 quarter, 1 nickel, 1 penny]

ACTIVITIES ***Congruent shapes.*** Take out the worksheet and ask the child to read the directions. Ask him to explain them.

Hold up the two 30-60 triangles in the position shown on the left below. Are these triangles congruent? [yes] Then flip 1 over as shown below on the right. Are these triangles congruent? [yes] Stress that even if a figure needs to be flipped over to fit exactly, it is still congruent.

Both sets of triangles are congruent. A flip may be necessary before matching.

The congruent shapes are shown below using patterns.

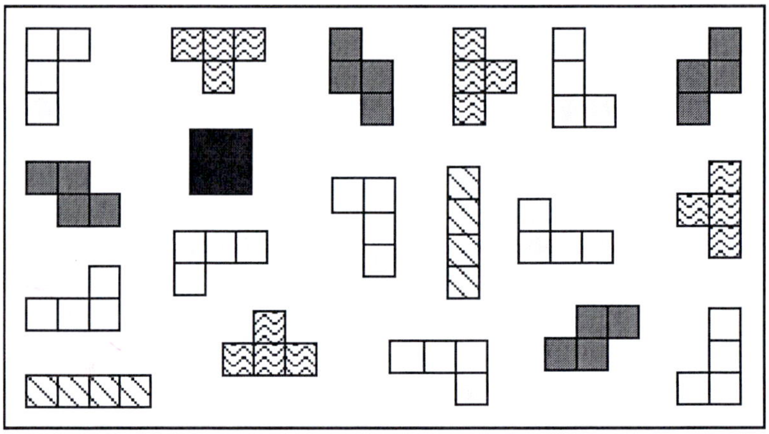

Note: The child might recognize these shapes as the pieces used in the computer game *Tetris.*

Then discuss the questions. How many different shapes are there? [**5**] How many different colors did you need? [5]

How many shapes of each color are there? [**1** (square), **2** (long), **4** each of *T* and *H,* and **8** of the *L*] What is special about these numbers? [**1, 2, 4, 8: each number is double the preceding number.**]

What are the areas in sq. cm of each shape? [**4 sq cm**] You might ask for the area of all the shapes. [19 × 4 = 76 sq. cm]

What are the perimeters in cm of each shape? [**8 cm** for the square and **10 cm** for the others] An easy way to count the perimeters is to first count the horizontal sides and then the vertical sides.

Are there any other shapes you could make with four squares with the sides touching? [I doubt it.]

Lesson 137

Combining Five Squares

OBJECTIVE 1. To find the 12 ways of combining 5 squares (pentominoes)

MATERIALS Worksheet 146, "Combining Five Squares"
1-inch tiles

WARM-UP What bills and coins do you need to make $2.57? [two $1 dollar bills, 2 quarters, 1 nickel, 2 pennies]

Ask the child to subtract $264.92 – $89.47 in her journal. [$175.45]

ACTIVITIES ***Review.*** Give the tiles to the child and ask her to use 4 squares at a time and to find all the different ways to combine them with the sides touching. [5] See the figures below.

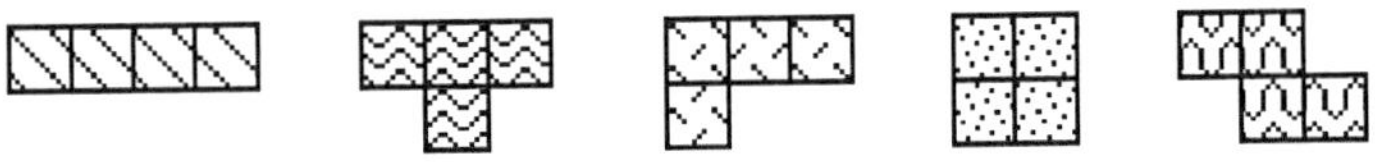

The five ways to combine four squares.

Combining 5 squares. Now ask to find all the different ways to combine 5 squares. Take out the worksheet, where she can draw the results freehand. The 12 possibilities are shown below.

Note: These arrangements are sometimes called pentominoes.

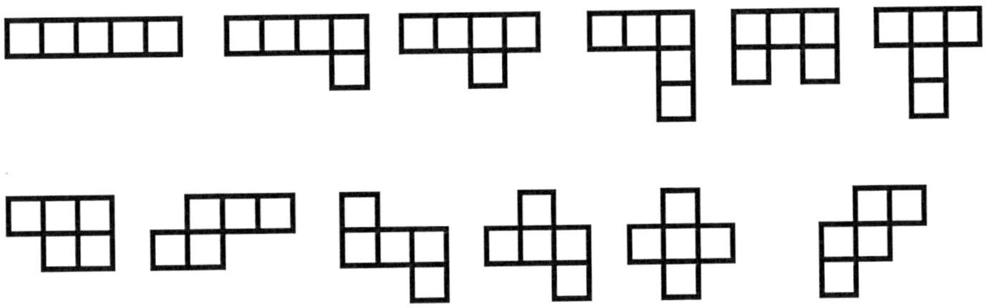

The 12 ways to combine 5 squares.

Note: The child will need to refer to these arrangements for the next lesson.

Remind her that two shapes are the same if they are congruent.

Areas and perimeters. If time remains, ask the child to find the areas [5 sq cm] and the perimeters. [12 cm for all except the first one in the second row, which is 10 cm]

Lesson 138

Building More Boxes

OBJECTIVES
1. To learn the term *scalene triangle*
2. To make boxes from different square arrangements
3. To practice visualizing which arrangements will make boxes

MATERIALS
Worksheet 147, "Building More Boxes"
A drawing board (The child could tape his own worksheet to the board.)
A T-square and 45 triangle 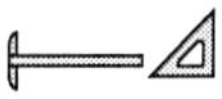
Scissors and removable tape
Ruler

WARM-UP
Draw an isosceles triangle and a triangle with no sides congruent as shown below. Tell the child that a triangle with no sides congruent is called a *scalene* triangle. Ask the following questions.

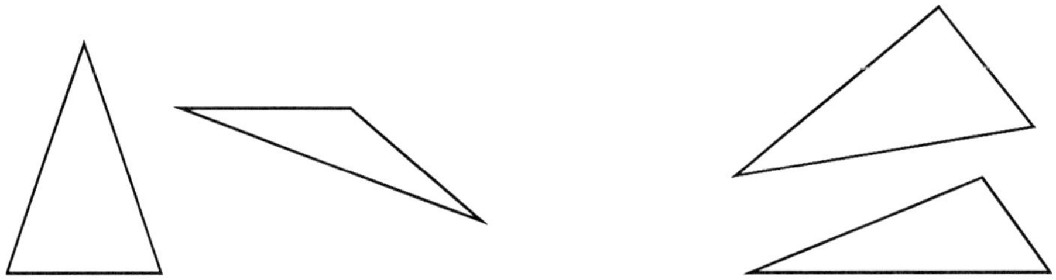

Isosceles triangles (two sides congruent).

Scalene trianglesa (no sides congruent).

Which triangle has 2 sides equal? [isosceles]

Which triangle has no sides equal? [scalene]

Which triangle has a line of symmetry? [isosceles]

ACTIVITIES
Constructing the box outlines. Explain to the child that today he will use the shapes he made in the previous lesson and try to make boxes from them. He will make 1 (or more if he is interested) of the 12 shapes from the previous lesson.

Then ask him to construct the shape on his paper with the sides of the squares 2 inches long. Ask him to plan ahead how it will fit on the paper.

Building the boxes. Stress that before the child cuts them out, he is to answer question 2 on his paper, Do you think it will make a box? [8 shapes will and 4 will not] Next he cuts it out, folds on the lines, and makes the box or partial box.

Note: The figure with a 2 by 2 square in it, as shown below, will need an extra cut shown by the dotted line.

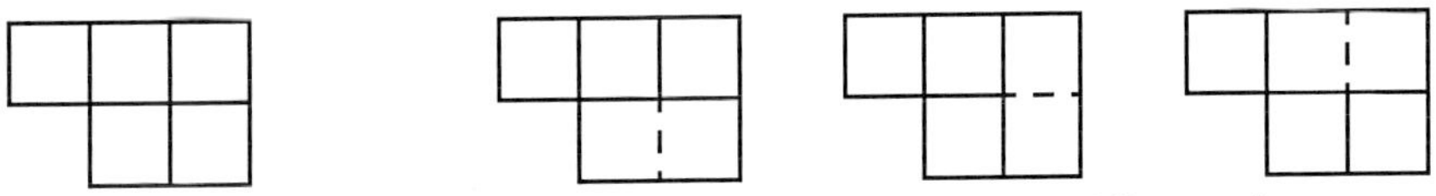

This figure also needs to be cut on *one* of the dotted lines shown.

The figures that will make a box are shown below.

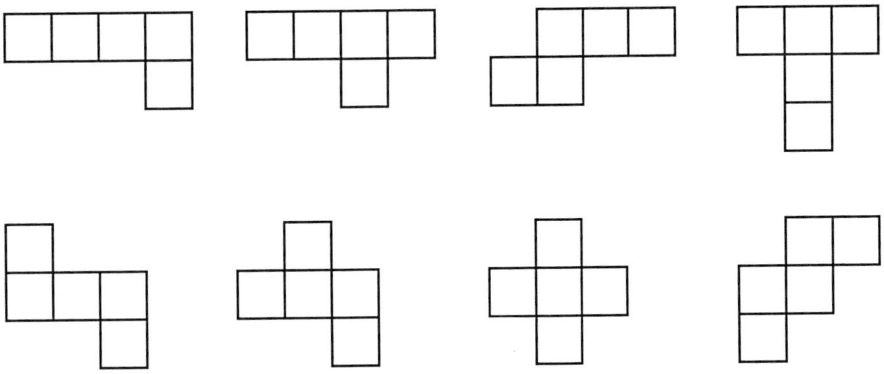

These shapes will fold into a box.

The figures that will *not* make a box are shown below.

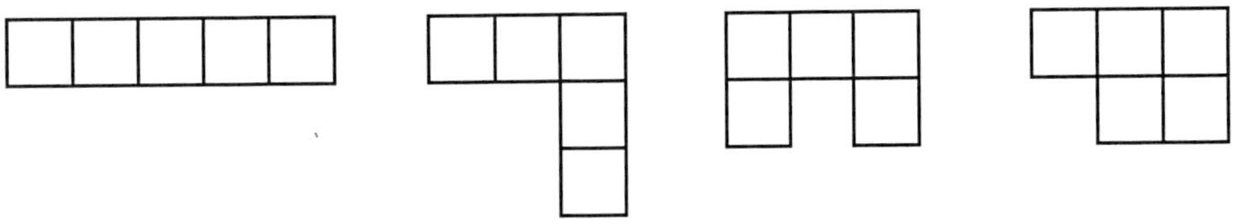

These shapes will *not* fold into a box.

> **Note:** Visualization is an extremely important skill for the child to learn. Many topics in advanced math require it.

Visualization. This is the important part of the lesson. Show the child each shape that he made in the previous lesson. Ask if it will make a box; allow him time to think. Then ask him to vote yes or no if he thinks it will make a box. Ask him how he would fold it.

If he is uncertain about any of the shapes, you might ask him to construct them, cut them out, and fold them

Lesson 139

Building a 4 × 3 × 1 Box in Inches

OBJECTIVE 1. To work with volume through constructing a box

MATERIALS Worksheet 148, "Building a 4 × 3 × 1 Box"
A drawing board, T-square, and 45 triangle
Scissors and removable tape
At least twelve 1-inch cubes

WARM-UP Tell the child you are going to give definitions of math terms and she is to write down the word that is described.

A mathematical statement that two things are the same. [equation]

A number that is part of another number. [fraction]

The answer after dividing. [quotient]

The answer after multiplying. [product]

Groups of the same number for example, 3, 6, and 9. [multiple]

A square that measures 1 inch on each side. [square inch]

Describes 2 figures that are exactly the same shape and size. [congruent]

Note: This dictionary project is a review of the year's work. The words are arranged by topic.

ACTIVITIES ***Constructing the 4 × 3 × 1 box.*** Arrange the 1-inch cubes in a 4 × 3 × 1 arrangement as shown below on the left. Explain she is to draw a box on the worksheet that these cubes will fit into exactly. She is to cut it out, fold it into the box, and fasten it with tape.

Note: Other box patterns are possible, although some of them will not fit on the paper.

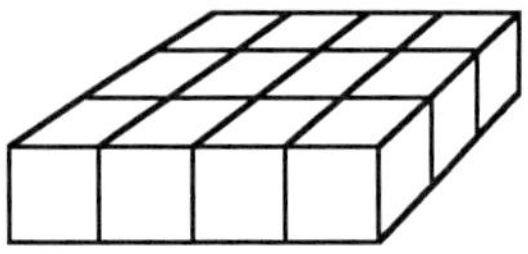

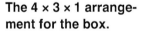

The 4 × 3 × 1 arrangement for the box.

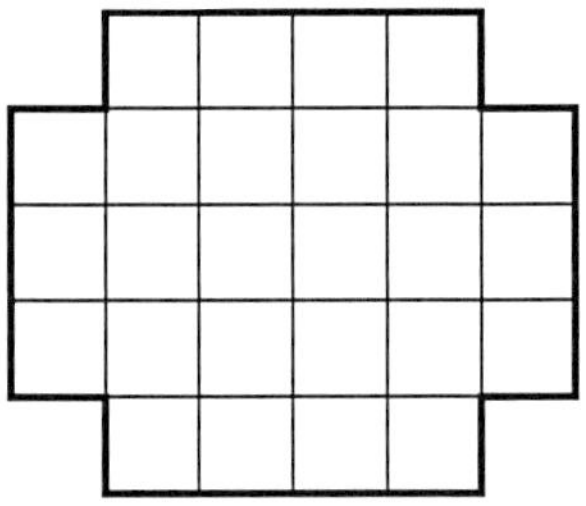

The box pattern.

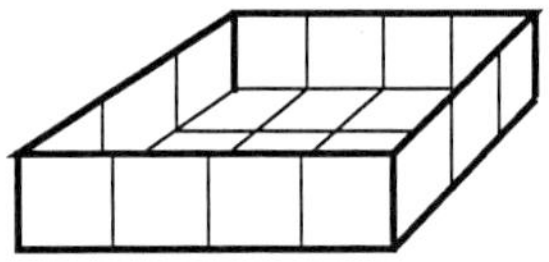

The completed box.

After the box is complete ask, How many cubes will fit in it? [12] What is the measurement of 1 cube? [1 inch wide, long, and high] What is the measurement of the box? [4 in. by 3 in. × 1 in.]

What is a cubic inch? [a cube that measures 1 in. × 1 in. × 1 in.] What does the box hold in cubic inches? [12 cu in.] Write out the words and the abbreviation.

Lesson 140

Building a 3 × 2 × 2 Box in Inches

OBJECTIVE 1. To work with volume through constructing a box

MATERIALS Worksheet 149, "Building a 3 × 2 × 2 Box"
A drawing board, T-square, and 45 triangle
Scissors and removable tape
At least twelve 1-inch cubes

WARM-UP Tell the child you are going to give definitions of math terms and he is to write down the word that is described.

The answer after adding. [sum]

A whole number when grouped by 2s has none left over. [even number]

A whole number that has 1 left over when grouped by 2s. [odd number]

A triangle with 2 sides congruent. [isosceles]

A triangle with 3 sides congruent. [equilateral]

A closed figure drawn with straight lines. [polygon]

A polygon with 4 sides. [quadrilateral]

ACTIVITIES ***Constructing the 3 × 2 × 2 box.*** Tell the child to follow the directions on the worksheet. For the child needing a physical arrangement, lay out the 1-inch cubes in a 3 × 2 × 2 as shown below on the left.

Note: Other box patterns are possible, although some of them will not fit on the paper.

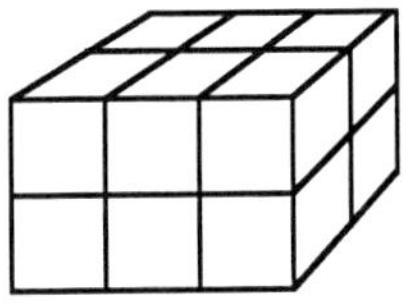

The 3 × 2 × 2 arrangement for the box.

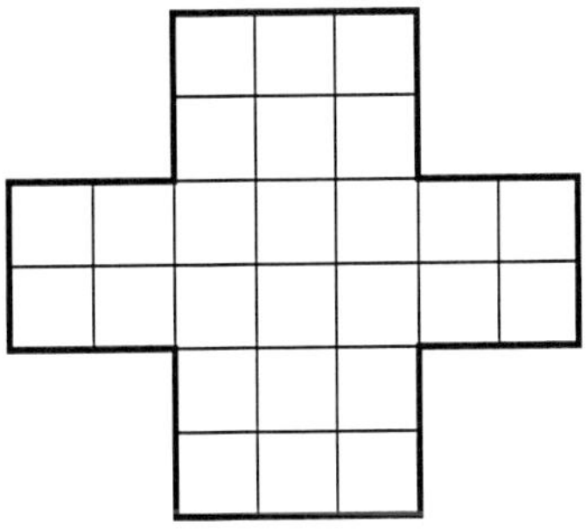

The box pattern.

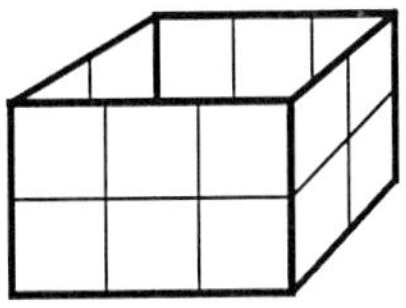

The completed box.

After the box is complete ask, How many cubes will fit in it? [12] What equation tells us how many? [3 × 2 × 2 = 12]

What is a cubic inch? [a cube that measures 1 in. × 1 in. × 1 in.] What does the box hold in cubic inches? [12 cu in.] Ask the child to write out the words and the abbreviation.

Lesson 141

Scaling

OBJECTIVES
1. To recognize parallel lines in a drawing
2. To duplicate a drawing at a different scale

MATERIALS Worksheet 150, "Scaling"
A drawing board with Worksheet 150 taped on it
A T-square and 45 triangle 

WARM-UP What is an isosceles triangle? [a triangle with 2 congruent sides]

What is an equilateral triangle? [a triangle with 3 congruent sides]

Give the child the following problem to solve: Robin is buying 36 packages of napkins. Each package has 200 napkins. How many napkins did Robin buy? [36 × 200 = 7200 napkins]

ACTIVITIES ***Drawing parallel lines.*** Take out the drawing board with the worksheet attached. What kind of lines can we draw with only a T-square? [parallel lines that are horizontal] How else can we draw parallel lines? [with a T-square and triangles]

Explain that sometimes we need to draw parallel lines at angles that are not on the triangles. For this we use one triangle and the T-square upside down. The T-square is upside down so it will be flat on the drawing board. Ask the child to read the first problem on the worksheet.

Demonstrate the procedure as shown below. Arrange the triangle and upside down T-square so that the triangle is parallel to the line. See the figure below on the left.

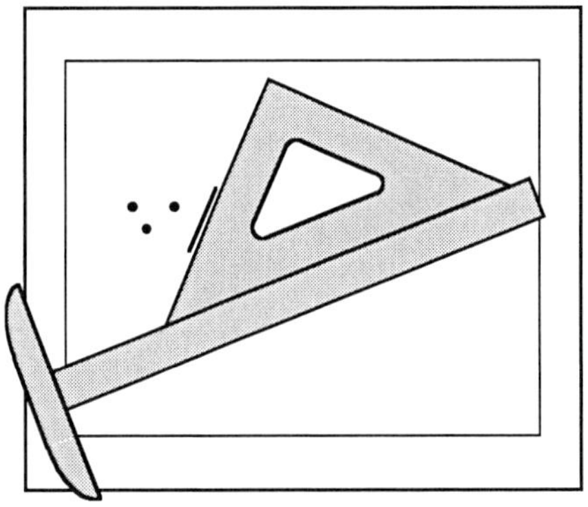

Arranging a triangle and T-square so the triangle is parallel to the line.

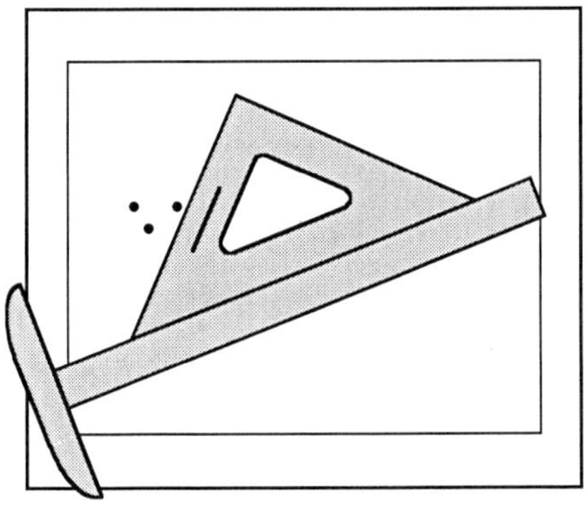

Slide the triangle on the T-square. Keep the T-square from moving.

Keep the T-square from moving while the triangle slides on the T-square. Then slide the triangle next to the dot as shown in the upper right figure and draw the line as shown in the left figure on top of the next page. Draw the remaining lines as shown in the right figure on the next page.

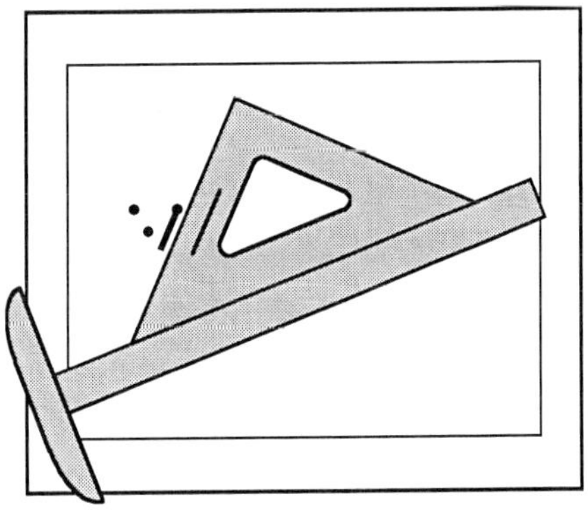

Drawing the line.

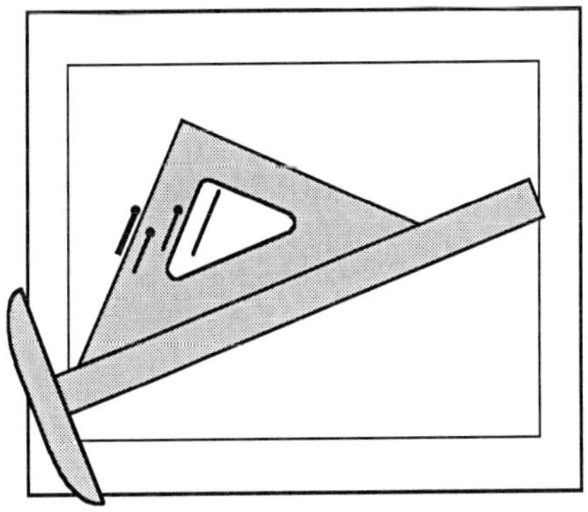

Slide the triangle without moving the T-square and draw the other lines.

Drawing the doghouse. Do you see any parallel lines? [four vertical lines; four lines including the roof, side of the house and inside the house; and two other roof lines]

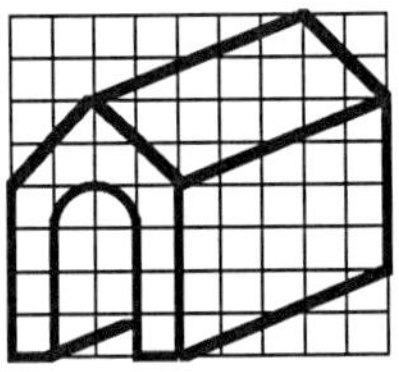

The doghouse, which is to be drawn to a larger scale.

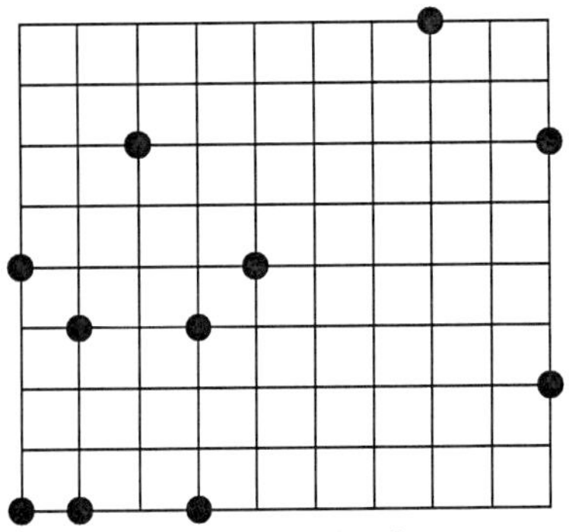

The dots for the doghouse.

Ask the child to draw dots to show where the lines begin and end. Discuss counting squares to find where the lines begin and end. Start from the lower left-hand corner. Count the spaces to the right and then count the spaces up. See the right figure above. Remind her to use the drawing tools for all the straight lines.

The three short roof lines can be drawn with the 45 triangle and T-square. For the other set of four diagonal parallel lines, draw the first line (best is the bottom outside line) with only one triangle and then draw the other lines as parallel lines, similar to the first exercise. Draw the curve freehand.

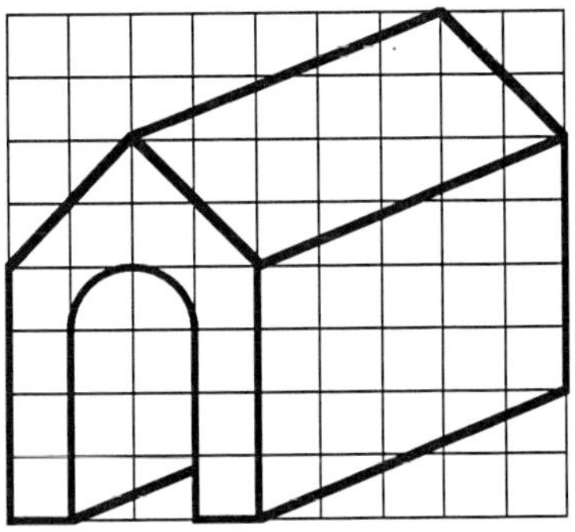

The doghouse drawn to the larger scale.

Lesson 142

Review

OBJECTIVE 1. To review and practice

MATERIALS Worksheet 151-A and 151-B, "Review" (2 versions)
Cards for playing games

ACTIVITIES ***Review worksheet.*** Give the child about 15 to 20 minutes to do the review sheet. The oral problems to be read are as follows:

$123 \times 10 =$ $\quad 1 \div 2 =$ $\quad 30 - ½ =$

Note: There is also a year-end test.

Review Sheet 151-A.

1. Write only the answers to the oral questions. **1230** $\frac{1}{2}$ $29\frac{1}{2}$

4. Write only the answers. $2\frac{3}{10} + 3\frac{7}{10} =$ **6** $\quad 151 - 68 =$ **83** $\quad 92 + 98 =$ **190**

7. Circle the groups of squares that are congruent with the shaded group on the left.

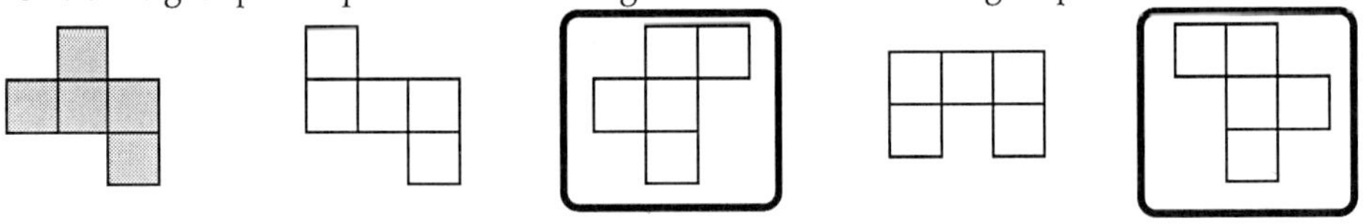

8. Circle the groups of squares that could make a box.

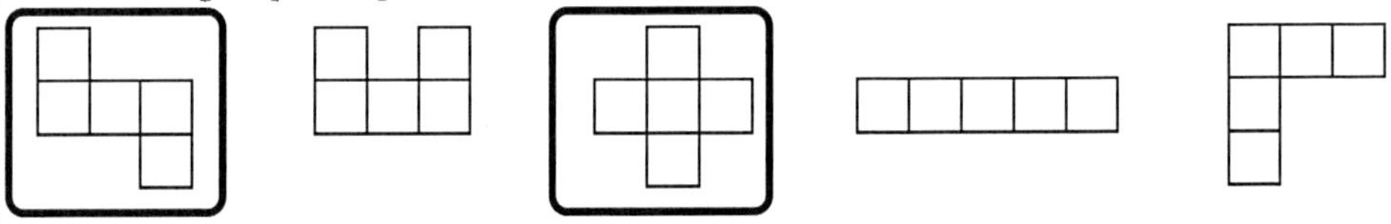

9. Draw lines to match the definitions with the words.

A triangle with 2 congruent sides.	equation
A triangle with 3 congruent sides.	equilateral triangle
A statement that two things are the same.	isosceles triangle
The answer after dividing.	quotient
A square that measures 1 inch on each side.	product
The answer after multiplying.	octagon
A polygon with 6 sides.	hexagon
A polygon with 8 sides.	square inch

18. While on a field trip, the teacher bought treats for 18 people. Each treat cost 59¢. What was the total cost? What was the change from $15? What bills and coins did the teacher receive?

```
 $0.59     $15.00     four $1 bills
  × 18   − 10.62     one quarter
  4 72      $4.38     one dime
  5 90                three pennies
$10.62
```

The oral problems to be read to the child are as follows:

$246 \times 10 =$ $1 \div 4 =$ $40 - ½ =$

Review Sheet 151-B.

1. Write only the answers to the oral questions. **2460** $\frac{1}{4}$ $39\frac{1}{2}$

4. Write only the answers. $4\frac{6}{10} + 3\frac{4}{10} =$ **8** $175 - 38 =$ **137** $83 + 87 =$ **170**

7. Circle the groups of squares that are congruent with the shaded group on the left.

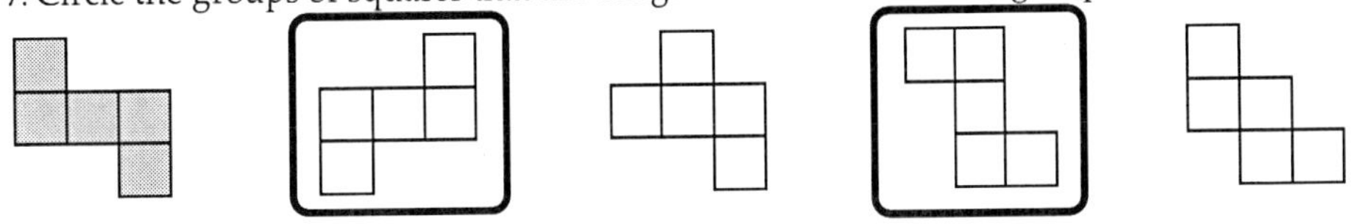

8. Circle the groups of squares that could make a box.

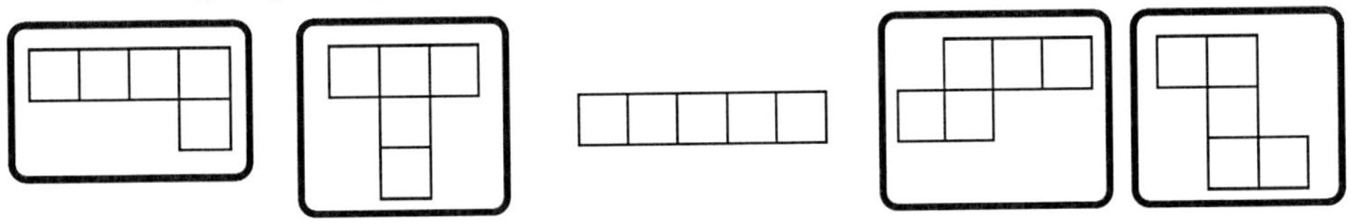

9. Draw lines to match the definitions with the words.

A polygon with 3 sides.	equation
A polygon with 4 sides.	triangle
A polygon with 6 sides.	quadrilateral
A polygon with 8 sides.	quotient
A statement that two things are the same.	product
The answer after dividing.	hexagon
The answer after multiplying.	octagon
A cube that measures 1 inch on all sides.	cubic inch

18. For a family party, Chris bought treats for 16 people. Each treat cost 79¢. What was the total cost? What was the change from $15? What bills and coins did Chris receive?

```
 $0.79     $15.00     two $1 bills
 × 16    – 12.64      one quarter
  4 74      $2.36     one dime
  7 90                one penny
$12.64
```

Games. Spend the remaining time playing games.

DIVISION. Mixed-Up Quotients (D3) *(Math Card Games).*

FRACTIONS. Fraction War (F7) and One (F6) *(Math Card Games).*

Name KEY

Date ______________________

Draw the hands.

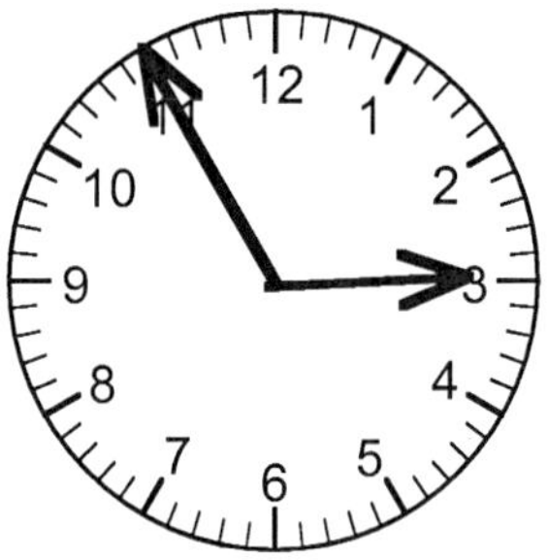

five before 3

11:05

Write the time two ways.

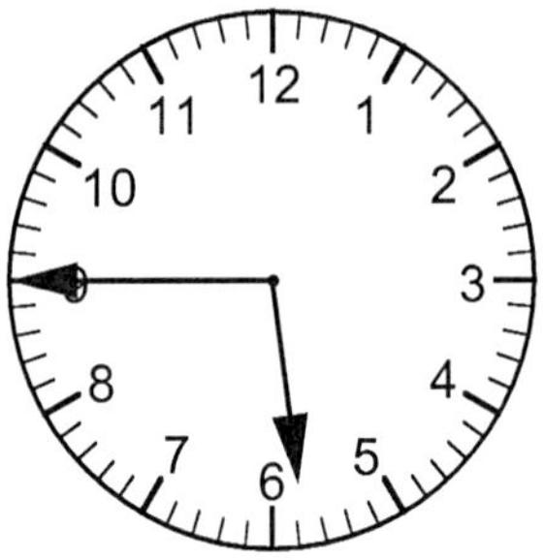

5:45

quarter to six

Show $4\frac{1}{2}$ inches on the ruler. Circle or shade it.

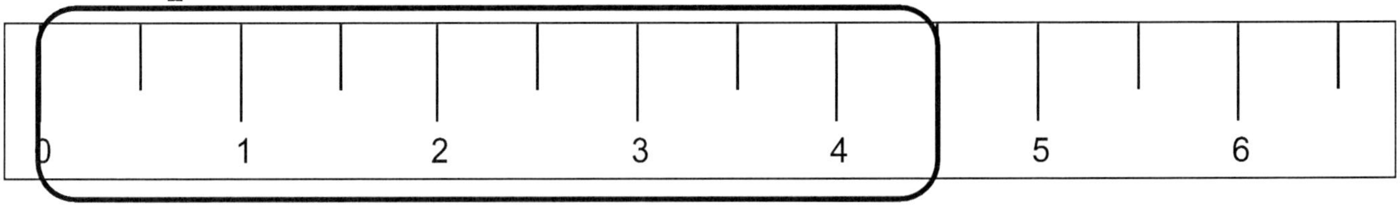

How long is this line in inches?
Circle your answer.

$3\frac{1}{2}$ (2) 4 $5\frac{1}{2}$

Add these. Write only the answers.

45 + 25 = 70

47 + 19 = 66

105 + 15 = 120

Subtract these. Write only the answers.

37 – 33 = 4

60 – 49 = 11

101 – 2 = 99

Write the multiples of 6.

6 12 18 24 30

36 42 48 54 60

Use the 6 table to find the answers.

6 x 4 = 24

6 x 10 = 60

6 x 6 = 36

What pattern do you see with the ones?

All even numbers. The ones in the second row are the same as the first row.

Name KEY

Alex earned $3.45 on each of five days. How much money did Alex earn?

$3.45
$3.45
$3.45
$3.45
$3.45
$17.25

Students measured a room that is in the shape of a rectangle. One side is 389" and the second side is 176". Draw a sketch. What is the perimeter?

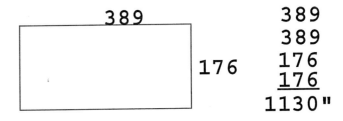

389
389
176
176
1130"

Write the missing amounts and the words in the rectangles below.

100¢ dollar									
50¢ half-dollar					50¢ half-dollar				
25¢ quarter			25¢ quarter		25¢ quarter			25¢ quarter	
10¢ dime	10¢ dime	10¢ dime	10¢ dime	10¢ dime	10¢ dime	10¢ dime	10¢ dime	10¢ dime	10¢ dime

Chris had five quarters and received two dollars as a gift. How much money does Chris have now?

5 quarters is $1.25 and 2 dollars is $2.00.
So Chris has $3.25.

Write the missing words and number of ounces in the rectangles below.

gallon 128 oz			
half gallon 64 oz		half gallon 64 oz	
quart 32 oz	quart 32 oz	quart 32 oz	quart 32 oz

Name **KEY**

Date ____________________

Draw the hands.

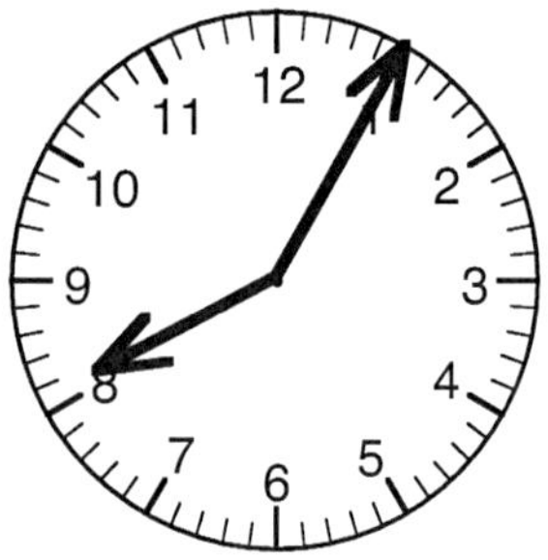

five after 8

7:55

Write the time two ways.

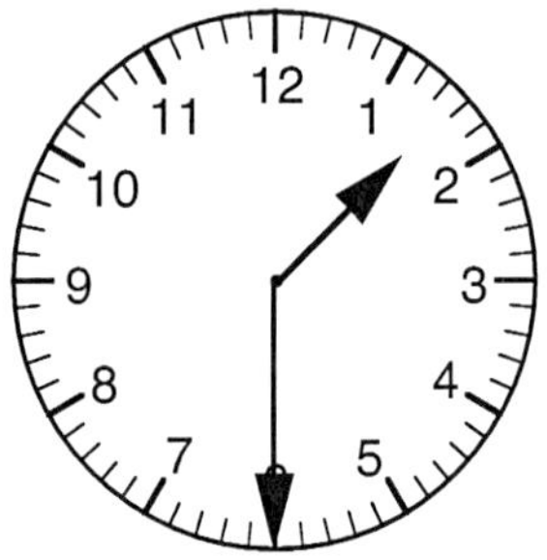

1:30

half past one

Show $3\frac{1}{2}$ inches on the ruler. Circle or shade it.

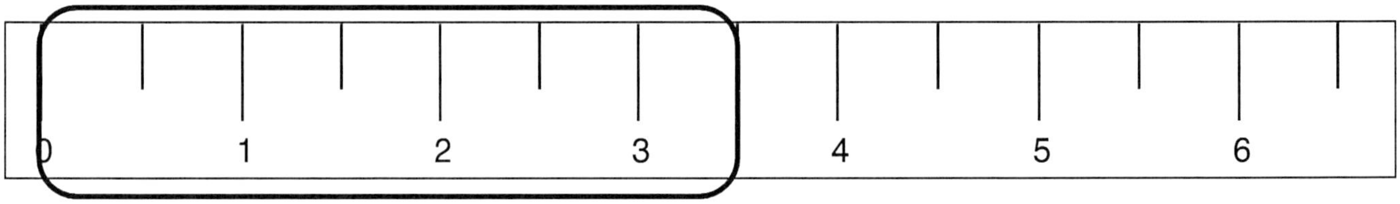

How long is this line in inches?
Circle your answer.

$6\frac{1}{2}$ 4 2 $2\frac{1}{2}$ (circled)

Add these. Write only the answers.

25 + 35 = **60**

47 + 28 = **75**

135 + 15 = **150**

Subtract these. Write only the answers.

28 – 25 = **3**

70 – 58 = **12**

100 – 3 = **97**

Write the multiples of 8.

8 **16** **24** **32** **40**

48 **56** **64** **72** **80**

What pattern do you see with the ones?

All even numbers. The ones are counting by 2s backwards. The ones in the second row are the same as the first row.

Use the 8 table to find the answers.

8 × 3 = **24**

8 × 9 = **72**

8 × 7 = **56**

Name **KEY**

Sammy bought four books. Each one cost $2.65. What was the total cost?

$2.65
$2.65
$2.65
$2.65
$10.60

Some third graders measured a room and found all the sides measured 254 inches. What is the shape of the room? What is the perimeter?

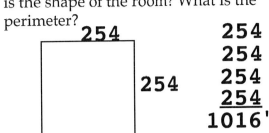

254

254

254
254
254
254
1016"

Write the missing fractions and the amounts in the rectangles below.

<table>
<tr><td colspan="20">1   100¢</td></tr>
<tr><td colspan="10">$\frac{1}{2}$ 50¢</td><td colspan="10">$\frac{1}{2}$ 50¢</td></tr>
<tr><td colspan="5">$\frac{1}{4}$ 25¢</td><td colspan="5">$\frac{1}{4}$ 25¢</td><td colspan="5">$\frac{1}{4}$ 25¢</td><td colspan="5">$\frac{1}{4}$ 25¢</td></tr>
<tr><td colspan="2">$\frac{1}{10}$ 10¢</td><td colspan="2">$\frac{1}{10}$ 10¢</td><td colspan="2">$\frac{1}{10}$ 10¢</td><td colspan="2">$\frac{1}{10}$ 10¢</td><td colspan="2">$\frac{1}{10}$ 10¢</td><td colspan="2">$\frac{1}{10}$ 10¢</td><td colspan="2">$\frac{1}{10}$ 10¢</td><td colspan="2">$\frac{1}{10}$ 10¢</td><td colspan="2">$\frac{1}{10}$ 10¢</td><td colspan="2">$\frac{1}{10}$ 10¢</td></tr>
</table>

Jamie had 11 dimes and received three dollars as a gift. How much money does Jamie have now?

11 dimes is $1.10 and 3 dollars is $3.00.
So Jamie has $4.10.

Write the missing fractions and the number of minutes in the rectangles below.

<table>
<tr><td colspan="4">1 hr   60 min</td></tr>
<tr><td colspan="2">$\frac{1}{2}$ hr   30 min</td><td colspan="2">$\frac{1}{2}$ hr   30 min</td></tr>
<tr><td>$\frac{1}{4}$ hr   15 min</td><td>$\frac{1}{4}$ hr   15 min</td><td>$\frac{1}{4}$ hr   15 min</td><td>$\frac{1}{4}$ hr   15 min</td></tr>
</table>

Name **KEY**

Date ______

[38 + 24, 101 - 25, 11 × 8]

1. Write only the answers to the oral questions. **62** **76** **88**

4. Write only the answers: 99 + 35 = **134** 81 – 56 = **25** 12 × 4 = **48**

7. Central Bank has 67 bags with money. Each bag has 1 thousand dollars. How much money is in the bags?

$67,000

8. Shade one half.

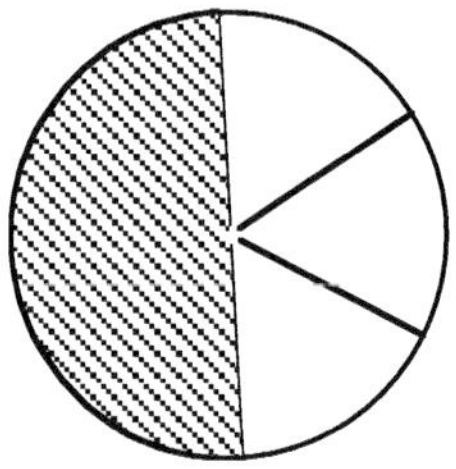

9. Draw and shade $1\frac{1}{2}$.

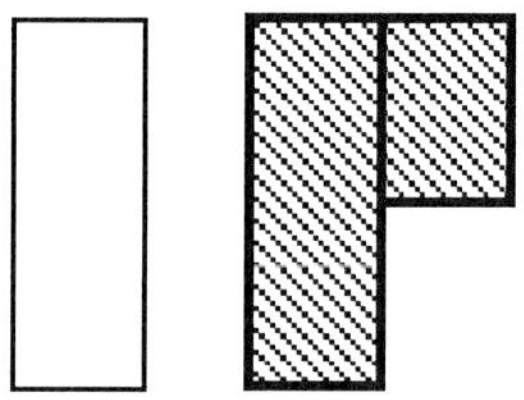

10. Apples cost 79¢ a pound and grapes cost $1.45 a pound. What is the total cost of 3 pounds of apples and 4 pounds of grapes?

.79	1.45	$2.37
× 3	× 4	+ $5.80
$2.37	$5.80	$8.17
Cost of apples	Cost of grapes	Total cost

11. Draw the fourth term and finish the table.

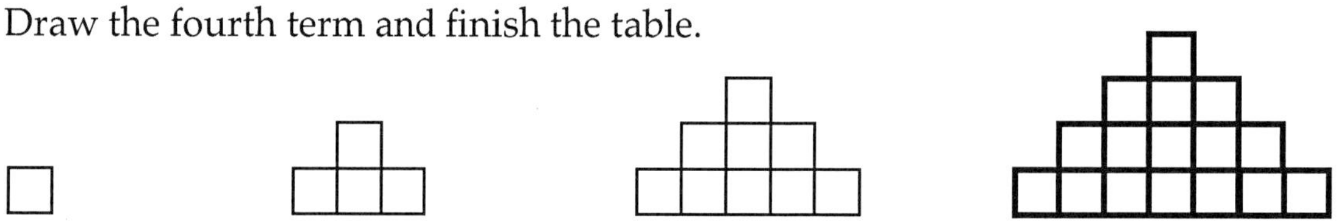

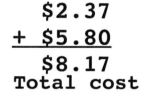

Term	1	2	3	4	5	6	10
Number of Squares	1	**4**	**9**	**16**	**25**	**36**	**100**

19. Write each division problem and answer three ways.

36 ÷ 4 = **9**	$\frac{36}{4} = 9$	$4\overline{)36}$ = 9
72 ÷ 9 = 8	$\frac{72}{9} = 8$	$9\overline{)72}$ = 8
55 ÷ 5 = 11	$\frac{55}{5} = 11$	$5\overline{)55}$ = 11

Name **KEY**

26. Write the next two terms for each pattern.

104	102	100	**98**	**96**
1	$1\frac{1}{2}$	2	**$2\frac{1}{2}$**	**3**
21	28	35	**42**	**49**
\$0.70	\$0.80	\$0.90	**\$1.00**	**\$1.10**
25¢	50¢	75¢	**\$1.00**	**\$1.25**

36. Jamey is buying candles for a birthday cake for an aunt who is 65 years old. There are 8 candles in a package. How many packages does Jamey need?

9 packages
(64 ÷ 8 = 8)

38. Draw the hands on the clock to show 5:12.

40. 12 × 17 = 204. How much is 12 × 18? **216** Mr. Coles has 204 eggs that he is packing in egg cartons. Twelve eggs fit in a carton. How many cartons does he need? Explain your work.

204 ÷ 12 = 17,
He needs 17 cartons.

43. What is the perimeter of a square with a side measuring 29 inches?

```
  29
× 4
 116
```

P = 116"

45. Add 693 + 5679 + 9650.

```
   693
  5679
+ 9650
 16,022
```

47. Subtract 48,537 – 9654.

```
  48,537
-  9,654
  38,883
```

49. Multiply \$6.85 × 37.

```
   $6.85
   × 37
   47 95
  205 50
 $253.45
```

Sets of Numbers for Finding Sums

0	0	0	0	0	0
9	9	9	9	9	9
8	8	8	8	8	8
7	7	7	7	7	7
6	6	6	6	6	6
6	6	6	6	6	6
5	5	5	5	5	5
5	5	5	5	5	5
2	2	2	2	2	2
1	1	1	1	1	1

Multiplication Table

×	1	2	3	4	5	6	7	8	9	10
1										
2										
3										
4										
5										
6										
7										
8										
9										
10										

Thousand Cube

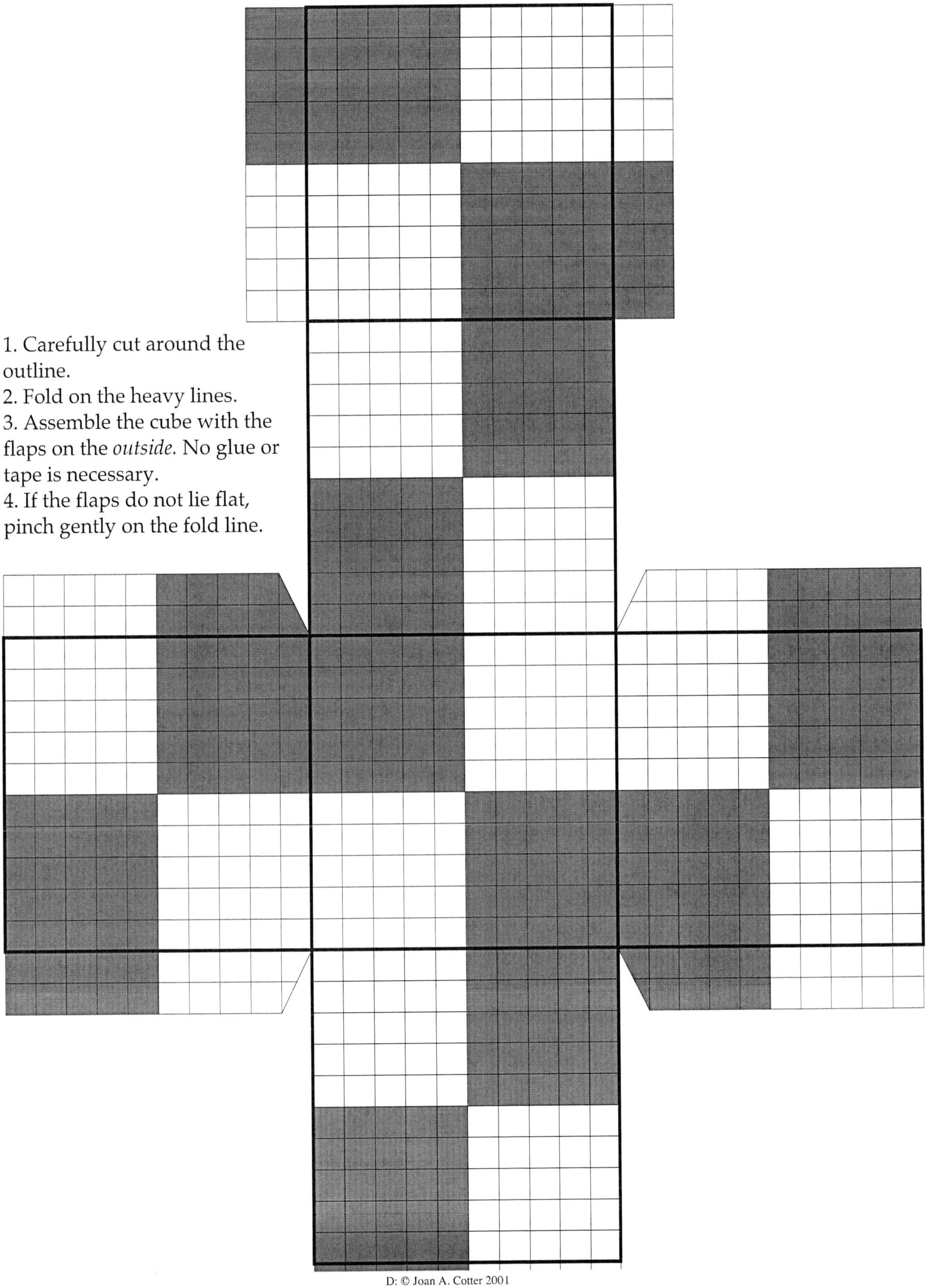

1. Carefully cut around the outline.
2. Fold on the heavy lines.
3. Assemble the cube with the flaps on the *outside.* No glue or tape is necessary.
4. If the flaps do not lie flat, pinch gently on the fold line.

Math Puzzles-1

Fill in the missing blanks. The sum of the first three numbers in a row equals the last number in that row. The sum of the first three numbers in a column equals the last number in that column. Check your puzzle when you are done by adding all the rows and columns.

1.

3	5	**1**	9
6	2	4	12
3	8	5	**16**
12	**15**	10	37

2.

6	5	1	12
7	**5**	**5**	17
6	**4**	3	**13**
19	14	9	42

3.

3	**4**	9	16
9	7	11	**27**
13	8	**12**	33
25	19	**32**	76

4.

11	**6**	8	**25**
14	7	9	30
9	**12**	7	28
34	25	**24**	83

5.

7	6	**4**	**17**
10	**9**	12	31
6	8	9	23
23	23	**25**	71

6.

8	**6**	8	22
10	12	**8**	30
6	**4**	**6**	16
24	22	22	68

Math Puzzles-2

Fill in the missing blanks. The sum of the first three numbers in a row equals the last number in that row. The sum of the first three numbers in a column equals the last number in that column. Check your puzzle when you are done by adding all the rows and columns.

1.

5	9	1	**15**
4	7	3	14
5	**4**	6	15
14	20	**10**	44

2.

4	4	**1**	9
10	6	**1**	17
3	4	12	19
17	14	14	**45**

3.

6	4	12	22
7	**7**	**4**	18
4	**1**	5	10
17	12	21	**44**

4.

6	10	1	**17**
12	3	**9**	24
3	**2**	**7**	12
21	15	17	53

5.

5	**7**	3	15
4	3	**1**	8
6	11	6	**23**
15	**21**	**10**	46

6.

6	12	12	30
8	**11**	6	25
4	10	**2**	16
18	33	20	71

Box Pattern

To make the box, cut along the outside and fold on the other lines. Then tape the sides together.

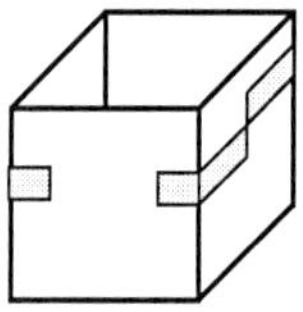

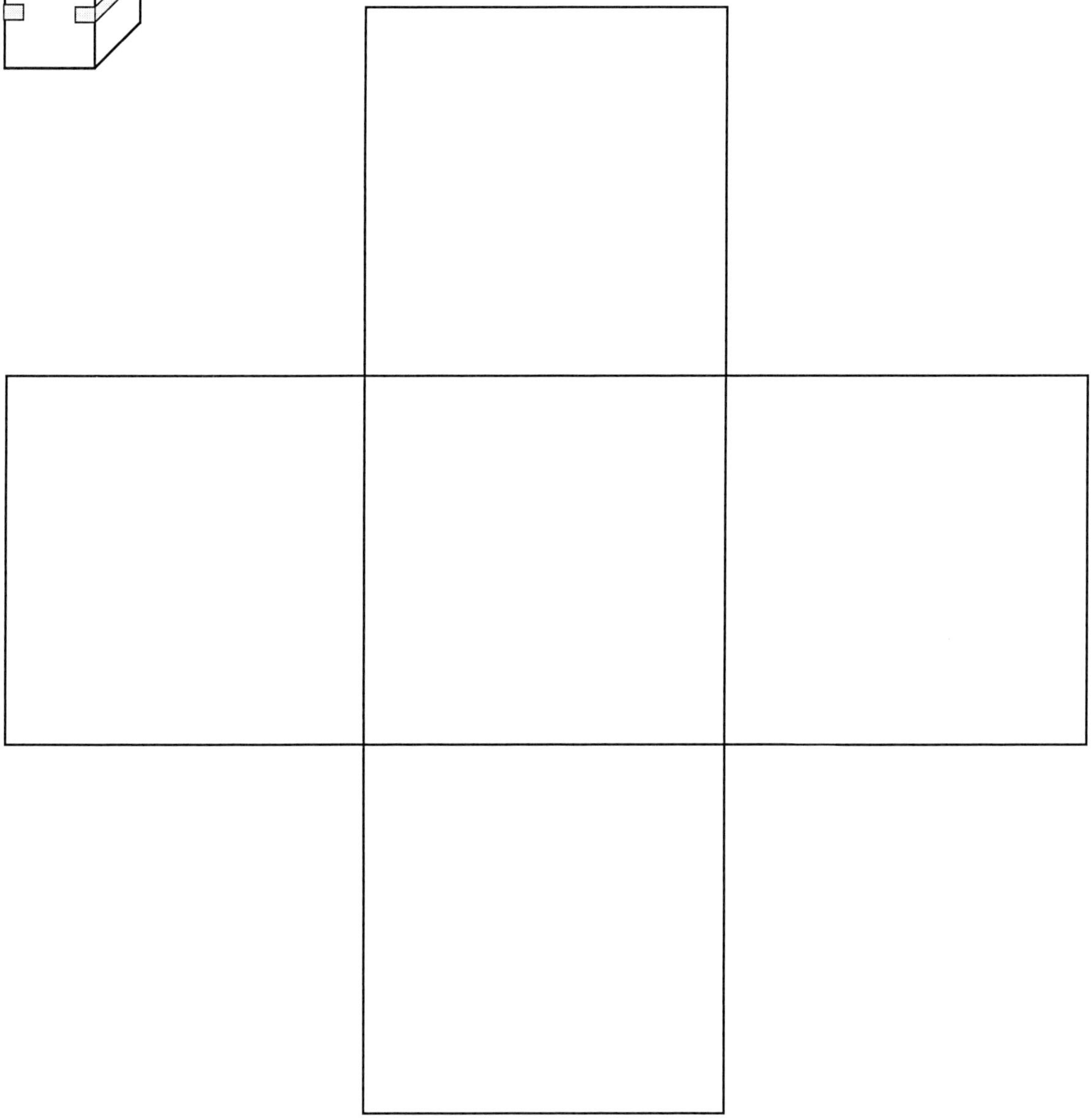

Games list for Level D

Game	Lesson	MCG
Between the Number's	56	C26
Check Number Patterns	50	N/A
Check Numbers	50	A63
Concentrating on One	113	F3
Continue my Rule	89	N/A
Corners	8	A9
Corners Solitaire with Stacks	18	A11
Corners Solitaire with Stock	18	A10
Corners Speed	18	A12
Corners Three Subtraction	50	N/A
Crazy Squares	90	P23
Division Memory	107	D2
Fraction War	116	F7
Minute Memory	57	C11
Minute Solitaire	57	C12
Missing Multiplication Cards	18	P5
Mixed Multiplication Cards	60	P7
Mixed-Up Quotients	114	D3
Multiple Authors	78	P16
Multiples Memory	18	P2
Multiples Solitaire	66	P19
Multiplication Card Speed	108	P31
Multiplication Memory	18	P10
Multiplication Old Main	66	P14
Mystery Multiplication Card	18	P4
Name That Time	57	C27
One	132	F6
Short Chain Subtraction	42	S18
Show your Product	78	P15
Slide-a-thon Solitaire	66	P9
Square Memory	90	P21
Subtraction Corners in the Thousands	48	N/A
Treasure Hunt	54	P6
What's my Rule	81	N/A
Zero Corners	33	S9